Vitamin-Binding Proteins

Functional Consequences

Vitamin-Binding Proteins
Functional Consequences

Edited by
Krishnamurti Dakshinamurti
Shyamala Dakshinamurti

CRC Press
Taylor & Francis Group
Boca Raton London New York

CRC Press is an imprint of the
Taylor & Francis Group, an **informa** business

CRC Press
Taylor & Francis Group
6000 Broken Sound Parkway NW, Suite 300
Boca Raton, FL 33487-2742

First issued in paperback 2019

© 2014 by Taylor & Francis Group, LLC
CRC Press is an imprint of Taylor & Francis Group, an Informa business

No claim to original U.S. Government works

ISBN-13: 978-1-4398-8016-6 (hbk)
ISBN-13: 978-0-367-37971-1 (pbk)

Visit the Taylor & Francis Web site at
http://www.taylorandfrancis.com

and the CRC Press Web site at
http://www.crcpress.com

Contents

Preface

Vitamins are essential micronutrients required by all protozoans and metazoans studied. They are of diverse chemical nature. We have the concept of a ligand (vitamin) reacting with a specific protein entity, the receptor, facilitating the absorption and transport of the ligand to the site of action, and further its participation in intracellular events that form part of the cellular signaling mechanisms, including, in many instances, the regulation of gene expression. Another significant aspect of the function of many vitamins is their role in cellular oxidation-reduction reactions and functioning as antioxidants.

The retinoids constitute a group of lipid-soluble morphogens related to retinol that play a vital role in various biological processes including embryonic pattern formation, organogenesis, cell proliferation and differentiation, apoptosis, and immunity. These functions, known to be conserved throughout the chordate phylum, are mediated by the binding of retinoic acid (RA) to two nuclear receptors, retinoic acid receptors (RAR) and retinoid X receptors (RXR). The interaction of the RA signaling network with other intracellular signaling cascades has been identified. Retinoid binding protein 4 has been implicated as an adipokine involved in insulin resistance. The protective role of RA in autoimmune diseases has been recognized.

Vitamin D, which functions as a steroid hormone, binds to a vitamin D receptor (VDR) expressed in most cells. Heterodimerization with RXR and binding to vitamin D response elements located in the promoter regions induces expressions of target genes involved in many cellular responses including immune responses. Through its effect on dendritic cell proliferation and maturation and T cell stimulation, vitamin D regulates adaptive and innate immunity and has a modulating effect on many inflammatory and autoimmune diseases.

Connective tissue in the vascular system plays an important role in maintenance of the intact vascular wall. Vitamins E and C influence the extracellular matrix (ECM) by their antioxidant functions. They bind to specific enzymes and act as cofactors and regulators with consequent modulation of vascular smooth muscle cell (VSMC) signal transduction and gene expression. The ability of vitamins E and C to influence VSMC proliferation, differentiation, and ECM production is important in the maintenance of intact vascular wall and in the repair of atherosclerotic lesions. Actively proliferating cells, but not quiescent cells, are susceptible to ascorbic acid, which inhibits cell division and promotes necrosis through its action on S-phase progression.

Apart from clotting factor proteins, vitamin K–dependent proteins include a variety of regulatory proteins. One of them, the matrix Gla-protein (MGP), through gamma carboxylation of glutamic acid residues, enables it to inhibit extraosseous calcification. Vascular calcification with its accompanying increased morbidity and mortality is associated with vitamin K deficiency, suggesting a favorable effect of vitamin K on vascular calcification.

Implication of advanced glycation end products (AGEs) and advanced lipoxidation end products (ALEs) in the pathogenesis of diabetes-mediated uremic complications as well as in aging and atherosclerosis are well recognized. Any chemical that would have an inhibitory effect on any of the steps in the process leading to vascular pathology has therapeutic potential. Benfotiamine, a lipid-soluble form of thiamine and pyridoxamine, a vitamin B_6 vitamer, have a significant place in the therapy of micro- and macrovascular defects associated with chronic diseases. Pyridoxamine is a potent scavenger of reactive carbonyls and inhibits formation of AGEs.

The crucial role played by vitamin B_6 in the nervous system and in neuroendocrinology is based on the fact that various putative neurotransmitters as well as taurine, sphingolipids, and polyamines are synthesized by pyridoxal phosphate (PLP)-dependent enzymes. There are numerous biological effects of vitamin B_6 unrelated to the role of PLP as a coenzyme. PLP is an antagonist of both the voltage-mediated and the ATP-mediated calcium transport systems. PLP modulates the activities of steroid hormone receptors and transcription factors. The preventive effect of vitamin B_6 on tumorigenesis might also derive from the antioxidant functions of this vitamin.

Biotin is central to the metabolism of carbohydrates, amino acids, and lipids as biotin is the prosthetic group of the carboxylases. In addition to this metabolic function, biotin influences transcription in organisms ranging from bacteria to humans. Biotin exerts complex effects on cell cycle and gene transcription through epigenetic mechanisms. Nuclear biotin holocarboxylase synthetase seems to interact with other chromatin proteins to form a multiprotein gene repressor complex.

Folate has a high affinity for folate receptors (FR) and retains its FR binding affinity and endocytosis properties even if it is covalently linked to a variety of molecules and particles. A strategy known as "active targeting" has been developed in which particles with ligands are designed to interact with tumor cells. This involves biomolecular-specific recognition. Intravenous injection of folate-linked particles is reported to exhibit antitumor effect. Rapidly dividing cells, including tumor cells, have a high demand for vitamin B_{12} cofactor than normal cells. Strategies have been developed to use vitamin B_{12} as a vehicle for the delivery of cytotoxic drugs or radioactive tracers to tumors.

This volume is devoted to a discussion of the function of vitamins based on their binding to specific proteins as well as their role as antioxidants, leading to effects on intracellular mechanisms that have a wide-ranging effect on a variety of cellular processes. Many of these effects have significant therapeutic potential in a wide spectrum of disease processes. The strategy of using particles bound to folate ligand in tumor targeting and the use of vitamin B_{12} as a vehicle to deliver cytotoxic drugs or radiotracers to tumors have immense therapeutic potential.

As editors, we have assembled the leading experts on the various vitamins to evaluate the current status of the role of vitamins and their binding proteins in the regulation of cellular processes with a view to their potential use in therapeutics. The material presented here has been designed to expand the horizon of researchers in these associated areas.

We express our gratitude to our colleagues who have spent their time and effort, in the midst of their busy schedules, to write their respective chapters. We thank our families for bearing with us during the preparation of this text.

Krishnamurti Dakshinamurti
Shyamala Dakshinamurti

Editors

Krishnamurti Dakshinamurti, emeritus professor of biochemistry and molecular biology, University of Manitoba, Winnipeg, Canada, received his PhD in 1957. After postdoctoral stints at the University of Illinois, Urbana, and at the Massachusetts Institute of Technology, Cambridge, he was appointed associate director of the Research Institute, Lancaster, Pennsylvania. He moved to the University of Manitoba, Faculty of Medicine, as associate professor of biochemistry and molecular biology in 1964, becoming a full professor in 1973. He spent 1974 and 1975 as a visiting professor of cell biology at the Rockefeller University, New York. He received the Borden Award of the Canadian Societies of Biological Sciences in 1973 and was elected Fellow of the Royal Society of Chemistry, United Kingdom in the same year. He is the author of more than two hundred peer-reviewed publications and five books in the areas of metabolic biochemistry and neuroscience. The monograph on "Vitamin Receptors," which he edited in 1994, was reprinted by Cambridge University Press in 2010 in their classics series. Dr. Dakshinamurti is a Fellow of the Royal Society of Medicine, London. He was elected to emeritus professorship of the University of Manitoba in 1998. He was codirector of the Centre for Health Policy Studies at the St. Boniface Hospital Research Centre. Currently he is the senior advisor of the Research Centre.

Shyamala Dakshinamurti is a neonatologist and biomedical researcher at the University of Manitoba. She received her MD and MSc from the University of Manitoba. Following residency in general pediatrics at the University of Chicago and a fellowship in neonatology at the University of Manitoba, she joined the University of Manitoba faculty in 2001. She started the Neonatal Pulmonary Biology Laboratory in 2003 as a clinician scientist and member of the Biology of Breathing research group in the Manitoba Institute of Child Health. Dr. Dakshinamurti was appointed to the Department of Pediatrics with concurrent appointment to the Department of Physiology in 2005, and became associate professor of pediatrics in 2008. Since 2005, she has been Coordinator of Neonatal Research for the eclectically disparate research streams within the section of neonatology, organizes the annual international Bowman Symposium in Neonatal Research, and is the research director for the University of Manitoba Neonatology Fellowship Program. She was appointed Dr. F.W. Du Val Clinical Research Professor in the Faculty of Medicine. Dr. Dakshinamurti's research interests are in pulmonary hemodynamics during circulatory transition and in the physiology and pharmacology of vascular smooth muscle.

Contributors

Wilbert S. Aronow
Department of Medicine
Division of Cardiology
New York Medical College
Valhalla, New York

Angelo Azzi
Vascular Biology Laboratory
Jean Mayer USDA Human Nutrition
    Research Center on Aging
Tufts University
Boston, Massachusetts

Diane Berry
Centre for Pediatric Epidemiology and
    Statistics
University College of London Institute
    of Child Health
London, United Kingdom

Pangala V. Bhat
Department of Medicine
University of Montreal
Montreal, Quebec, Canada

Daniel Camara Teixeira
Department of Nutrition and Health
    Sciences
University of Nebraska–Lincoln
Lincoln, Nebraska

João E. Carvalho
Villefranche-sur-Mer Developmental
    Biology Laboratory
Villefranche-sur-Mer, France

Russell W. Chesney
Department of Pediatrics
The University of Tennessee Health
    Science Center
Memphis, Tennessee

Inpyo Choi
Immunotherapy Research Center
Korea Research Institute for Bioscience
    and Biotechnology
Daejon, Republic of Korea

Eliane Fischer
Center for Radiopharmaceutical
    Sciences
Paul Scherrer Institute
Villigen, Switzerland

Michel Fontés
Therapy of Genetic Disorders
Faculty of Medicine–Timone Sector
University of Aix-Marseille
Marseille, France

Evelyne Furger
Center for Radiopharmaceutical
    Sciences
Paul Scherrer Institute
Villigen, Switzerland

Elina Hyppönen
School of Population Health
University of South Australia
Adelaide, Australia

Young Jun Park
Immunotherapy Research Center
Korea Research Institute for Bioscience
    and Biotechnology
Daejon, Republic of Korea

Haiyoung Jung
Immunotherapy Research Center
Korea Research Institute for Bioscience
    and Biotechnology
Daejon, Republic of Korea

Jagadish Khanagavi
Department of Medicine
Division of Cardiology
New York Medical College
Valhalla, New York

Ramesh C. Khanal
Utah State University
Logan, Utah

Dandan Liu
Department of Nutrition and Health
    Sciences
University of Nebraska–Lincoln
Lincoln, Nebraska

Yoshie Maitani
Fine Drug Targeting Research
    Laboratory
Institute of Medicinal Chemistry
Hoshi University
Tokyo, Japan

Daniel-Constantin Manolescu
Departments of Medicine and Nutrition
University of Montreal
Montreal, Quebec, Canada

Yu Meng
Utah State University
Logan, Utah

Mohsen Meydani
Vascular Biology Laboratory
Jean Mayer USDA Human Nutrition
Research Center on Aging
Tufts University
Boston, Massachusetts

Chie Miyabe
Department of Medicine and
    Rheumatology
Graduate School of Medical and Dental
    Sciences
Tokyo Medical and Dental University
Tokyo, Japan

Yoshishige Miyabe
Department of Medicine and
    Rheumatology
Graduate School of Medical and Dental
    Sciences
Tokyo Medical and Dental University
Tokyo, Japan

Toshihiro Nanki
Department of Clinical Research
    Medicine
Teikyo University
Tokyo, Japan

Ilka Nemere
Utah State University
Logan, Utah

Fryad Rahman
Therapy of Genetic Disorders
Faculty of Medicine–Timone Sector
University of Aix-Marseille
Marseille, France

Chandrasekar Palaniswamy
Department of Medicine
Division of Cardiology
New York Medical College
Valhalla, New York

Michael Schubert
Villefranche-sur-Mer Developmental
    Biology Laboratory
Villefranche-sur-Mer, France

Arunabh Sekhri
Department of Medicine
Division of Cardiology
New York Medical College
Valhalla, New York

Mahendra P. Singh
Department of Nutrition and Health
    Sciences
University of Nebraska–Lincoln
Lincoln, Nebraska

Tremaine M. Sterling
Utah State University
Logan, Utah

Hyun Woo Suh
Immunotherapy Research Center
Korea Research Institute for Bioscience
    and Biotechnology
Daejon, Republic of Korea

Janos Zempleni
Department of Nutrition and Health
    Sciences
University of Nebraska–Lincoln
Lincoln, Nebraska

Yang Zhang
Utah State University
Logan, Utah

Jean-Marc Zingg
Vascular Biology Laboratory
Jean Mayer USDA Human Nutrition
Research Center on Aging
Tufts University
Boston, Massachusetts

1 Retinoic Acid
Metabolism, Developmental Functions, and Evolution

João E. Carvalho and Michael Schubert

CONTENTS

1.1 INTRODUCTION

A large amount of data collected during the past one hundred years has established that vitamin A (also known as retinol) and its derivatives, the retinoids (Figure 1.1), play crucial roles during vertebrate development. The retinoids are fat-soluble morphogens that are involved in a variety of processes, such as pattern formation during early development as well as in organogenesis, cell proliferation, cell differentiation, apoptosis, vision, immune responses, and tissue homeostasis (Blomhoff and Blomhoff 2006, Campo-Paysaa et al. 2008, Collins and Mao 1999, Duester 2008, Glover et al. 2006, Hall et al. 2011, Mark et al. 2006, Niederreither and Dollé 2008,

Noy 2010, Theodosiou et al. 2010). Initial insights into retinol-dependent developmental processes were obtained from nutritional excess and deficiency studies carried out in different animal models including pigs and rats (Hale 1933, Wilson and Warkany 1949). In particular, the so-called vitamin A deficiency (VAD) models produced valuable information about the biological roles of retinol in different organs during embryonic development (Wilson et al. 1953). The subsequent emergence of genetic and molecular tools allowed a much more detailed dissection of the developmental functions of retinol and its derivatives including the identification of the molecular machinery underlying retinoid-dependent signaling (Giguere et al. 1987, Mark et al. 2006, Petkovich et al. 1987).

It is now generally accepted that all-*trans* retinoic acid (RA) is the main biologically active form of retinol, although a biological role for other retinoids, such as 9-*cis* RA, is still a matter of debate (Kane 2012, Theodosiou et al. 2010) (Figure 1.1). Moreover, although alternative mechanisms for triggering retinoid-dependent signaling have recently been proposed (Theodosiou et al. 2010), in vertebrates the main cellular mechanism of RA action occurs by binding of RA to heterodimers of two nuclear receptors: the retinoic acid receptor (RAR) and the retinoid X receptor (RXR) (Mark et al. 2006). This classical mode of retinoid action involves fixation of RA by the DNA-bound RXR/RAR heterodimer, the subsequent recruitment of co-activators, and the ultimate activation of transcription of target genes (Figure 1.2) (Chambon 1996).

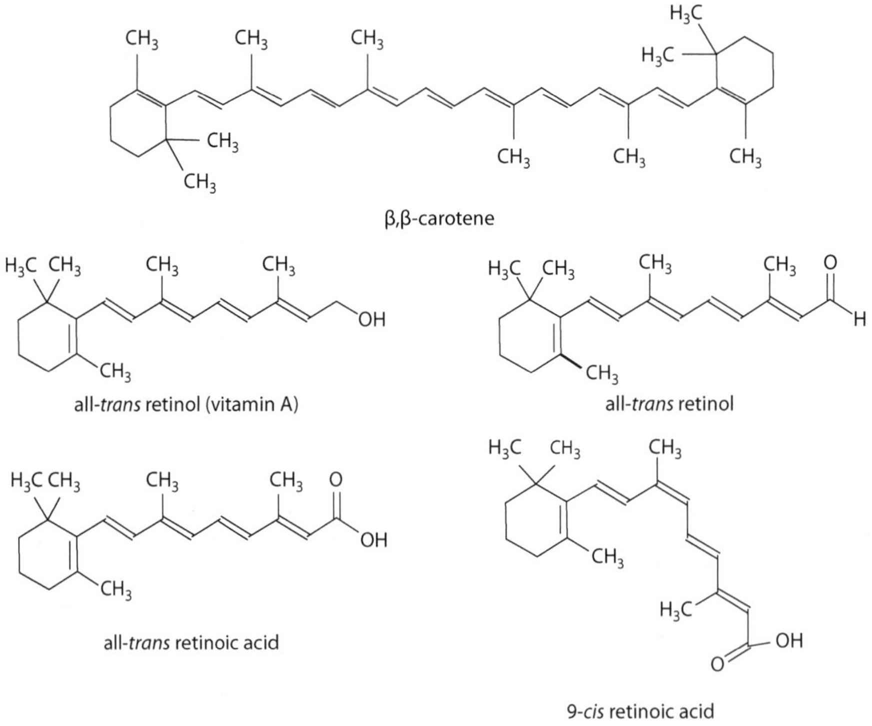

FIGURE 1.1   Chemical structures of different retinoids.

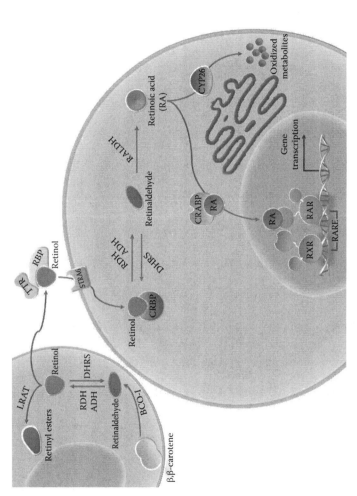

FIGURE 1.2 Retinoid metabolism. Conversion of β,β-carotene to retinol (vitamin A), synthesis of retinoic acid, the major biologically active metabolite, and activation of retinoid-dependent signaling. ADH, alcohol dehydrogenase; BCO-I, β,β-carotene-15,15'-monooxygenase 1; CRABP, cellular retinoic acid-binding protein; CRBP, cellular retinol-binding protein; CYP26, cytochrome P450 family 26; DHRS, dehydrogenase/reductase SDR family member; LRAT, lecithin:retinol acyltransferase; RA, retinoic acid; RALDH, retinaldehyde dehydrogenase; RAR, retinoic acid receptor; RARE, retinoic acid response element; RBP, retinol-binding protein; RDH, retinol dehydrogenase; RXR, retinoid X receptor; STRA6, stimulated by retinoic acid gene 6; TTR, transthyretin.

In this review, we first describe the canonical RA pathway in vertebrates, whose components include proteins responsible for RA production, degradation, and transport. Later, we discuss recently described alternative mechanisms of RA function, followed by a summary of the main developmental functions of RA signaling in vertebrates. We conclude by comparing the available data on RA in vertebrates with the latest information on RA signaling obtained from non-vertebrate animal models.

1.2 RETINOID METABOLISM

Due to their inability to synthesize retinol *de novo*, animals need to obtain retinoids through the diet, mainly in the form of carotenoids, such as β,β-carotene (Figure 1.1). Carotenoids are isoprenoids that are widespread in nature and are typically associated with the yellow, orange, red, or purple colors found in vegetables, fruits, flowers, birds, and crustaceans (Fraser and Bramley 2004). Alternatively, retinol can also be taken up directly in the form of pro-vitamin A carotenoids (Blomhoff and Blomhoff 2006).

The conversion of ingested carotenoids, such as β,β-carotene, into retinaldehyde is the first step of the classical retinoid metabolism (Figure 1.2). This process is driven by the enzyme β,β-carotene-15,15′-monooxygenase 1 (BCO-I) (von Lintig et al. 2005). This enzyme is crucial for proper embryonic development, at least in diets with β,β-carotene as the only source of retinol. For example, BCO-I knockout mice accumulate high levels of β,β-carotene (Hessel et al. 2007), a condition that is similar in zebrafish embryos (Lampert et al. 2003). Additional to BCO-I, β,β-carotene-9,10-dioxygenase 2 (BCO-II) has been described as a second carotenoid oxygenase in vertebrates (Kiefer et al. 2001). BCO-II mainly functions in regulating cell survival, proliferation, and preventing oxidative stress in accordance with its mitochondrial location (Lobo et al. 2012). Additionally, BCO-II might also function as the key enzyme for a non-canonical pathway for RA production (Simões-Costa et al. 2008), but this involvement of BCO-II in retinol metabolism remains to be confirmed *in vivo* (Spiegler et al. 2012).

Retinaldehyde is subsequently hydrolyzed to retinol, which can be used for two different purposes: esterification and tissue storage or oxidative metabolism leading to synthesis of RA, the main biologically active retinoid. The process of esterification is conducted by the enzyme lecithin:retinol acyltransferase (LRAT), which produces retinyl esters (Figure 1.2). Under retinoid oversupply, LRAT has been proposed to play a crucial role in maintaining retinol homeostasis (Ross 2003). For RA synthesis, retinol is transported by retinol-binding protein (RBP) to target tissues (D'Ambrosio et al. 2011). RBP is a specific transport protein circulating in the blood and is associated one to one with another serum protein, transthyretin (TTR) (Figure 1.2) (Richardson 2009, Yamauchi and Ishihara 2009).

In mice, RBP, which cannot cross the placenta (Quadro et al. 2004), has been detected at the junction of the uterine wall and the placenta at E13.5 (embryonic day 13.5) and during subsequent development in the visceral endoderm of the yolk sac (Spiegler et al. 2012). Similarly, RBP has been detected in developing chick and zebrafish embryos (Li et al. 2007, Quadro et al. 2005). Given that RBP is the only known specific carrier protein of retinol, it is rather surprising that mice lacking

RBP (RBP$^{-/-}$) are viable, showing no major malformations when exposed to a retinol-sufficient diet (Barron et al. 1998). Moreover, using an elegant genetic approach, it was shown that, in a BCO-I$^{-/-}$/RBP$^{-/-}$ background, BCO-I$^{+/-}$/RBP$^{-/-}$ offspring can develop normally and maintain retinoid homeostasis, when BCO-I$^{-/-}$/RBP$^{-/-}$ pregnant mice are maintained under a VAD diet with β,β-carotene being the only source of retinoids administrated via intra peritoneal injection (Kim et al. 2011). These results also suggest that β,β-carotene alone can be an adequate source of maternal retinoids (Kim et al. 2011).

The cellular uptake of retinol is carried out by an RBP membrane receptor, called stimulated by retinoic acid gene 6 (STRA6) (Kawaguchi et al. 2007) (Figure 1.2). STRA6 works as a bidirectional retinol transporter, and its activity is controlled by intracellular retinoid levels (D'Ambrosio et al. 2011). In addition, expression of STRA6 is also regulated by RA (D'Ambrosio et al. 2011). Interestingly, STRA6 does not seem to function like other known membrane receptors, transporters, or channels, and there are, at present, no data available about the molecular mechanisms of STRA6-dependent substrate uptake (Sun 2012).

Once inside the cell, RA is synthesized from retinol in two oxidation steps. The first step consists of the reversible oxidation of retinol into retinaldehyde, which is carried out by enzymes of two different classes: the cytosolic alcohol dehydrogenases (ADHs), belonging to the medium-chain dehydrogenase/reductase family, and the microsomal retinol dehydrogenases (RDHs), which are members of the short-chain dehydrogenase/reductase family (Parés et al. 2008) (Figure 1.2). Several members of the ADH family, including ADH1, ADH3, and ADH4, have been shown to catalyze the oxidation of retinol into retinaldehyde *in vitro* (Duester 2008, Parés et al. 2008). Mice carrying null mutations for ADH1, ADH3, and ADH4 have revealed partially redundant functions of these three ADHs (Parés et al. 2008). Moreover, while both ADH1 and ADH4 seem to be required specifically in conditions of retinoid deficiency or excess, ADH3 can probably be regarded as a ubiquitous ADH (Theodosiou et al. 2010). Like for the ADHs, multiple RDHs, such as RDH10, have been shown to be capable of synthesizing retinaldehyde from retinol *in vitro* (Parés et al. 2008). However, recent experimental evidence suggests that the most important player in retinaldehyde synthesis during embryonic development is RDH10 (Farjo et al. 2011). Loss of RDH10 functions in mice is lethal between E10.5 and E14.5 with embryos showing severe defects in embryonic patterning and morphogenesis (Sandell et al. 2007). It has recently been suggested that the regulation of endogenous retinaldehyde levels represents an additional mechanism for controlling RA signaling activity in developing embryos: within a given cell, RA production is thus dependent on the balance between retinaldehyde synthesis from retinol and retinaldehyde reduction to retinol. While in developing embryos the former reaction, as mentioned previously, is catalyzed by ADH or RDH10, the latter reaction can be carried out, for example, by DHRS3a (dehydrogenase/reductase SDR family member 3a), which, like RDH10, is a member of the short-chain dehydrogenase/reductase family (Feng et al. 2010).

The second oxidation step is the conversion of retinaldehyde into RA, an irreversible reaction carried out by retinaldehyde dehydrogenases (RALDHs) (Figure 1.2). Three RALDHs of the ALDH1 class (RALDH1, 2, and 3 also called ALDH1A1, ALDH1A2, and ALDH1A3, respectively) and one RALDH of the ALDH8 class

(RALDH4 also called ALDH8A1) have been implicated in endogenous RA production with each enzyme presenting a distinct and tissue-specific distribution (Theodosiou et al. 2010). RALDH2 is the earliest RALDH enzyme to be expressed during development, and its expression is crucial for early embryogenesis, since mice lacking RALDH2 exhibit severe abnormalities and die before mid-gestation (Niederreither et al. 1999). These phenotypes can be almost completely rescued by maternal administration of RA, suggesting that RALDH2 is indeed required for endogenous RA synthesis during development (Mic et al. 2002, Niederreither et al. 1999). RALDH2 homologs in the frog *Xenopus laevis* and in zebrafish have been shown to have similar roles during early embryonic development (Begemann et al. 2001, Chen et al. 2001). RALDH1, RALDH3, and RALDH4 are expressed at later developmental stages, with RALDH1 and RALDH3 contributing to the patterning of the respiratory and visual systems (Duester et al. 2003, Rhinn and Dollé 2012) and RALDH4 being detectable in the fetal liver (Lin et al. 2003).

Some reports have suggested that RA synthesis during embryonic development may also occur independently of RALDHs, possibly through the action of members of the cytochrome P450 family of monooxygenases (Collins and Mao 1999). Functional studies point to CYP1B1 as a potential candidate (Chambers et al. 2007), because this enzyme can efficiently oxidize retinol to retinaldehyde and subsequently to RA and exhibits an expression pattern consistent with RA synthesis patterns in the developing embryo (Chambers et al. 2007).

Catabolism of endogenous RA is driven by enzymes of the cytochrome P450 subfamily 26 (CYP26A1, CYP26B1, and CYP26C1), which are responsible for the oxidation of RA into metabolites (Figure 1.2), such as 4-hydroxy-RA, 4-oxo-RA, or 18-hydroxy-RA (Bempong et al. 1995). All three CYP26 enzymes function in embryonic development and exhibit dynamic expression patterns both in time and space (Reijntjes et al. 2004, Ross and Zolfaghari 2011). CYP26A1 expression starts at the gastrula stage in mouse endoderm, in chick it is initially expressed in ectoderm and mesoderm close to Hensen's node, whereas in frogs and zebrafish the gene is found in the anterior ectoderm during early gastrulation (White and Schilling 2008).

Mice devoid of CYP26A1 die before the first postnatal day presenting a posteriorized hindbrain and vertebral column as well as severe caudal truncations (Abu-Abed et al. 2001). In contrast, CYP26B1 seems to be fundamentally involved in limb patterning and outgrowth (Yashiro et al. 2004) and is required for maintenance of germ cells during testes development (MacLean et al. 2007). Like CYP26A1, CYP26C1 is expressed in the hindbrain, but mice deficient for CYP26C1 are viable and do not show severe anatomical abnormalities, suggesting some degree of functional redundancy between CYP26A1 and CYP26C1 (Uehara et al. 2007). Double mutants for CYP26A1 and CYP26C1 have a much stronger phenotype when compared to both CYP26A1 or CYP26C1 single mutants. Simultaneous loss of RALDH2 partially rescues these phenotypes, confirming that the malformations observed after loss of both CYP26A1 and CYP26C1 are due to elevated levels of RA (Uehara et al. 2007).

Retinoids present within a given cell are bound by intracellular-binding proteins, such as cellular retinol-binding proteins (CRBPs) or cellular retinoic acid-binding proteins (CRABPs) (Figure 1.2), all of which exhibit extremely high affinities for their substrates (Noy 2000). CRBP-I, for example, protects retinol from oxidation and isomerization by limiting the interaction with enzymes capable of recognizing retinol (Napoli 2000). During development, CRBP-I is expressed in several tissues including motor neurons, spinal cord, lung, liver, and placenta (Ghyselinck et al. 1999). Mice lacking CRBP-I do not exhibit any morphological abnormalities, suggesting that RA metabolism is not dependent on CRBP-I function (Ghyselinck et al. 1999). Nonetheless, in the absence of CRBP-I, retinol stocks are severely depleted (Matt et al. 2005). Compared to CRBP-I, CRBP-II has a more restricted expression pattern, being mainly localized in fetal intestine and liver, but also in the yolk sac from E10.5 to E15.5 (Noy 2000). Its expression suggests a role as binding protein for recently absorbed retinol (Suruga et al. 1997). Interestingly, developmental expression of CRBP-I overlaps significantly with that of CRABP-II, while the patterns of CRBP-I and CRABP-I are mutually exclusive (Napoli 1999).

Several functions have been proposed for the two CRABP proteins, but their antagonistic effect on RA transport seems to be the most important: while CRABP-I seems to mediate CYP26-dependent RA degradation (Napoli 1999), CRABP-II is probably responsible for the transport of RA to its receptors in the nucleus, a fact supported by the increase of RA signaling when CRABP-II is overexpressed in cell lines and *Xenopus* embryos (Dong et al. 1999, Noy 2000). Although CRABP-I$^{-/-}$ and CRABP-II$^{-/-}$ single mutant as well as CRABP-I$^{-/-}$/CRABP-II$^{-/-}$ double mutant mice failed to reveal the biological functions of CRABPs (Napoli 2000), a recent analysis in zebrafish has highlighted the importance of CRABPs in stabilizing the RA morphogen gradient in the developing hindbrain by delivering RA both to its receptors and to CYP26 for degradation (Cai et al. 2012).

1.3 CLASSICAL RETINOID SIGNAL TRANSDUCTION: THE RXR/RAR HETERODIMER

Until very recently, the main dogma of retinoid research stated that RA functions are mediated by heterodimers of two nuclear receptors: RAR and RXR (Chambon 1996, Gronemeyer et al. 2004). There are usually three RAR genes encoded in mammalian genomes (RARα, RARβ, and RARγ) (Linney et al. 2011), while the zebrafish, for example, has lost RARβ but duplicated both RARα (yielding RARαa and RARαb) and RARγ (yielding RARγa and RARγb) (Waxman and Yelon 2007). Congruent with the overall RAR toolkit, mammalian genomes generally also possess three RXR genes (RXRα, RXRβ, and RXRγ). In contrast, the zebrafish genome, for instance, encodes six RXR genes (RARαa, RARαb, RARβa, RARβb, RARγa, and RARγb) (Waxman and Yelon 2007). The RXR/RAR heterodimer acts as ligand-dependent transcription factor by binding to RA response elements (RAREs) located within the regulatory DNA of RA target genes. *In vitro* studies have shown that RAR is able to bind and to be activated by both all-*trans* RA and 9-*cis* RA, whereas RXR binds and is activated only by 9-*cis* RA. However, it remains to be determined whether 9-*cis* RA is present in vertebrate embryos and if its interaction with RXR effectively triggers a biological process (Mic et al. 2003).

RXR/RAR typically associates with two DNA sequence stretches composed of direct repeats (DRs) with the conserved nucleotide sequence (A/G)G(G/T)(G/T)(G/C)A. The spacing between the two DRs is typically one, two, or five nucleotides long (Balmer and Blomhoff 2005, Chambon 1996, Ross et al. 2000). According to the canonical model, RXR/RAR heterodimers bind to DNA even in the absence of ligand, thereby recruiting a co-repressor complex that mediates chromatin compaction and target gene repression (Vilhais-Neto and Pourquié 2008). In the presence of ligand, due to a conformational change of the receptor induced by the ligand, there is dissociation of the co-repressor complex and recruitment of a co-activator complex. This binding of co-activators mediates chromatin decondensation and assembly of the transcriptional pre-initiation complex ultimately leading to target gene transcription (Rhinn and Dollé 2012).

Genetic studies have established that the three mammalian RAR paralogs have both specific and partially redundant functions during embryogenesis (Matt et al. 2003). Interestingly, RARα$^{-/-}$, RARβ$^{-/-}$, and RARγ$^{-/-}$ single mutant mice are able to reach adulthood, presenting only minor morphological phenotypes indicative of VAD syndromes (Mark et al. 2009). In contrast, RARα$^{-/-}$/RARβ$^{-/-}$, RARα$^{-/-}$/RARγ$^{-/-}$, and RARβ$^{-/-}$/RARγ$^{-/-}$ double mutant mice die during embryonic development exhibiting severe malformations typical of VAD (Mark et al. 2009). RXRα$^{-/-}$ mice present several severe abnormalities and die during development, with most of the abnormalities equally found in the different RAR$^{-/-}$ compound mutants (Mark et al. 2009). In contrast, RXRβ$^{-/-}$/RXRγ$^{-/-}$ double mutant mice do not show any evident morphogenetic defects, indicating that RXRα is the most important RXR implicated in morphogenetic patterning during development (Krezel et al. 1996). RAR$^{-/-}$ and RXRα$^{-/-}$ mutations seem to act in synergy, since null mutations for individual RAR paralogs combined with null mutations of RXRα reveal potentially redundant functions of the different RAR paralogs. These different compound mutants thus reveal the fundamental functions of the RXRα/RARα, RXRα/RARβ, and RXRα/RARγ heterodimers in transducing the RA signal during embryogenesis (Samarut and Rochette-Egly 2012).

1.4 NON-CLASSICAL RETINOID SIGNAL TRANSDUCTION

In addition to the classical mode of retinoid action, mediated by RXR/RAR heterodimers, several alternative mechanisms for triggering retinoid-dependent signaling cascades in vertebrates have recently been proposed. For example, RA has been shown to specifically activate the orphan receptor peroxisome proliferator-activated receptor β/δ (PPARβ/δ), but neither one of its paralogs, PPARα and PPARγ (Shaw et al. 2003). Interestingly, CRABP does not play a role in the delivery of RA to this receptor, as this function is carried out by FABP5 (Schug et al. 2007, Tan et al. 2002). Despite the evident existence of this alternative pathway, the affinity of RA for association with FABP5/PPARβ/δ is much lower than its affinity for CRABP/RAR (Schug et al. 2007). Importantly, the ratio of the proteins involved in intracellular RA delivery (FABP5 and CRABP) has been shown to be a crucial component for triggering alternative versus classical RA responses (Schug et al. 2007). Given that the activation of the FABP5/PPARβ/δ pathway leads to a very different biological output both in terms of cell survival (Schug et al. 2007) and differentiation (Berry and Noy 2009), this alternative pathway thus significantly increases the complexity of the RA signaling response.

A number of other nuclear receptors have also been shown to bind and be activated by retinoids. For example, testicular receptor 4 (TR4) has recently been suggested to be a retinoid receptor, because the binding of both retinol and RA can induce conformational changes of TR4 leading to the activation of this receptor (Zhou et al. 2011). TR4 can either function as homodimer or heterodimerize with testicular receptor 2 (TR2) to function in spermatogenesis, lipid and lipoprotein regulation, and central nervous system development (Zhou et al. 2011). However, additional experiments are required to reveal the physiological and developmental implications of the activation of TR4 by retinoids, such as RA (Zhou et al. 2011).

Like TR4, COUP-TFII (chicken ovalbumin upstream promoter-transcription factor II) is capable of recruiting co-activators and thus of activating reporter construct expression in the presence of retinoids (Kruse et al. 2008). Both all-*trans* RA and 9-*cis* RA can associate with, and activate, COUP-TFII. However, the biological relevance of this interaction is questionable because the retinoid concentration needed to trigger a response might be significantly higher than endogenous retinoid levels (Kruse et al. 2008). Future work will thus have to assess the potential role of retinoids in mediating COUP-TFII-specific functions, for example, in neuronal development, cell fate determination, and circadian rhythm (Kruse et al. 2008).

RA is also able to bind to RA receptor-related orphan receptor β (RORβ) *in vitro*. This receptor functions in the regulation of the circadian rhythm, and its activity has been shown to be regulated by RA *in vivo* (Stehlin-Gaon et al. 2003). Given that RAR has also been shown to inhibit a fundamental protein of the circadian clock, CLOCK/BMAL1, at least *in vitro* (McNamara et al. 2001, Shirai et al. 2006), the modification of RORβ functions by RA might represent an alternative pathway for integrating the circadian clock. Further studies are evidently required to understand how RORβ acts in association with RA.

Members of completely different protein families have also been suggested to convey retinoid signals. For instance, the association of retinol-bound RBP with STRA6 has been shown to trigger a signaling cascade mediated by the Janus kinase JAK2 and its associated transcription factors, the signal transducers and activators of transcription (STATs) (Berry and Noy 2012). Activation of the JAK/STAT cascade by retinol-bound RBP also leads to an increase of expression of STAT target genes, such as inhibitors of insulin signaling and regulators of lipid homeostasis (Berry and Noy 2012).

Moreover, RA has been shown to modulate the activity of protein kinase C α (PKCα). Thus, all-*trans* RA and acidic phospholipids compete for binding to PKCα, and this competitive binding is responsible for controlling PKCα activity (Ochoa et al. 2003). RA can also activate cAMP response element-binding protein (CREB) in cell lines, where RXR/RAR heterodimers have been silenced (Aggarwal et al. 2006). RA rapidly activates CREB, which leads to a concomitant increase of DNA binding by activated CREB (Aggarwal et al. 2006). RA is also responsible for an increase in the activity of both extracellular-signal-regulated kinases 1/2 (ERK1/2) and ribosomal s6 kinase (RSK), an important kinase acting upstream of CREB (Aggarwal et al. 2006). It has been suggested that PKC is involved in this RA-dependent activation (Aggarwal et al. 2006). Another study has shown that RA activates ERK1/2 through induction of mitogen-activated protein kinase kinase 1/2 (MEK1/2), and that this specific activation leads to caspase-3-dependent apoptosis (Zanotto-Filho et al. 2008).

The PI3K/Akt (phosphatidylinositol 3-kinase/v-akt murine thymoma viral oncogene homolog) signaling pathway has also been shown to be activated by RA in a transient process essential for neural differentiation (López-Carballo et al. 2002). Interestingly, an atypical orphan nuclear receptor, DAX1 (dosage-sensitive sex reversal-adrenal hypoplasia critical region on the chromosome X protein 1), has been implicated in this process (Nagl et al. 2009): acting through PI3K/Akt signaling, RA increases DAX1 expression, which leads to the modification of nitric oxide signaling levels, hence modulating neural differentiation (Nagl et al. 2009).

In addition to the aforementioned classical mechanisms of RA signaling, which involve binding of RXR/RAR heterodimers to DNA, RXR/RAR-dependent RA activity may also occur independently of DNA binding, a feature referred to as non-genomic signaling. One example of a non-genomic signaling response to RA is the RAR-dependent activation of p38MAPK (p38 mitogen-activated protein kinase) and of the downstream mitogen and stress-activated kinase 1 (MSK1) (Duong and Rochette-Egly 2011, Piskunov and Rochette-Egly 2011). This non-genomic signaling mechanism that ultimately leads to the activation of cdk7 (cyclin-dependent kinase 7) is mediated by a fraction of cellular RAR anchored in lipid rafts at the cell membrane, where RAR forms a complex with G protein alpha q (Gαq) (Piskunov and Rochette-Egly 2011). A fraction of RAR proteins thus functions in an RA-dependent, non-genomic manner to integrate processes occurring at the cell membrane. These roles played by RA and RAR might serve as fine-tune mechanisms for the rapid integration of extracellular cues. Future studies will certainly be needed to address the biological roles of these non-genomic signaling cascades during development.

1.5 RETINOID FUNCTIONS DURING DEVELOPMENT

During vertebrate development, RA regulates a significant number of pleiotropic effects, most of which, however, are not exclusively dependent on RA (Gutierrez-Mazariegos et al. 2011). These processes also rely on signals from other intercellular signaling pathways, such as the fibroblast growth factor (FGF) (Yasuda et al. 1992), sonic hedgehog (SHH) (Ribes et al. 2009), WNT, NODAL (Engberg et al. 2010), and bone morphogenetic protein (BMP) (Sheng et al. 2010) cascades. This combined action of RA and other signaling pathways can have very different architectures. Interactions can be synergistic or antagonistic (Gutierrez-Mazariegos et al. 2011), they may act directly or indirectly on a given target gene (Gutierrez-Mazariegos et al. 2011), they may also regulate upstream or downstream effectors of a particular cascade (Gutierrez-Mazariegos et al. 2011), or be involved in one or several regulatory loops (Gutierrez-Mazariegos et al. 2011). In the following section, we will discuss various tissue-specific roles of RA signaling during development (Figure 1.3) and highlight examples of some of the interactions of the RA pathway with other signaling cascades.

1.5.1 CENTRAL NERVOUS SYSTEM PATTERNING DRIVEN BY RA

There is an extensive amount of experimental evidence describing roles of RA signaling in the developing central nervous system (CNS). For example, it has been

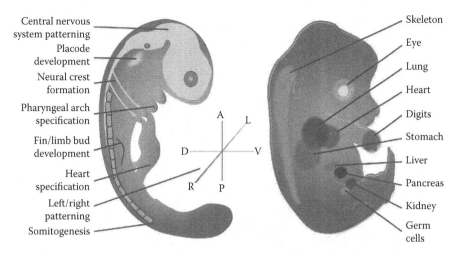

FIGURE 1.3 Summary of retinoic acid (RA) signaling functions during development. The schematic drawings of two vertebrate embryos chiefly correspond to early somitogenesis (left) and organogenesis (right) stages. RA signaling is involved in a multitude of developmental processes (Duester 2008, Kam et al. 2012, Rhinn and Dollé 2012) and the illustration highlights the roles of this signaling cascade detailed in the text.

known for a long time that treatment of vertebrate embryos with exogenous RA results in the loss of forebrain/midbrain territories with a significant expansion of the hindbrain (Avantaggiato et al. 1996, Pourquié 2003, White et al. 2000, Zhang et al. 1996). Thus, the precise regulation of RA function during development is fundamental for the proper establishment of anteroposterior (AP) polarity in the vertebrate CNS (Avantaggiato et al. 1996, Durston et al. 1989, White et al. 2000).

The activity of RA in the hindbrain is regulated by dynamic expression of RAR and RXR (i.e., of the RA receptors), of RALDH2 (i.e., the RA source), and of the CYP26s (i.e., the RA sinks) (Hernandez et al. 2007, Sirbu et al. 2005, White and Schilling 2008). In particular, the expression of RAR paralogs and the activity of different CYP26 enzymes regulate RA signaling in the hindbrain leading to rhombomere-specific activation of *hox* genes (Rijli et al. 1998, White and Schilling 2008).

The AP regionalization of the hindbrain is thus dependent on RA signaling, and this function of RA is directly mediated by *hox* genes (Glover et al. 2006, Rijli et al. 1998): within the hindbrain, the direct regulation by RA of the overlapping expression of different *hox* genes, the so-called *hox* code, ensures the proper patterning of this CNS region (Marshall et al. 1996, Rijli et al. 1998). Interestingly, in addition to its role in transcriptional activation of *hox* genes, RAR and RXR may also promote *hox* gene expression through interactions with chromatin-remodeling enzymes (del Corral and Storey 2004).

The anterior limit of *hox* expression in the CNS is at least in part defined by diffusible signals emanating from the midbrain-hindbrain boundary region, such as FGF8, working antagonistically to the RA signal (Irving and Mason 2000). RA/FGF antagonism seems to be a recurrent scheme for the regulation of *hox* genes, given that such a crosstalk has also been described for the regulation of *hox* genes along the AP axis of the vertebrate CNS (White and Schilling 2008). RA thus preferentially activates anteriorly expressed *hox* genes, while FGF induces progressively more posteriorly expressed *hox* genes (Irving and Mason 2000, Liu et al. 2001).

In addition to AP regionalization of the CNS, RA signaling also controls neuronal differentiation and specification and is also involved in the patterning of the neural tube along the dorsoventral (DV) axis (Maden 2002). In the spinal cord, RA interacts with SHH to establish ventral determinants that later will give rise to motor neurons. This specification consists of the induction of genes, such as *pax6*, *nkx6.1,* or *olig2*, that are responsible for mediating DV patterning of the spinal cord (Novitch et al. 2003). When RA signaling is not present, the ventral neurons are not induced to differentiate from the spinal cord neuroectodermal progenitor cells (Molotkova et al. 2005, Novitch et al. 2003). Finally, neurite outgrowth during development is also regulated by RA, since lack of RA signaling results in abnormal axonal projections (Corcoran et al. 2000, White et al. 2000).

1.5.2   RA Signaling in Neural Crest Cells (NCCs) and Placode Development

NCCs are a transient population of cells in vertebrates that originate at the dorsal neural tube from where they delaminate and subsequently migrate through the embryo (Krispin et al. 2010, Martínez-Morales et al. 2011). NCCs ultimately differentiate into a wide variety of cell types, such as sensory neurons, Schwann cells, myoblasts, chondrocytes, osteocytes, or melanocytes, among others (Hall 2009). Migrating NCCs are already specified to a particular cell lineage according to their place of origin along the AP axis of the neural tube and hence actively participate in the regional patterning of the embryo (Minoux and Rijli 2010, Trainor and Krumlauf 2001). It has been shown that exogenous RA triggers abnormal migration of NCCs from rhombomere 4 into the first pharyngeal arch, as opposed to their normal migration to the second pharyngeal arch (Minoux and Rijli 2010, Plant et al. 2000). RA signaling has also been implicated in controlling differentiation and survival of NCCs at different developmental stages (Begemann et al. 2001, Martínez-Morales et al. 2011). One of the most important processes NCCs need to undergo to become migratory is an epithelial-mesenchymal transition, a process that is regulated by interaction of RA and FGF signaling (Martínez-Morales et al. 2011). Indeed, alterations of RA or FGF signaling activity disrupt the timing of epithelial-mesenchymal transitions of NCCs along the AP axis of the CNS, which ultimately leads to changes in the cell fates of the NCCs (Krispin et al. 2010).

Placodes are discrete, specialized ectodermal areas located at stereotyped positions in the embryonic head (Schlosser 2010). They give rise to a wide range of cell types, including sensory neurons that contribute to cranial ganglia (Schlosser 2010). RA signaling has been implicated in the development of the otic (Dupé et al. 1999), lens (Matt et al. 2005), olfactory (Song et al. 2004), and lateral line placodes (Gibbs and Northcutt 2004), but the precise functions of RA in the different placodes remain elusive. Eye morphogenesis and retina differentiation are also controlled by

RA, although the RA-dependent signal seems to act primarily on the NCC-derived periocular mesenchyme (Matt et al. 2005, McCaffery et al. 1999).

1.5.3   RA SIGNALING AND LEFT/RIGHT (LR) AXIAL PATTERNING

In addition to patterning along the AP and DV body axes, RA signaling is also implicated in the early determination of LR asymmetry during vertebrate development. Indeed, RA seems to be involved in the side-specific activation of genes involved in LR specification, such as *lefty*, *pitx*, and *nodal* (Wasiak and Lohnes 1999). It has initially been suggested that the RA-dependent establishment of LR asymmetry is controlled by vesicles loaded with RA and SHH that are delivered to the future left side of the embryo by rotating cilia in an FGF-dependent process (Tanaka et al. 2005). This hypothesis was subsequently challenged by the finding that an antagonism between FGF8 and RA in the ectoderm is required for generating symmetry. Failure of this antagonism to be established generates excessive FGF8 signaling in adjacent mesoderm that leads to reduction of somite size and LR asymmetry (Sirbu and Duester 2006). In addition, it has been shown that RA action during somitogenesis is associated with the maintenance of symmetric somite formation along the LR axis (Vermot and Pourquié 2005). RA thus works as a buffer to stabilize and balance the LR signaling cues in the presomitic mesoderm (PSM) (Vermot and Pourquié 2005).

1.5.4   RA SIGNALING IN SOMITOGENESIS

Somites are metameric structures of mesodermal origin that give rise to vertebrae and their associated muscles and tendons, the skeletal muscles of the body wall and limbs, and to the dermis of the back (Brent and Tabin 2002, Dubrulle and Pourquié 2004). The biological process of forming new somites appears to be highly conserved during vertebrate embryogenesis and is controlled by rhythmic gene transcription within the PSM, involving mainly NOTCH and WNT signaling components that establish the so-called segmentation clock (Pourquié 2003). In addition to its role in the maintenance of LR asymmetry, RA produced by RALDH2 in anterior somites is involved in repressing FGF8 signals emanating from the posterior ectoderm (Duester 2007). In this way, reciprocal inhibition of FGF8 signals by RA and of RA signals by FGF8 is crucial for positioning segment boundaries in the PSM (Moreno and Kintner 2004). This interaction is responsible for the translation of pulsations of the segmentation clock into the periodic establishment of segment boundaries, with RA canceling the effects of FGF8 and hence triggering the expression of segmentation genes responsible for somite formation (Duester 2007, Moreno and Kintner 2004). Furthermore, through the control of *hox* gene expression, the RA/FGF antagonism also has a very important function in defining AP positional information in structures derived from the PSM (del Corral and Storey 2004).

1.5.5   RA IN HEART FIELD SPECIFICATION

RA signaling also plays important roles in patterning the developing heart, in particular during cardiac field specification, during AP regionalization of the heart tube, and in the establishment of LR asymmetry during heart looping (Chazaud et al. 1999, Niederreither et al. 1999, Rosenthal and Xavier-Neto 2000, Xavier-Neto et al. 2001). Interestingly, at least the role of RA in AP patterning of the heart is well conserved within vertebrates, given that exogenous RA leads to similar heart regionalization defects in mice (Lin et al. 2010), chicken (Hochgreb et al. 2003), and zebrafish (Stainier and Fishman 1992). In vertebrates, RALDH2 is expressed early in the heart mesoderm in an AP wave, hence defining cardiac precursors and delimitating the heart field along the AP axis (Hochgreb et al. 2003). RA signaling exerts its effect on AP regionalization of the heart, at least in part, by downregulating *isl1* expression, probably by reducing FGF signaling levels (Sirbu et al. 2008).

1.5.6 RA Signaling and Kidney Development

The kidney is another target tissue of RA signaling during development. The kidney field is derived from intermediate mesoderm, and recent studies have shown that RA signaling is required for early specification of this kidney field (Cartry et al. 2006, Serluca and Fishman 2001, Wingert et al. 2007). In the frog *Xenopus laevis*, for example, early determinants of a pronephric fate, such as *pax8* and *lhx1*, are controlled by RA signaling (Cartry et al. 2006). In zebrafish, RA is required for the positioning and segmentation of the pronephric kidney (Wingert et al. 2007). Moreover, the frog *pteg* and zebrafish *wt1* genes, both of which play important roles during kidney development in the respective animal, are direct targets of RA signaling (Bollig et al. 2009, Lee et al. 2010). In mice, it has further been shown that RA signaling, by controlling *ret* expression in ureteric bud cells, is crucial for ureteric bud formation and for branching morphogenesis within the developing collecting duct system (Rosselot et al. 2010). Interestingly, RA signaling activity in ureteric bud cells depends mainly on RA generated by RALDH2 in adjacent stromal mesenchyme (Rosselot et al. 2010).

1.5.7 Relevance of RA Signaling for Body Appendage Development

Based on pharmacological experiments that had shown, for example, that exogenous RA induces mirror-image duplications in chick wing buds (Tickle et al. 1982), it had initially been suggested that RA signaling plays crucial roles in limb bud patterning of tetrapod vertebrates (Campo-Paysaa et al. 2008). However, more recent work suggests that RA signaling might be less important for tetrapod limb development than initially expected. Albeit required during forelimb induction, RA might not be necessary at all for hind limb budding and patterning (Zhao et al. 2009). Moreover, the involvement of RA in forelimb formation seems to be indirect, by creating a permissive environment for FGF signaling to develop a limb bud (Zhao et al. 2009). Conversely, in zebrafish, RA is indispensable for pectoral fin induction (Gibert et al. 2006, Grandel and Brand 2011), which is suggestive of a divergence in RA signaling functions between fish fin buds and tetrapod limb buds (Campo-Paysaa et al. 2008). Later in development, RA signaling likely controls tissue remodeling and apoptosis at the digit-interdigit junction, with RALDH2 being expressed in interdigital mesenchyme and CYP26B1 in digits, hence restricting RA action to the interdigital zones (Zhao et al. 2010).

1.5.8 ENDODERM SPECIFICATION AND RA

RA signaling is also essential for patterning the developing endoderm. RA is required for the determination of endodermal fields along the AP axis, each of which will give rise to a particular organ (Bayha et al. 2009). In particular, exposure to exogenous RA prevents the expression of genes normally present in the anterior-most endoderm and, at the same time, activates and shifts anteriorly the expression of genes normally expressed at more posterior endoderm levels (Bayha et al. 2009).

In the anterior endoderm, RA signaling is required for the regional patterning of the pharyngeal arches. The pharyngeal arches are structures that develop along the embryonic foregut endoderm in a very complex process that involves the interaction between cells from all germ layers (ectoderm, mesoderm, endoderm, and the neuroectoderm-derived NCCs). RA is produced locally in the pharyngeal arches by RALDH2 in the mesoderm adjacent to the endoderm (Bayha et al. 2009, Niederreither et al. 2003). When RA signaling is disrupted, the first and second pharyngeal arches develop normally, while severe defects are observed in structures derived from more posterior pharyngeal arches (Niederreither et al. 2003). Interestingly, RA seems dispensable for the specification of pharyngeal endoderm, but required for morphogenesis and segmentation of the posterior pharyngeal arches (Kopinke et al. 2006).

Some organs derived from endoderm in the posterior foregut also require RA for proper development (Wang et al. 2006). These organs include the lungs, stomach, liver, and pancreas (Wang et al. 2006). In the lung, for example, loss of RA signaling causes a disruption of lung bud outgrowth (Wang et al. 2006). This defect is due to a reduction of FGF10 levels, which, together with BMP4, is required for mediating proper branching morphogenesis (Wang et al. 2006). Activation of FGF signaling by RA is also required for early stomach and liver development (Wang et al. 2006), while, during pancreas development, mesodermal RALDH2 is required for specification of the dorsal, but not the ventral, pancreatic bud (Molotkov et al. 2005, Stafford et al. 2004).

1.6 EVOLUTION OF RA SIGNALING

Retinoid signaling was long thought to be vertebrate specific, but recent studies have revealed major roles for RA signaling that are conserved in all chordates (Figure 1.4), which, in addition to the vertebrates, include two invertebrate groups: the cephalochordates and the urochordates (also called tunicates) (Campo-Paysaa et al. 2008, Cañestro et al. 2006, Marletaz et al. 2006). While the tunicates constitute the sister group of vertebrates, the cephalochordates are located at the base of the chordate phylum (Schubert et al. 2006). Nonetheless, of all invertebrates, the cephalochordates, such as the lancelet or amphioxus (for example *Branchiostoma floridae,* the Florida amphioxus), are characterized by the most vertebrate-like retinoid signaling system, both in terms of molecular composition and biological functions (Campo-Paysaa et al. 2008, Koop and Holland 2008, Schubert et al. 2006, Theodosiou et al. 2010).

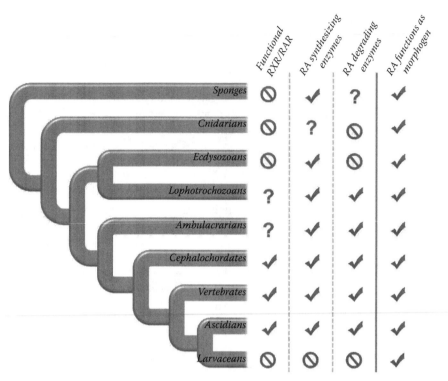

FIGURE 1.4 Evolution of RA signaling components and functions. A simplified phyloge-netic tree of metazoan animals is shown and the presence (✓) or absence (⊘) of functional RXR/RAR heterodimers, of RA synthesizing (RALDH), and of RA degrading (CYP26) enzymes is highlighted. Uncertainty over presence or absence is illustrated by (?). The phy-logenetic tree also indicates the distribution of intercellular signaling functions of RA (as morphogen) in the different lineages. RA, retinoic acid; RAR, retinoic acid receptor; RXR, retinoid X receptor.

1.6.1 CHORDATE RA SIGNALING

When considering the key players of RA metabolism, some conserved and divergent features are observable when comparing vertebrates with invertebrate chordates (Figure 1.4). For example, while vertebrate genomes encode multiple RAR and RXR paralogs, amphioxus has only one RAR and one RXR, forming a single heterodimer (Escriva et al. 2002). Like its vertebrate counterparts, this amphioxus RXR/RAR heterodimer binds to and is activated by RA (Escriva et al. 2002). Using ligands specific for each of the three mammalian RAR paralogs (RARα, RARβ, and RARγ), it was shown that the ligand-binding capacities of amphioxus RAR resemble those of mammalian RARβ, suggesting that the ancestral RAR of vertebrates was of the RARβ type (Escriva et al. 2006). Comparison of developmental gene expression patterns between amphioxus RAR and the three vertebrate RAR paralogs further supported this notion and sug-gested that, while RARβ kept the ancestral features in terms of ligand specificity and development expression, after duplication, RARα and RARγ might have acquired novel functions in the course of vertebrate diversification (Escriva et al. 2006).

Although the amphioxus genome encodes orthologs of the enzymes required in vertebrates for RA synthesis (RALDH) and degradation (CYP26), it remains to be determined whether these cephalochordate RALDH and CYP26 are indeed capable of, respectively, RA synthesis and degradation (Albalat et al. 2011, Sobreira et al. 2011). Regarding the endogenous production of retinaldehyde from retinol, a bona fide ADH has previously been characterized biochemically in amphioxus (Cañestro et al. 2000, Cañestro et al. 2003, Dalfó et al. 2007), but the functional capacities of the RDH orthologs that have been identified in cephalochordate genomes still need to be assessed (Albalat et al. 2011, Holland et al. 2008). Interestingly, although a candidate for an intracellular retinol-binding protein (CRBP) has previously been described in amphioxus, there is no convincing evidence for the existence of an amphioxus intracellular RA-binding protein (CRABP) (Albalat et al. 2011, Holland et al. 2008, Jackman et al. 2004), and the concerted absence in amphioxus of the main components for retinol storage, transport, and cellular uptake further suggest that these systems might be functional innovations of vertebrates (Albalat et al. 2011).

Several studies have assessed the roles of RA signaling during amphioxus development (Campo-Paysaa et al. 2008, Escriva et al. 2002, Holland and Holland 1996, Koop and Holland 2008). For example, in the amphioxus CNS, it was shown that RA regulates both AP regional patterning and neuronal specification in a *hox*-dependent manner (Schubert et al. 2006). Likewise, the amphioxus ectoderm is also patterned by RA: disruption of endogenous RA signaling levels lead to alteration of an ectodermal *hox* code, which in turn results in AP regionalization defects and changes in the combinatorial code defining specific populations of ectodermal sensory neurons (Schubert et al. 2004). Recently, it has also been shown that, at later developmental stages, RA signaling is involved in the formation of the amphioxus tail fin, which arises from the posterior ectoderm (Koop et al. 2011).

In the endoderm, RA signaling via *hox1* has been shown to be required for the definition of the posterior limit of the amphioxus pharynx by limiting expression of pro-pharyngeal genes to the anteriormost (thus pharyngeal) endoderm (Schubert et al. 2005). Moreover, in the posterior endoderm, RA signaling controls the expression of two *parahox* genes, *xlox* and *cdx*, hence mediating AP patterning of the developing amphioxus hindgut (Osborne et al. 2009). While it is very likely that RA signaling directly activates expression of amphioxus *hox* genes (Koop et al. 2010), further work is needed to assess in more detail the regulation of amphioxus *parahox* genes by RA (Osborne et al. 2009).

Contrasting the situation in amphioxus, RA signaling in tunicates has secondarily been modified in different lineages (Campo-Paysaa et al. 2008, Cañestro et al. 2006, Koop and Holland 2008). For example, although the genome of the ascidian tunicate *Ciona intestinalis* contains the basic molecular components of the RA signaling cascade, that is a single receptor heterodimer (RXR/RAR) and enzymes potentially capable of RA synthesis (RALDH) and degradation (CYP26), the regulation of at least some of these components in this tunicate has been modified (Campo-Paysaa et al. 2008, Cañestro et al. 2007, Koop and Holland 2008). Transcription of RAR in this ascidian tunicate is not regulated by RA, contrasting with amphioxus and vertebrates, where RAR expression is directly regulated by RA (Nagatomo et al. 2003). Moreover, the larvacean tunicate *Oikopleura dioica* has lost most of the molecular components defining the canonical RA pathway, including RAR, RALDH, and CYP26 (Campo-Paysaa et al. 2008, Cañestro and Postlethwait 2007).

Along these lines, the roles for RA signaling during tunicate development seem to be limited. For example, although RA signaling might function during development of the ascidian tunicate CNS (De Bernardi et al. 1994, Nagatomo et al. 2003), RA probably only has a limited role in the regulation of *hox* gene expression (Nagatomo and Fujiwara 2003, Natale et al. 2011, Wada et al. 2006). Moreover, given that tunicates in general have lost several *hox* genes and that both cluster organization and collinear expression have at least partially been lost (Ikuta et al. 2004, Seo et al. 2004), the contribution of an RA-regulated *hox* code to tunicate CNS development is very limited (Ikuta et al. 2010).

Interestingly, exogenous RA has been shown to activate and shift anteriorly the expression of ascidian *hox1* in both the CNS and ectoderm (Kanda et al. 2009, Katsuyama et al. 1995, Katsuyama and Saiga 1998, Nagatomo and Fujiwara 2003) and, at least in the ectoderm, this activation is directly mediated by the ascidian RXR/RAR heterodimer, which represents a very rare example for RA-dependent gene expression in ascidians (Kanda et al. 2009). This RA-*hox1* network in the ectoderm is required for formation of the ascidian atrial siphon placode, which is considered homologous to the vertebrate otic placode (Sasakura et al. 2012). The *hox1* gene is also expressed in the ascidian endoderm (Ikuta et al. 2004), where exogenous RA leads to a loss of expression of pharyngeal markers and to a posteriorization of the endoderm leading to a complete loss of the branchial basket (Hinman and Degnan 1998, 2000; Ogasawara et al. 1999). Intriguingly, these effects of RA are reminiscent of the phenotypes observed in amphioxus and vertebrates following treatment with exogenous RA (Campo-Paysaa et al. 2008).

Taken together, at the base of the chordate lineage the basic molecular components constituting the RA signaling cascade were most likely already present. Moreover, the RA pathway was probably involved in the specification of neuronal cell populations in both CNS and ectoderm, and RA signals also contributed to the regional patterning along the AP axis of the CNS, ectoderm, and endoderm. In all three tissue layers, this RA-dependent AP regionalization process was probably directly mediated by *hox* genes (Campo-Paysaa et al. 2008, Marletaz et al. 2006). In the tunicate lineage, the functions of the RA signaling cascade have been secondarily modified leading to diverged functions, such as the partial loss of RA-dependent regulation of the *hox* code in ascidians, and culminating in the loss of key components of the RA pathway, including RAR, RALDH, and CYP26 in larvaceans (Cañestro et al. 2007).

1.6.2 Non-Chordate RA Signaling

Outside the chordate lineage, evidence for functional roles of retinoids is scarce (Figure 1.4). Although genes encoding orthologs of vertebrate RAR and RXR as well as orthologs of the basic vertebrate components for endogenous RA synthesis (RALDH) and degradation (CYP26) are present in ambulacrarians (such as hemichordates and echinoderms) and lophotrochozoans (such as annelids and mollusks) (Albalat and Cañestro 2009, Campo-Paysaa et al. 2008, Theodosiou et al. 2010), extensive experimental evidence for retinoid and/or RXR/RAR heterodimer functions in these lineages is unfortunately still lacking. It has been suggested that RA treatments result in a delay of embryonic development in echinoderms, and, in mollusks, retinoids (in particular 9-*cis* RA) disrupt embryogenesis and are also involved in neuronal differentiation, neuron outgrowth, and growth cone guidance (Farrar et al. 2009). This role of retinoids in mollusk neurogenesis is apparently independent of a functional RXR/RAR heterodimer, but instead requires a cytoplasmic localization of RXR (Carter et al. 2010). Also, in mollusks, exogenous RA affects eye formation and leads to arrested development at the trochophore larval stage (Créton et al. 1993).

Intriguingly, the available *in silico* data suggest that ecdysozoans (which include insects, crustaceans, and nematodes) might have secondarily lost both the canonical retinoid receptor RAR and a CYP26-based RA degrading machinery (Albalat and Cañestro 2009, Campo-Paysaa et al. 2008, Theodosiou et al. 2010), although retinoid functions have been described in tissue regeneration of both insects and crustaceans (Halme et al. 2010, Hopkins 2001). Other functions of retinoids in animals, whose genomes do, *a priori*, not encode the RAR receptor include, for example, hydroid specification, cell proliferation, and neuronal differentiation in cnidarians (Estephane and Anctil 2010, Müller 1984), and tissue regression and spicule formation in sponges (Müller et al. 2011, Wiens et al. 2003). Collectively, these data suggest that in the course of metazoan evolution the functions for retinoids might have originated before a functional RXR/RAR heterodimer and that the presence of both RAR and RXR in a genome does not automatically imply a functional RXR/RAR heterodimer capable of binding and being activated by retinoids.

1.7 CONCLUSION

Considerable time and effort have been invested by the scientific community in the analysis of metabolism and functions of RA during development. It is clear that the signaling cascade controlled by this morphogen intervenes in various tissues at different time points during embryogenesis. Because of this multitude of biological functions, we are still lacking a comprehensive understanding of the developmental roles played by the RA pathway. Moreover, we are only starting to understand the complex regulation during development of the RA signaling activity, which depends not only on a very fine regulation of endogenous RA availability, but also on the complex expression of receptors and cofactors, all of which are controlled by intricate interactions of RA signaling with other signaling cascades creating regulatory loops of various sorts.

Further work is thus needed to unite the various facets of the RA signaling network from metabolism to signal transduction and from interactions with other signaling cascades to developmental functions into a coherent ensemble representing the regulatory network as a whole. In this context, model organisms with representative, vertebrate-like RA signaling cascades that lack genetic redundancy and secondary modifications, such as the cephalochordate amphioxus, will be an important asset for providing the blueprint of an ancestral RA network that was elaborated in vertebrates following the genomic expansion by two rounds of whole genome duplications (Van de Peer et al. 2009).

The usefulness of expanding the scope of RA-centered research to non-vertebrate models is not only limited to instructing the complexity of vertebrate systems. As a matter of fact, data from various organisms might have the potential to reveal the evolutionary origins of the RA signaling network and the ancestral functions that retinoids might have assumed during development of the first metazoan animals. The results that have already been obtained in several invertebrate taxa suggest that retinoids might not necessarily require the canonical retinoid receptors RAR and RXR to elicit a biological function and that the presence in a genome of RAR and RXR does thus not automatically imply that the heterodimer formed by these receptors is reactive to retinoids. In this context, it is interesting to note that retinoids have very ancient roles in light reception and in the visual system, which represents a pool of retinoids that, in the course of evolution, might have been co-opted for roles in morphogen-dependent signaling and that has ultimately been elaborated into one of the most important intercellular signaling cascades in development.

ACKNOWLEDGEMENTS

The authors are indebted to Ricardo Lara-Ramírez, Florent Campo-Paysaa, Juliana Gutierrez-Mazariegos, and Eric Samarut for critical reading of the manuscript, and to Vincent Laudet for having hosted our group in his laboratory. This work was supported by funds from ANR (ANR-09-BLAN-0262-02 and ANR-11-JSV2-002-01) and CNRS to Michael Schubert.

REFERENCES

Abu-Abed, S., P. Dollé, D. Metzger, et al. 2001. The Retinoic Acid-Metabolizing Enzyme, Cyp26a1, Is Essential for Normal Hindbrain Patterning, Vertebral Identity, and Development of Posterior Structures. *Genes Dev* 15, no 2: 226–40.

Aggarwal, S., S. W. Kim, K. Cheon, et al. 2006. Nonclassical Action of Retinoic Acid on the Activation of the Camp Response Element-Binding Protein in Normal Human Bronchial Epithelial Cells. *Mol Biol Cell* 17, no 2: 566–75.

Albalat, R., F. Brunet, V. Laudet, and M. Schubert. 2011. Evolution of Retinoid and Steroid Signaling: Vertebrate Diversification from an Amphioxus Perspective. *Genome Biol Evol* 3: 985–1005.

Albalat, R., and C. Cañestro. 2009. Identification of Aldh1a, Cyp26 and Rar Orthologs in Protostomes Pushes Back the Retinoic Acid Genetic Machinery in Evolutionary Time to the Bilaterian Ancestor. *Chem-Biol Interact* 178, no 1-3: 188–96.

Avantaggiato, V., D. Acampora, F. Tuorto, and A. Simeone. 1996. Retinoic Acid Induces Stage-Specific Repatterning of the Rostral Central Nervous System. *Dev Biol* 175, no 2: 347–57.

Balmer, J. E., and R. Blomhoff. 2005. A Robust Characterization of Retinoic Acid Response Elements Based on a Comparison of Sites in Three Species. *J Steroid Biochem* 96, no 5: 347–54.

Barron, M., D. McAllister, S. M. Smith, and J. Lough. 1998. Expression of Retinol Binding Protein and Transthyretin During Early Embryogenesis. *Dev Dynam* 212, no 3: 413–22.

Bayha, E., M. C. Jørgensen, P. Serup, and A. Grapin-Botton. 2009. Retinoic Acid Signaling Organizes Endodermal Organ Specification Along the Entire Antero-Posterior Axis. *Plos One* 4, no 6: e5845.

Begemann, G., T. F. Schilling, G. J. Rauch, R. Geisler, and P. W. Ingham. 2001. The Zebrafish Neckless Mutation Reveals a Requirement for Raldh2 in Mesodermal Signals That Pattern the Hindbrain. *Development* 128, no 16: 3081–94.

Bempong, D. K., I. L. Honigberg, and N. M. Meltzer. 1995. Normal-Phase Lc-Ms Determination of Retinoic Acid Degradation Products. *J Pharmaceut Biomed* 13, no 3: 285–91.

Berry, D. C., and N. Noy. 2009. All-*trans*-Retinoic Acid Represses Obesity and Insulin Resistance by Activating Both Peroxisome Proliferation-Activated Receptor β/δ and Retinoic Acid Receptor. *Mol Cell Biol* 29, no 12: 3286–96.

Berry, D. C., and N. Noy. 2012. Signaling by Vitamin A and Retinol-Binding Protein in Regulation of Insulin Responses and Lipid Homeostasis. *Biochim Biophys Acta* 1821, no 1: 168–76.

Blomhoff, R., and H. K. Blomhoff. 2006. Overview of Retinoid Metabolism and Function. *J Neurobiol* 66, no 7: 606–30.

Bollig, F., B. Perner, B. Besenbeck, et al. 2009. A Highly Conserved Retinoic Acid Responsive Element Controls *Wt1a* Expression in the Zebrafish Pronephros. *Development* 136, no 17: 2883–92.

Brent, A. E., and C. J. Tabin. 2002. Developmental Regulation of Somite Derivatives: Muscle, Cartilage and Tendon. *Curr Opin Genet Dev* 12, no 5: 548–57.

Cai, A. Q., K. Radtke, A. Linville, et al. 2012. Cellular Retinoic Acid-Binding Proteins Are Essential for Hindbrain Patterning and Signal Robustness in Zebrafish. *Development* 139, no 12: 2150–55.

Campo-Paysaa, F., F. Marlétaz, V. Laudet, and M. Schubert. 2008. Retinoic Acid Signaling in Development: Tissue-Specific Functions and Evolutionary Origins. *Genesis* 46, no 11: 640–56.

Cañestro, C., L. Godoy, R. Gonzàlez-Duarte, and R. Albalat. 2003. Comparative Expression Analysis of *Adh3* During Arthropod, Urochordate, Cephalochordate, and Vertebrate Development Challenges Its Predicted Housekeeping Role. *Evol Dev* 5, no 2: 157–62.

Cañestro, C., L. Hjelmqvist, R. Albalat, et al. 2000. Amphioxus Alcohol Dehydrogenase Is a Class 3 Form of Single Type and of Structural Conservation but with Unique Developmental Expression. *Eur J Biochem* 267, no 22: 6511–18.

Cañestro, C., and J. H. Postlethwait. 2007. Development of a Chordate Anterior-Posterior Axis without Classical Retinoic Acid Signaling. *Dev Biol* 305, no 2: 522–38.

Cañestro, C., J. H. Postlethwait, R. Gonzàlez-Duarte, and R. Albalat. 2006. Is Retinoic Acid Genetic Machinery a Chordate Innovation? *Evol Dev* 8, no 5: 394–406.

Cañestro, C., H. Yokoi, and J. H. Postlethwait. 2007. Evolutionary Developmental Biology and Genomics. *Nat Rev Genet* 8, no 12: 932–42.

Carter, C. J., N. Farrar, R. L. Carlone, and G. E. Spencer. 2010. Developmental Expression of a Molluscan Rxr and Evidence for Its Novel, Nongenomic Role in Growth Cone Guidance. *Dev Biol* 343, no 1-2: 124–37.

Cartry, J., M. Nichane, V. Ribes, et al. 2006. Retinoic Acid Signalling Is Required for Specification of Pronephric Cell Fate. *Dev Biol* 299, no 1: 35–51.

Chambers, D., L. Wilson, M. Maden, and A. Lumsden. 2007. Raldh-Independent Generation of Retinoic Acid During Vertebrate Embryogenesis by Cyp1b1. *Development* 134, no 7: 1369–83.

Chambon, P. 1996. A Decade of Molecular Biology of Retinoic Acid Receptors. *Faseb J* 10, no 9: 940–54.

Chazaud, C., P. Chambon, and P. Dollé. 1999. Retinoic Acid Is Required in the Mouse Embryo for Left-Right Asymmetry Determination and Heart Morphogenesis. *Development* 126, no 12: 2589–96.

Chen, Y. L., N. Pollet, C. Niehrs, and T. Pieler. 2001. Increased Xraldh2 Activity Has a Posteriorizing Effect on the Central Nervous System of *Xenopus* Embryos. *Mech Develop* 101, no 1-2: 91–103.

Collins, M. D., and G. E. Mao. 1999. Teratology of Retinoids. *Annu Rev Pharmacol* 39: 399–430.

Corcoran, J., B. Shroot, J. Pizzey, and M. Maden. 2000. The Role of Retinoic Acid Receptors in Neurite Outgrowth from Different Populations of Embryonic Mouse Dorsal Root Ganglia. *J Cell Sci* 113: 2567–74.

Créton, R., G. Zwaan, and R. Dohmen. 1993. Specific Developmental Defects in Mollusks after Treatment with Retinoic Acid During Gastrulation. *Dev Growth Differ* 35, no 3: 357–64.

Dalfó, D., N. Marqués, and R. Albalat. 2007. Analysis of the Nadh-Dependent Retinaldehyde Reductase Activity of Amphioxus Retinol Dehydrogenase Enzymes Enhances Our Understanding of the Evolution of the Retinol Dehydrogenase Family. *Febs J* 274, no 14: 3739–52.

D'Ambrosio, D. N., R. D. Clugston, and W. S. Blaner. 2011. Vitamin A Metabolism: An Update. *Nutrients* 3, no 1: 63–103.

De Bernardi, F., C. Sotgia, and G. Ortolani. 1994. Retinoic Acid Treatment of Ascidian Embryos: Effects on Larvae and Metamorphosis. *Anim Dev* 3: 75–81.

del Corral, R. D., and K. G. Storey. 2004. Opposing Fgf and Retinoid Pathways: A Signalling Switch That Controls Differentiation and Patterning Onset in the Extending Vertebrate Body Axis. *Bioessays* 26, no 8: 857–69.

Dong, D., S. E. Ruuska, D. J. Levinthal, and N. Noy. 1999. Distinct Roles for Cellular Retinoic Acid-Binding Proteins I and Ii in Regulating Signaling by Retinoic Acid. *J Biol Chem* 274, no 34: 23695–98.

Dubrulle, J., and O. Pourquié. 2004. Coupling Segmentation to Axis Formation. *Development* 131, no 23: 5783–93.

Duester, G. 2007. Retinoic Acid Regulation of the Somitogenesis Clock. *Birth Defects Res C Embryo Today* 81, no 2: 84–92.

Duester, G. 2008. Retinoic Acid Synthesis and Signaling During Early Organogenesis. *Cell* 134, no 6: 921–31.

Duester, G., F. A. Mic, and A. Molotkov. 2003. Cytosolic Retinoid Dehydrogenases Govern Ubiquitous Metabolism of Retinol to Retinaldehyde Followed by Tissue-Specific Metabolism to Retinoic Acid. *Chem-Biol Interact* 143: 201–10.

Duong, V., and C. Rochette-Egly. 2011. The Molecular Physiology of Nuclear Retinoic Acid Receptors. From Health to Disease. *Biochem Biophys Acta* 1812, no 8: 1023–31.

Dupé, V., N. B. Ghyselinck, O. Wendling, P. Chambon, and M. Mark. 1999. Key Roles of Retinoic Acid Receptors Alpha and Beta in the Patterning of the Caudal Hindbrain, Pharyngeal Arches and Otocyst in the Mouse. *Development* 126, no 22: 5051–59.

Durston, A. J., J. P. M. Timmermans, W. J. Hage, et al. 1989. Retinoic Acid Causes an Anteroposterior Transformation in the Developing Central Nervous System. *Nature* 340, no 6229: 140–44.

Engberg, N., M. Kahn, D. R. Petersen, M. Hansson, and P. Serup. 2010. Retinoic Acid Synthesis Promotes Development of Neural Progenitors from Mouse Embryonic Stem Cells by Suppressing Endogenous, Wnt-Dependent Nodal Signaling. *Stem Cells* 28, no 9: 1498–509.

Escriva, H., S. Bertrand, P. Germain, et al. 2006. Neofunctionalization in Vertebrates: The Example of Retinoic Acid Receptors. *Plos Genet* 2, no 7: 955–65.

Escriva, H., N. D. Holland, H. Gronemeyer, C. Laudet, and L. Z. Holland. 2002. The Retinoic Acid Signaling Pathway Regulates Anterior/Posterior Patterning in the Nerve Cord and Pharynx of Amphioxus, a Chordate Lacking Neural Crest. *Development* 129, no 12: 2905–16.

Estephane, D., and M. Anctil. 2010. Retinoic Acid and Nitric Oxide Promote Cell Proliferation and Differentially Induce Neuronal Differentiation *in vitro* in the Cnidarian *Renilla koellikeri*. *Dev Neurobiol* 70, no 12: 842–52.

Farjo, K. M., G. Moiseyev, O. Nikolaeva, et al. 2011. Rdh10 Is the Primary Enzyme Responsible for the First Step of Embryonic Vitamin A Metabolism and Retinoic Acid Synthesis. *Dev Biol* 357, no 2: 347–55.

Farrar, N. R., J. M. Dmetrichuk, R. L. Carlone, and G. E. Spencer. 2009. A Novel, Nongenomic Mechanism Underlies Retinoic Acid-Induced Growth Cone Turning. *J Neurosci* 29, no 45: 14136–42.

Feng, L., R. E. Hernandez, J. S. Waxman, D. Yelon, and C. B. Moens. 2010. Dhrs3a Regulates Retinoic Acid Biosynthesis through a Feedback Inhibition Mechanism. *Dev Biol* 338, no 1: 1–14.

Fraser, P. D., and P. M. Bramley. 2004. The Biosynthesis and Nutritional Uses of Carotenoids. *Prog Lipid Res* 43, no 3: 228–65.

Ghyselinck, N. B., C. Båvik, V. Sapin, et al. 1999. Cellular Retinol-Binding Protein I Is Essential for Vitamin A Homeostasis. *Embo J* 18, no 18: 4903–14.

Gibbs, M. A., and R. G. Northcutt. 2004. Retinoic Acid Repatterns Axolotl Lateral Line Receptors. *Int J Dev Biol* 48, no 1: 63–6.

Gibert, Y., A. Gajewski, A. Meyer, and G. Begemann. 2006. Induction and Prepatterning of the Zebrafish Pectoral Fin Bud Requires Axial Retinoic Acid Signaling. *Development* 133, no 14: 2649–59.

Giguere, V., E. S. Ong, P. Segui, and R. M. Evans. 1987. Identification of a Receptor for the Morphogen Retinoic Acid. *Nature* 330, no 6149: 624–29.

Glover, J. C., J. S. Renaud, and F. M. Rijli. 2006. Retinoic Acid and Hindbrain Patterning. *J Neurobiol* 66, no 7: 705–25.

Grandel, H., and M. Brand. 2011. Zebrafish Limb Development Is Triggered by a Retinoic Acid Signal During Gastrulation. *Dev Dynam* 240, no 5: 1116–26.

Gronemeyer, H., J. A. Gustafsson, and V. Laudet. 2004. Principles for Modulation of the Nuclear Receptor Superfamily. *Nat Rev Drug Discov* 3, no 11: 950–64.

Gutierrez-Mazariegos, J., M. Theodosiou, F. Campo-Paysaa, and M. Schubert. 2011. Vitamin A: A Multifunctional Tool for Development. *Semin Cell Dev Biol* 22, no 6: 603–10.

Hale, F. 1933. Pigs Born Without Eye Balls. *J Heredity* 24, no 3: 105–06.

Hall, B. K. 2009. *The Neural Crest and Neural Crest Cells on Vertebrate Development and Evolution*. 2nd ed. ed: Springer.

Hall, J. A., J. R. Grainger, S. P. Spencer, and Y. Belkaid. 2011. The Role of Retinoic Acid in Tolerance and Immunity. *Immunity* 35, no 1: 13–22.

Halme, A., M. Cheng, and I. K. Hariharan. 2010. Retinoids Regulate a Developmental Checkpoint for Tissue Regeneration in *Drosophila*. *Curr Biol* 20, no 5: 458–63.

Hernandez, R. E., A. P. Putzke, J. P. Myers, L. Margaretha, and C. B. Moens. 2007. Cyp26 Enzymes Generate the Retinoic Acid Response Pattern Necessary for Hindbrain Development. *Development* 134, no 1: 177–87.

Hessel, S., A. Eichinger, A. Isken, et al. 2007. Cmo1 Deficiency Abolishes Vitamin A Production from β-Carotene and Alters Lipid Metabolism in Mice. *J Biol Chem* 282, no 46: 33553–61.

Hinman, V. F., and B. M. Degnan. 1998. Retinoic Acid Disrupts Anterior Ectodermal and Endodermal Development in Ascidian Larvae and Postlarvae. *Dev Genes Evol* 208, no 6: 336–45.

Hinman, V. F., and B. M. Degnan. 2000. Retinoic Acid Perturbs *Otx* Gene Expression in the Ascidian Pharynx. *Dev Genes Evol* 210, no 3: 129–39.

Hochgreb, T., V. L. Linhares, D. C. Menezes, et al. 2003. A Caudorostral Wave of Raldh2 Conveys Anteroposterior Information to the Cardiac Field. *Development* 130, no 22: 5363–74.

Holland, L. Z., R. Albalat, K. Azumi, et al. 2008. The Amphioxus Genome Illuminates Vertebrate Origins and Cephalochordate Biology. *Genome Res* 18, no 7: 1100–11.

Holland, L. Z., and N. D. Holland. 1996. Expression of *Amphihox-1* and *Amphipax-1* in Amphioxus Embryos Treated with Retinoic Acid: Insights into Evolution and Patterning of the Chordate Nerve Cord and Pharynx. *Development* 122, no 6: 1829–38.

Hopkins, P. M. 2001. Limb Regeneration in the Fiddler Crab, *Uca pugilator*: Hormonal and Growth Factor Control. *Am Zool* 41, no 3: 389–98.

Ikuta, T., N. Satoh, and H. Saiga. 2010. Limited Functions of Hox Genes in the Larval Development of the Ascidian *Ciona intestinalis. Development* 137, no 9: 1505–13.

Ikuta, T., N. Yoshida, N. Satoh, and H. Saiga. 2004. *Ciona intestinalis* Hox Gene Cluster: Its Dispersed Structure and Residual Colinear Expression in Development. *P Natl Acad Sci USA* 101, no 42: 15118–23.

Irving, C., and I. Mason. 2000. Signalling by Fgf8 from the Isthmus Patterns Anterior Hindbrain and Establishes the Anterior Limit of Hox Gene Expression. *Development* 127, no 1: 177–86.

Jackman, W. R., J. M. Mougey, G. D. Panopoulou, and C. B. Kimmel. 2004. *Crabp* and *Maf* Highlight the Novelty of the Amphioxus Club-Shaped Gland. *Acta Zool-Stockholm* 85, no 2: 91–99.

Kam, R. K., Y. Deng, Y. Chen, and H. Zhao. 2012. Retinoic Acid Synthesis and Functions in Early Embryonic Development. *Cell Biosci* 2, no 1: 11.

Kanda, M., H. Wada, and S. Fujiwara. 2009. Epidermal Expression of *Hox1* Is Directly Activated by Retinoic Acid in the *Ciona intestinalis* Embryo. *Dev Biol* 335, no 2: 454–63.

Kane, M. A. 2012. Analysis, Occurrence, and Function of 9-*cis*-Retinoic Acid. *Biochem Biophys Acta* 1821, no 1: 10–20.

Katsuyama, Y., and H. Saiga. 1998. Retinoic Acid Affects Patterning Along the Anterior-Posterior Axis of the Ascidian Embryo. *Dev Growth Differ* 40, no 4: 413–22.

Katsuyama, Y., S. Wada, S. Yasugi, and H. Saiga. 1995. Expression of the *Labial* Group Hox Gene *Hrhox-1* and Its Alteration Induced by Retinoic Acid in Development of the Ascidian *Halocynthia roretzi. Development* 121, no 10: 3197–205.

Kawaguchi, R., J. M. Yu, J. Honda, et al. 2007. A Membrane Receptor for Retinol Binding Protein Mediates Cellular Uptake of Vitamin A. *Science* 315, no 5813: 820–25.

Kiefer, C., S. Hessel, J. M. Lampert, et al. 2001. Identification and Characterization of a Mammalian Enzyme Catalyzing the Asymmetric Oxidative Cleavage of Provitamin A. *J Biol Chem* 276, no 17: 14110–16.

Kim, Y. K., L. Wassef, S. Chung, et al. 2011. β-Carotene and Its Cleavage Enzyme β-Carotene-15,15′-Oxygenase (CmoI) Affect Retinoid Metabolism in Developing Tissues. *Faseb J* 25, no 5: 1641–52.

Koop, D., and L. Z. Holland. 2008. The Basal Chordate Amphioxus as a Simple Model for Elucidating Developmental Mechanisms in Vertebrates. *Birth Defects Res C Embryo Today* 84, no 3: 175–87.

Koop, D., L. Z. Holland, D. Setiamarga, M. Schubert, and N. D. Holland. 2011. Tail Regression Induced by Elevated Retinoic Acid Signaling in Amphioxus Larvae Occurs by Tissue Remodeling, Not Cell Death. *Evol Dev* 13, no 5: 427–35.

Koop, D., N. D. Holland, M. Sémon, et al. 2010. Retinoic Acid Signaling Targets *Hox* Genes During the Amphioxus Gastrula Stage: Insights into Early Anterior-Posterior Patterning of the Chordate Body Plan. *Dev Biol* 338, no 1: 98–106.

Kopinke, D., J. Sasine, J. Swift, W. Z. Stephens, and T. Piotrowski. 2006. Retinoic Acid Is Required for Endodermal Pouch Morphogenesis and Not for Pharyngeal Endoderm Specification. *Dev Dynam* 235, no 10: 2695–709.

Krezel, W., V. Dupé, M. Mark, et al. 1996. Rxrγ Null Mice Are Apparently Normal and Compound Rxrα$^{+/-}$/Rxrβ$^{-/-}$/Rxrγ$^{-/-}$ Mutant Mice Are Viable. *P Natl Acad Sci USA* 93, no 17: 9010–4.

Krispin, S., E. Nitzan, Y. Kassem, and C. Kalcheim. 2010. Evidence for a Dynamic Spatiotemporal Fate Map and Early Fate Restrictions of Premigratory Avian Neural Crest. *Development* 137, no 4: 585–95.

Kruse, S. W., K. Suino-Powell, X. E. Zhou, et al. 2008. Identification of Coup-Tfii Orphan Nuclear Receptor as a Retinoic Acid-Activated Receptor. *Plos Biol* 6, no 9: 2002–15.

Lampert, J. M., J. Holzschuh, S. Hessel, et al. 2003. Provitamin A Conversion to Retinal Via the β,β-Carotene-15,15′-Oxygenase (*Bcox*) Is Essential for Pattern Formation and Differentiation During Zebrafish Embryogenesis. *Development* 130, no 10: 2173–86.

Lee, S. J., S. Kim, S. C. Choi, and J. K. Han. 2010. *Xpteg (Xenopus Proximal Tubules-Expressed Gene)* Is Essential for Pronephric Mesoderm Specification and Tubulogenesis. *Mech Develop* 127, no 1-2: 49–61.

Li, Z., V. Korzh, and Z. Y. Gong. 2007. Localized *Rbp4* Expression in the Yolk Syncytial Layer Plays a Role in Yolk Cell Extension and Early Liver Development. *Bmc Dev Biol* 7: 117.

Lin, M., M. Zhang, M. Abraham, S. M. Smith, and J. L. Napoli. 2003. Mouse Retinal Dehydrogenase 4 (Raldh4), Molecular Cloning, Cellular Expression, and Activity in 9-*cis*-Retinoic Acid Biosynthesis in Intact Cells. *J Biol Chem* 278, no 11: 9856–61.

Lin, S. C., P. Dollé, L. Ryckebusch, et al. 2010. Endogenous Retinoic Acid Regulates Cardiac Progenitor Differentiation. *P Natl Acad Sci USA* 107, no 20: 9234–39.

Linney, E., S. Donerly, L. Mackey, and B. Dobbs-McAuliffe. 2011. The Negative Side of Retinoic Acid Receptors. *Neurotoxicol Teratol* 33, no 6: 631–40.

Liu, J. P., E. Laufer, and T. M. Jessell. 2001. Assigning the Positional Identity of Spinal Motor Neurons: Rostrocaudal Patterning of Hox-C Expression by Fgfs, Gdf11, and Retinoids. *Neuron* 32, no 6: 997–1012.

Lobo, G. P., J. Amengual, G. Palczewski, D. Babino, and J. von Lintig. 2012. Mammalian Carotenoid-Oxygenases: Key Players for Carotenoid Function and Homeostasis. *Biochem Biophys Acta* 1821, no 1: 78–87.

López-Carballo, G., L. Moreno, S. Masiá, P. Pérez, and D. Barettino. 2002. Activation of the Phosphatidylinositol 3-Kinase/Akt Signaling Pathway by Retinoic Acid Is Required for Neural Differentiation of Sh-Sy5y Human Neuroblastoma Cells. *J Biol Chem* 277, no 28: 25297–304.

MacLean, G., H. Li, D. Metzger, P. Chambon, and M. Petkovich. 2007. Apoptotic Extinction of Germ Cells in Testes of *Cyp26b1* Knockout Mice. *Endocrinology* 148, no 10: 4560–67.

Maden, M. 2002. Retinoid Signalling in the Development of the Central Nervous System. *Nat Rev Neurosci* 3, no 11: 843–53.

Mark, M., N. B. Ghyselinck, and P. Chambon. 2006. Function of Retinoid Nuclear Receptors. Lessons from Genetic and Pharmacological Dissections of the Retinoic Acid Signaling Pathway During Mouse Embryogenesis. *Annu Rev Pharmacol* 46: 451–80.

Mark, M., N. B. Ghyselinck, and P. Chambon. 2009. Function of Retinoic Acid Receptors During Embryonic Development. *Nucl Recept Signal* 7: e002.

Marletaz, F., L. Z. Holland, V. Laudet, and M. Schubert. 2006. Retinoic Acid Signaling and the Evolution of Chordates. *Int J Biol Sci* 2, no 2: 38–47.

Marshall, H., A. Morrison, M. Studer, H. Pöpperl, and R. Krumlauf. 1996. Retinoids and Hox Genes. *Faseb J* 10, no 9: 969–78.

Martínez-Morales, P. L., R. D. del Corral, I. Olivera-Martínez, et al. 2011. Fgf and Retinoic Acid Activity Gradients Control the Timing of Neural Crest Cell Emigration in the Trunk. *J Cell Biol* 194, no 3: 489–503.

Matt, N., V. Dupé, J. M. Garnier, et al. 2005. Retinoic Acid-Dependent Eye Morphogenesis Is Orchestrated by Neural Crest Cells. *Development* 132, no 21: 4789–800.

Matt, N., N. B. Ghyselinck, O. Wendling, P. Chambon, and M. Mark. 2003. Retinoic Acid-Induced Developmental Defects Are Mediated by Rarβ/Rxr Heterodimers in the Pharyngeal Endoderm. *Development* 130, no 10: 2083–93.

Matt, N., C. K. Schmidt, V. Dupé, et al. 2005. Contribution of Cellular Retinol-Binding Protein Type 1 to Retinol Metabolism During Mouse Development. *Dev Dynam* 233, no 1: 167–76.

McCaffery, P., E. Wagner, J. O'Neil, M. Petkovich, and U. C. Dräger. 1999. Dorsal and Ventral Retinal Territories Defined by Retinoic Acid Synthesis, Break-Down and Nuclear Receptor Expression. *Mech Develop* 82, no 1-2: 119–30.

McNamara, P., S. B. Seo, R. D. Rudic, et al. 2001. Regulation of Clock and Mop4 by Nuclear Hormone Receptors in the Vasculature: A Humoral Mechanism to Reset a Peripheral Clock. *Cell* 105, no 7: 877–89.

Mic, F. A., R. J. Haselbeck, A. E. Cuenca, and G. Duester. 2002. Novel Retinoic Acid Generating Activities in the Neural Tube and Heart Identified by Conditional Rescue of *Raldh2* Null Mutant Mice. *Development* 129, no 9: 2271–82.

Mic, F. A., A. Molotkov, D. M. Benbrook, and G. Duester. 2003. Retinoid Activation of Retinoic Acid Receptor but Not Retinoid X Receptor Is Sufficient to Rescue Lethal Defect in Retinoic Acid Synthesis. *P Natl Acad Sci USA* 100, no 12: 7135–40.

Minoux, M., and F. M. Rijli. 2010. Molecular Mechanisms of Cranial Neural Crest Cell Migration and Patterning in Craniofacial Development. *Development* 137, no 16: 2605–21.

Molotkov, A., N. Molotkova, and G. Duester. 2005. Retinoic Acid Generated by Raldh2 in Mesoderm Is Required for Mouse Dorsal Endodermal Pancreas Development. *Dev Dynam* 232, no 4: 950–57.

Molotkova, N., A. Molotkov, I. O. Sirbu, and G. Duester. 2005. Requirement of Mesodermal Retinoic Acid Generated by *Raldh2* for Posterior Neural Transformation. *Mech Develop* 122, no 2: 145–55.

Moreno, T. A., and C. Kintner. 2004. Regulation of Segmental Patterning by Retinoic Acid Signaling During *Xenopus* Somitogenesis. *Dev Cell* 6, no 2: 205–18.

Müller, W. A. 1984. Retinoids and Pattern Formation in a Hydroid. *J Embryol Exp Morph* 81, no Jun: 253–71.

Müller, W. E. G., M. Binder, J. von Lintig, et al. 2011. Interaction of the Retinoic Acid Signaling Pathway with Spicule Formation in the Marine Sponge *Suberites domuncula* Through Activation of Bone Morphogenetic Protein-1. *Biochem Biophys Acta* 1810, no 12: 1178–94.

Nagatomo, K., and S. Fujiwara. 2003. Expression of *Raldh2*, *Cyp26* and *Hox-1* in Normal and Retinoic Acid-Treated *Ciona intestinalis* Embryos. *Gene Expr Patterns* 3, no 3: 273–77.

Nagatomo, K., T. Ishibashi, Y. Satou, N. Satoh, and S. Fujiwara. 2003. Retinoic Acid Affects Gene Expression and Morphogenesis Without Upregulating the Retinoic Acid Receptor in the Ascidian *Ciona intestinalis*. *Mech Develop* 120, no 3: 363–72.

Nagl, F., K. Schönhofer, B. Seidler, et al. 2009. Retinoic Acid-Induced Nnos Expression Depends on a Novel Pi3k/Akt/Dax1 Pathway in Human Tgw-Nu-I Neuroblastoma Cells. *Am J Physiol Cell Physiol* 297, no 5: C1146–56.

Napoli, J. L. 1999. Interactions of Retinoid Binding Proteins and Enzymes in Retinoid Metabolism. *Biochem Biophys Acta* 1440, no 2-3: 139–62.

Napoli, J. L. 2000. A Gene Knockout Corroborates the Integral Function of Cellular Retinol-Binding Protein in Retinoid Metabolism. *Nutr Rev* 58, no 8: 230–6.

Natale, A., C. Sims, M. L. Chiusano, et al. 2011. Evolution of Anterior *Hox* Regulatory Elements among Chordates. *Bmc Evol Biol* 11: 330.

Niederreither, K., and P. Dollé. 2008. Retinoic Acid in Development: Towards an Integrated View. *Nat Rev Genet* 9, no 7: 541–53.

Niederreither, K., V. Subbarayan, P. Dollé, and P. Chambon. 1999. Embryonic Retinoic Acid Synthesis Is Essential for Early Mouse Post-Implantation Development. *Nat Genet* 21, no 4: 444–48.

Niederreither, K., J. Vermot, I. Le Roux, et al. 2003. The Regional Pattern of Retinoic Acid Synthesis by Raldh2 Is Essential for the Development of Posterior Pharyngeal Arches and the Enteric Nervous System. *Development* 130, no 11: 2525–34.

Novitch, B. G., H. Wichterle, T. M. Jessell, and S. Sockanathan. 2003. A Requirement for Retinoic Acid-Mediated Transcriptional Activation in Ventral Neural Patterning and Motor Neuron Specification. *Neuron* 40, no 1: 81–95.

Noy, N. 2000. Retinoid-Binding Proteins: Mediators of Retinoid Action. *Biochem J* 348: 481–95.

Noy, N. 2010. Between Death and Survival: Retinoic Acid in Regulation of Apoptosis. *Annu Rev Nutr* 30: 201–17.

Ochoa, W. F., A. Torrecillas, I. Fita, et al. 2003. Retinoic Acid Binds to the C2-Domain of Protein Kinase Cα. *Biochemistry* 42, no 29: 8774–79.

Ogasawara, M., H. Wada, H. Peters, and N. Satoh. 1999. Developmental Expression of *Pax1/9* Genes in Urochordate and Hemichordate Gills: Insight into Function and Evolution of the Pharyngeal Epithelium. *Development* 126, no 11: 2539–50.

Osborne, P. W., G. Benoit, V. Laudet, M. Schubert, and D. E. K. Ferrier. 2009. Differential Regulation of Parahox Genes by Retinoic Acid in the Invertebrate Chordate Amphioxus (*Branchiostoma floridae*). *Dev Biol* 327, no 1: 252–62.

Parés, X., J. Farrés, N. Kedishvili, and G. Duester. 2008. Medium-Chain and Short-Chain Dehydrogenases/Reductases in Retinoid Metabolism. *Cell Mol Life Sci* 65, no 24: 3936–49.

Petkovich, M., N. J. Brand, A. Krust, and P. Chambon. 1987. A Human Retinoic Acid Receptor Which Belongs to the Family of Nuclear Receptors. *Nature* 330, no 6147: 444–50.

Piskunov, A., and C. Rochette-Egly. 2011. A Retinoic Acid Receptor Rarα Pool Present in Membrane Lipid Rafts Forms Complexes with G Protein Aq to Activate P38mapk. *Oncogene 31, no 28: 3333–45.*

Plant, M. R., M. E. MacDonald, L. I. Grad, S. J. Ritchie, and J. M. Richman. 2000. Locally Released Retinoic Acid Repatterns the First Branchial Arch Cartilages *in vivo*. *Dev Biol* 222, no 1: 12–26.

Pourquié, O. 2003. The Segmentation Clock: Converting Embryonic Time into Spatial Pattern. *Science* 301, no 5631: 328–30.

Quadro, L., L. Hamberger, M. E. Gottesman, et al. 2004. Transplacental Delivery of Retinoid: The Role of Retinol-Binding Protein and Lipoprotein Retinyl Ester. *Am J Physiol-Endoc M* 286, no 5: E844–E51.

Quadro, L., L. Hamberger, M. E. Gottesman, et al. 2005. Pathways of Vitamin A Delivery to the Embryo: Insights from a New Tunable Model of Embryonic Vitamin A Deficiency. *Endocrinology* 146, no 10: 4479–90.

Reijntjes, S., E. Gale, and M. Maden. 2004. Generating Gradients of Retinoic Acid in the Chick Embryo: *Cyp26c1* Expression and a Comparative Analysis of the Cyp26 Enzymes. *Dev Dynam* 230, no 3: 509–17.

Rhinn, M., and P. Dollé. 2012. Retinoic Acid Signalling During Development. *Development* 139, no 5: 843–58.

Ribes, V., I. Le Roux, M. Rhinn, B. Schuhbaur, and P. Dollé. 2009. Early Mouse Caudal Development Relies on Crosstalk Between Retinoic Acid, Shh and Fgf Signalling Pathways. *Development* 136, no 4: 665–76.

Richardson, S. J. 2009. Evolutionary Changes to Transthyretin: Evolution of Transthyretin Biosynthesis. *Febs J* 276, no 19: 5342–56.

Rijli, F. M., A. Gavalas, and P. Chambon. 1998. Segmentation and Specification in the Branchial Region of the Head: The Role of the *Hox* Selector Genes. *Int J Dev Biol* 42, no 3: 393–401.

Rosenthal, N., and J. Xavier-Neto. 2000. From the Bottom of the Heart: Anteroposterior Decisions in Cardiac Muscle Differentiation. *Curr Opin Cell Biol* 12, no 6: 742–46.

Ross, A. C. 2003. Retinoid Production and Catabolism: Role of Diet in Regulating Retinol Esterification and Retinoic Acid Oxidation. *J Nutr* 133, no 1: 291s–96s.

Ross, A. C., and R. Zolfaghari. 2011. Cytochrome P450s in the Regulation of Cellular Retinoic Acid Metabolism. *Annu Rev Nutr* 31, no 1: 65–87.

Ross, S. A., P. J. McCaffery, U. C. Drager, and L. M. De Luca. 2000. Retinoids in Embryonal Development. *Physiol Rev* 80, no 3: 1021–54.

Rosselot, C., L. Spraggon, I. Chia, et al. 2010. Non-Cell-Autonomous Retinoid Signaling Is Crucial for Renal Development. *Development* 137, no 2: 283–92.

Samarut, E., and C. Rochette-Egly. 2012. Nuclear Retinoic Acid Receptors: Conductors of the Retinoic Acid Symphony During Development. *Mol Cell Endocrinol* 348, no 2: 348–60.

Sandell, L. L., B. W. Sanderson, G. Moiseyev, et al. 2007. Rdh10 Is Essential for Synthesis of Embryonic Retinoic Acid and Is Required for Limb, Craniofacial, and Organ Development. *Genes Dev* 21, no 9: 1113–24.

Sasakura, Y., M. Kanda, T. Ikeda, et al. 2012. Retinoic Acid-Driven *Hox1* Is Required in the Epidermis for Forming the Otic/Atrial Placodes During Ascidian Metamorphosis. *Development* 139, no 12: 2156–60.

Schlosser, G. 2010. Making Senses: Development of Vertebrate Cranial Placodes. *Int Rev Cell Mol Biol* 283: 129–234.

Schubert, M., H. Escriva, J. Xavier-Neto, and V. Laudet. 2006. Amphioxus and Tunicates as Evolutionary Model Systems. *Trends Ecol Evol* 21, no 5: 269–77.

Schubert, M., N. D. Holland, H. Escriva, L. Z. Holland, and V. Laudet. 2004. Retinoic Acid Influences Anteroposterior Positioning of Epidermal Sensory Neurons and Their Gene Expression in a Developing Chordate (Amphioxus). *P Natl Acad Sci USA* 101, no 28: 10320–25.

Schubert, M., N. D. Holland, V. Laudet, and L. Z. Holland. 2006. A Retinoic Acid-*Hox* Hierarchy Controls Both Anterior/Posterior Patterning and Neuronal Specification in the Developing Central Nervous System of the Cephalochordate Amphioxus. *Dev Biol* 296, no 1: 190–202.

Schubert, M., J. K. Yu, N. D. Holland, et al. 2005. Retinoic Acid Signaling Acts Via Hox1 to Establish the Posterior Limit of the Pharynx in the Chordate Amphioxus. *Development* 132, no 1: 61–73.

Schug, T. T., D. C. Berry, N. S. Shaw, S. N. Travis, and N. Noy. 2007. Opposing Effects of Retinoic Acid on Cell Growth Result from Alternate Activation of Two Different Nuclear Receptors. *Cell* 129, no 4: 723–33.

Seo, H. C., R. B. Edvardsen, A. D. Maeland, et al. 2004. Hox Cluster Disintegration with Persistent Anteroposterior Order of Expression in *Oikopleura dioica*. *Nature* 431, no 7004: 67–71.

Serluca, F. C., and M. C. Fishman. 2001. Pre-Pattern in the Pronephric Kidney Field of Zebrafish. *Development* 128, no 12: 2233–41.

Shaw, N., M. Elholm, and N. Noy. 2003. Retinoic Acid Is a High Affinity Selective Ligand for the Peroxisome Proliferator-Activated Receptor β/δ. *J Biol Chem* 278, no 43: 41589–92.

Sheng, N. Y., Z. H. Xie, C. Wang, et al. 2010. Retinoic Acid Regulates Bone Morphogenic Protein Signal Duration by Promoting the Degradation of Phosphorylated Smad1. *P Natl Acad Sci USA* 107, no 44: 18886–91.

Shirai, H., K. Oishi, and N. Ishida. 2006. Bidirectional Clock/Bmal1-Dependent Circadian Gene Regulation by Retinoic Acid *in vitro*. *Biochem Bioph Res Co* 351, no 2: 387–91.

Simões-Costa, M. S., A. P. Azambuja, and J. Xavier-Neto. 2008. The Search for Non-Chordate Retinoic Acid Signaling: Lessons from Chordates. *J Exp Zool Part B* 310B, no 1: 54–72.

Sirbu, I. O., and G. Duester. 2006. Retinoic-Acid Signalling in Node Ectoderm and Posterior Neural Plate Directs Left-Right Patterning of Somitic Mesoderm. *Nat Cell Biol* 8, no 3: 271–77.

Sirbu, I. O., L. Gresh, J. Barra, and G. Duester. 2005. Shifting Boundaries of Retinoic Acid Activity Control Hindbrain Segmental Gene Expression. *Development* 132, no 11: 2611–22.

Sirbu, I. O., X. L. Zhao, and G. Duester. 2008. Retinoic Acid Controls Heart Anteroposterior Patterning by Down-Regulating *Isl1* Through the *Fgf8* Pathway. *Dev Dynam* 237, no 6: 1627–35.

Sobreira, T. J. P., F. Marlétaz, M. Simões-Costa, et al. 2011. Structural Shifts of Aldehyde Dehydrogenase Enzymes Were Instrumental for the Early Evolution of Retinoid-Dependent Axial Patterning in Metazoans. *P Natl Acad Sci USA* 108, no 1: 226–31.

Song, Y., J. N. Hui, K. K. Fu, and J. M. Richman. 2004. Control of Retinoic Acid Synthesis and Fgf Expression in the Nasal Pit Is Required to Pattern the Craniofacial Skeleton. *Dev Biol* 276, no 2: 313–29.

Spiegler, E., Y. K. Kim, L. Wassef, V. Shete, and L. Quadro. 2012. Maternal-Fetal Transfer and Metabolism of Vitamin A and Its Precursor β-Carotene in the Developing Tissues. *Biochem Biophys Acta* 1821, no 1: 88–98.

Stafford, D., A. Hornbruch, P. R. Mueller, and V. E. Prince. 2004. A Conserved Role for Retinoid Signaling in Vertebrate Pancreas Development. *Dev Genes Evol* 214, no 9: 432–41.

Stainier, D. Y. R., and M. C. Fishman. 1992. Patterning the Zebrafish Heart Tube: Acquisition of Anteroposterior Polarity. *Dev Biol* 153, no 1: 91–101.

Stehlin-Gaon, C., D. Willmann, D. Zeyer, et al. 2003. All-*trans* Retinoic Acid Is a Ligand for the Orphan Nuclear Receptor Rorβ. *Nat Struct Biol* 10, no 10: 820–25.

Sun, H. 2012. Membrane Receptors and Transporters Involved in the Function and Transport of Vitamin A and Its Derivatives. *Biochem Biophys Acta* 1821, no 1: 99–112.

Suruga, K., T. Goda, M. Igarashi, et al. 1997. Cloning of Chick Cellular Retinol-Binding Protein, Type II and Comparison to That of Some Mammals: Expression of the Gene at Different Developmental Stages, and Possible Involvement of Rxrs and Ppar. *Comp Biochem Physiol A Physiol* 118, no 3: 859–69.

Tan, N. S., N. S. Shaw, N. Vinckenbosch, et al. 2002. Selective Cooperation Between Fatty Acid Binding Proteins and Peroxisome Proliferator-Activated Receptors in Regulating Transcription. *Mol Cell Biol* 22, no 14: 5114–27.

Tanaka, Y., Y. Okada, and N. Hirokawa. 2005. Fgf-Induced Vesicular Release of Sonic Hedgehog and Retinoic Acid in Leftward Nodal Flow Is Critical for Left-Right Determination. *Nature* 435, no 7039: 172–77.

Theodosiou, M., V. Laudet, and M. Schubert. 2010. From Carrot to Clinic: An Overview of the Retinoic Acid Signaling Pathway. *Cell Mol Life Sci* 67, no 9: 1423–45.

Tickle, C., B. Alberts, L. Wolpert, and J. Lee. 1982. Local Application of Retinoic Acid to the Limb Bond Mimics the Action of the Polarizing Region. *Nature* 296, no 5857: 564–66.

Trainor, P. A., and R. Krumlauf. 2001. *Hox* Genes, Neural Crest Cells and Branchial Arch Patterning. *Curr Opin Cell Biol* 13, no 6: 698–705.

Uehara, M., K. Yashiro, S. Mamiya, et al. 2007. Cyp26a1 and Cyp26c1 Cooperatively Regulate Anterior-Posterior Patterning of the Developing Brain and the Production of Migratory Cranial Neural Crest Cells in the Mouse. *Dev Biol* 302, no 2: 399–411.

Van de Peer, Y., S. Maere, and A. Meyer. 2009. The Evolutionary Significance of Ancient Genome Duplications. *Nat Rev Genet* 10, no 10: 725–32.

Vermot, J., and O. Pourquié. 2005. Retinoic Acid Coordinates Somitogenesis and Left-Right Patterning in Vertebrate Embryos. *Nature* 435, no 7039: 215–20.

Vilhais-Neto, G. C., and O. Pourquié. 2008. Retinoic Acid. *Curr Biol* 18, no 7: 550–52.

von Lintig, J., S. Hessel, A. Isken, et al. 2005. Towards a Better Understanding of Carotenoid Metabolism in Animals. *Biochem Biophys Acta* 1740, no 2: 122–31.

Wada, H., H. Escriva, S. C. Zhang, and V. Laudet. 2006. Conserved Rare Localization in Amphioxus *Hox* Clusters and Implications for *Hox* Code Evolution in the Vertebrate Neural Crest. *Dev Dynam* 235, no 6: 1522–31.

Wang, Z. X., P. Dollé, W. V. Cardoso, and K. Niederreither. 2006. Retinoic Acid Regulates Morphogenesis and Patterning of Posterior Foregut Derivatives. *Dev Biol* 297, no 2: 433–45.

Wasiak, S., and D. Lohnes. 1999. Retinoic Acid Affects Left-Right Patterning. *Dev Biol* 215, no 2: 332–42.

Waxman, J. S., and D. Yelon. 2007. Comparison of the Expression Patterns of Newly Identified Zebrafish Retinoic Acid and Retinoid X Receptors. *Dev Dynam* 236, no 2: 587–95.

White, J. C., M. Highland, M. Kaiser, and M. Clagett-Dame. 2000. Vitamin A Deficiency Results in the Dose-Dependent Acquisition of Anterior Character and Shortening of the Caudal Hindbrain of the Rat Embryo. *Dev Biol* 220, no 2: 263–84.

White, R. J., and T. F. Schilling. 2008. How Degrading: Cyp26s in Hindbrain Development. *Dev Dynam* 237, no 10: 2775–90.

Wiens, M., R. Batel, M. Korzhev, and W. E. G. Müller. 2003. Retinoid X Receptor and Retinoic Acid Response in the Marine Sponge *Suberites domuncula*. *J Exp Biol* 206, no 18: 3261–71.

Wilson, J. G., C. B. Roth, and J. Warkany. 1953. An Analysis of the Syndrome of Malformations Induced by Maternal Vitamin A Deficiency. Effects of Restoration of Vitamin A at Various Times During Gestation. *Am J Anat* 92, no 2: 189–217.

Wilson, J. G., and J. Warkany. 1949. Aortic-Arch and Cardiac Anomalies in the Offspring of Vitamin A Deficient Rats. *Am J Anat* 85, no 1: 113–55.

Wingert, R. A., R. Selleck, J. Yu, et al. 2007. The *Cdx* Genes and Retinoic Acid Control the Positioning and Segmentation of the Zebrafish Pronephros. *Plos Genet* 3, no 10: 1922–38.

Xavier-Neto, J., N. Rosenthal, F. A. Silva, et al. 2001. Retinoid Signaling and Cardiac Anteroposterior Segmentation. *Genesis* 31, no 3: 97–104.

Yamauchi, K., and A. Ishihara. 2009. Evolutionary Changes to Transthyretin: Developmentally Regulated and Tissue-Specific Gene Expression. *Febs J* 276, no 19: 5357–66.

Yashiro, K., X. L. Zhao, M. Uehara, et al. 2004. Regulation of Retinoic Acid Distribution Is Required for Proximodistal Patterning and Outgrowth of the Developing Mouse Limb. *Dev Cell* 6, no 3: 411–22.

Yasuda, Y., N. Nishi, J. A. Takahashi, et al. 1992. Induction of Avascular Yolk Sac Due to Reduction of Basic Fibroblast Growth Factor by Retinoic Acid in Mice. *Dev Biol* 150, no 2: 397–413.

Zanotto-Filho, A., M. Cammarota, D. P. Gelain, et al. 2008. Retinoic Acid Induces Apoptosis by a Non-Classical Mechanism of Erk1/2 Activation. *Toxicol in vitro* 22, no 5: 1205–12.

Zhang, Z. Y., J. E. Balmer, A. Lovlie, S. H. Fromm, and R. Blomhoff. 1996. Specific Teratogenic Effects of Different Retinoic Acid Isomers and Analogs in the Developing Anterior Central Nervous System of Zebrafish. *Dev Dynam* 206, no 1: 73–86.

Zhao, X. L., T. Brade, T. J. Cunningham, and G. Duester. 2010. Retinoic Acid Controls Expression of Tissue Remodeling Genes *Hmgn1* and *Fgf18* at the Digit–Interdigit Junction. *Dev Dynam* 239, no 2: 665–71.

Zhao, X. L., I. O. Sirbu, F. A. Mic, et al. 2009. Retinoic Acid Promotes Limb Induction Through Effects on Body Axis Extension but Is Unnecessary for Limb Patterning. *Curr Biol* 19, no 12: 1050–57.

Zhou, X. E., K. M. Suino-Powell, Y. Xu, et al. 2011. The Orphan Nuclear Receptor Tr4 Is a Vitamin A-Activated Nuclear Receptor. *J Biol Chem* 286, no 4: 2877–85.

2 Serum Retinol-Binding Protein, Obesity, and Insulin Resistance

Pangala V. Bhat and Daniel-Constantin Manolescu

CONTENTS

2.1 INTRODUCTION

Increased adipose mass, which is associated with insulin resistance and type 2 diabetes mellitus (T2DM) (Kahn and Flier 2000), is characterized by the impaired ability of insulin to reduce glucose output in the liver and to augment glucose uptake in adipose and muscle tissues (Saltiel and Khan 2001). The excess energy stored in adipose tissue plays a major role in the maintenance of whole-body energy homeostasis. It is well established that adipose tissues not only serve as inert energy storage depots but also secrete a number of signaling factors into the circulation, such as C-reactive protein, leptin, resistin, adiponectin, tumor necrosis factor-α (TNFα), and others (Rajala and Scherer 2003). These factors are believed to be involved in the development of insulin resistance. Another such signaling factor, retinol-binding protein-4 (RBP4), a protein whose only function was thought to be the delivery of retinol (vitamin A) to tissues, was recently implicated as an adipokine in insulin resistance (Yang et al. 2005). This chapter summarizes the evidence gathered from studies in mice and humans on the linkage between RBP4, obesity, and insulin resistance.

2.2   SERUM RBP4 AND RETINOID METABOLISM

In 1968, Goodman and colleagues (Kanai et al. 1968) first reported the existence of a 21-kDa-transport protein for vitamin A in blood and called it RBP (retinol-binding protein). Later, in 2005, this protein was renamed RBP4 (Yang et al. 2005). Since 1968, extensive research has been performed to characterize it (Peterson 1971; Goodman 1974 and 1982; Blaner and Goodman 1990). A vast amount of literature exists on the molecular structure, distribution, synthesis, and regulation of RBP4 (Blaner 1989 and references therein). The liver is the major site where RBP4 is produced, and retinol is mobilized from liver vitamin A storage depots bound to newly synthesized RBP4 in a 1:1 molar ratio. In the circulation, RBP4 occurs as a protein complex with the high molecular weight protein transthyretin (TTR), which prevents its glomerular filtration and reduces its renal clearance (Peterson 1971). RBP4 ablation is not lethal to embryos. However, RBP4 deficiency in mice results in impaired vision (Vogel et al. 2002). Although hepatocytes are major sites of RBP4 synthesis and secretion, other organs and tissues, including adipose tissues, express RBP4 (Makeover et al. 1989; Tsutsumi et al. 1992). Adipocytes synthesize about 20% of the RBP4 amount in the liver and can release it from adipose tissues into the circulation (Tsutsumi et al. 1992). Recent studies have shown that RBP4 is not required for intrahepatic transport and storage of vitamin A (Quadro et al. 2002 and 2004). Little is known about the role of extra-hepatically synthesized RBP4. In normal physiological conditions, circulating retinol-RBP4 levels remain constant, changing only in response to extremes in the nutritional intake of vitamin A, protein, calories, zinc, and to hormonal factors, stress, or some disease states (Soprono and Blaner 1994).

Circulating holo-RBP4 delivers retinol to retinoid-responsive cells. Recent experiments have demonstrated the occurrence of cell surface receptors for retinol-RBP4 that facilitates the uptake of retinol from RBP4 into cells (Kawaguchi et al. 2007; Pasutto et al. 2007). Once inside target cells, retinol undergoes metabolism to metabolites, such as retinyl esters (storage), oxidation to the active compound retinoic acid (RA), and catabolism to 4-oxo and 5,6 epoxy derivatives (Frolik et al. 1979; Bhat and Lacroix 1983; Labrecque et al. 1995). Enzymes involved in the formation of these metabolites are well characterized (Napoli 2000; Bhat 2005), and their specific functions in vitamin A homeostasis have been established by gene knockout in mice (Gottesman et al. 2001). The active metabolite RA (*all-trans* and 9-*cis*) binds to specific nuclear retinoic acid receptors (RARα, β, and γ, and RXRs), which are ligand-dependent transcriptional factors that regulate gene expression (Chambon 1996). For efficient gene transcription, RXRs form homodimers and heterodimers with RARs, thyroid hormone receptor, vitamin D3 receptor, and peroxisome proliferator-activated receptors (PPARs) (Zang et al. 1992; Mader et al. 1993; Rosen and Spiegelman 2001).

2.2.1 Vitamin A and Adipogenesis

Adipose tissues are targets of RA action (Villarroya et al. 1999) and regulate differentiation via RA-nuclear receptors RARs and RXRs (Aleman et al. 2004). Vitamin A status influences the development and function of adipose tissues in whole animals, with vitamin A deficiency favoring increased fat deposition. Adipocytes are dynamically involved in retinoid storage and metabolism (Tsutsumi et al. 1992). Retinol dehydrogenase type 1, an enzyme that participates in RA production, is highly expressed in adipose tissues (Sima et al. 2011) and has recently been shown to play a major role in the regulation of fat depots in adipose tissues (Reichert et al. 2011). The transcriptional factor PPARγ is abundantly expressed in adipocytes and forms a heterodimeric partner with RXRα, β, and γ, which also are expressed in adipose tissues (Metzer et al. 2005). Heterodimers PPARγ/RXRs regulate the transcription of genes in insulin action, adipocyte differentiation, lipid metabolism, and inflammation (Lenhard 2001). Therefore, PPARγ/RXRs are molecular targets in diabetes, and several drugs that activate these heterodimers have been developed to improve insulin action in T2DM (Mukherjee et al. 1997; Lebovitz and Banerji 2001).

2.3 OBESITY, INSULIN RESISTANCE, AND T2DM

Obesity is associated with a global increase in metabolic syndrome and T2DM. Among several essential features of these diseases, such as insulin resistance, impaired insulin secretion, hepatic steatosis, dyslipidemia, and atherosclerosis, insulin resistance is a common denominator (Kahn and Flier 2000). Under such conditions, hyperglycemia develops when increased insulin secretion no longer compensates for insulin resistance. In obesity-associated insulin resistance, multiple endocrine, inflammatory, and neural pathways are affected, which leads to disturbed signaling that is cell intrinsic (Qatanani and Lazar 2007). In obese subjects, plasma fatty acid elevation can facilitate insulin resistance by activating protein kinases (Petersen and Shulman 2006). Obesity-related changes in the secretion of adipokines, such as leptin, adiponectin, resistin, TNFα, and interleukin-6, impair insulin sensitivity (Kershaw and Flier 2004). Macrophage accumulation increases in the white adipose tissues of obese individuals, resulting in augmented adipose tissue production of inflammatory cytokines that affects insulin signaling (Weisberg et al. 2003). Furthermore, obesity-associated changes in brain responses to hormonal and metabolic signals alter peripheral insulin sensitivity (Pocai et al. 2005).

The liver, skeletal muscles, and adipose tissues are major targets of metabolic insulin action. Insulin regulates glucose homeostasis by decreasing hepatic glucose output and by augmenting the rate of glucose uptake by skeletal muscles and adipose tissues. Under normal physiological conditions, skeletal muscles are regarded as predominant sites of insulin-stimulated glucose uptake, and much less glucose is taken up by adipose tissues. Insulin-stimulated glucose uptake into muscle cells and adipocytes depends largely on translocation of the insulin-regulated glucose transporter GLUT4 from intracellular compartments to the cell surface (Shepherd and Khan 1999). GLUT4 expression is diminished in adipocytes but not in skeletal muscles of animals and humans with obesity and T2DM (DeFronzo 1997; Shepherd and Khan 1999). Since skeletal muscles are major sinks for glucose disposal, unaltered GLUT4 expression by the diabetogenic effect of reduced GLUT4 in adipocytes is unexpected. To explain this observation, Abel et al. (2001) hypothesized that a factor secreted into the circulation from adipose tissues induces insulin resistance in liver and muscle tissues.

2.4   ASSOCIATION BETWEEN RBP4, OBESITY, AND INSULIN RESISTANCE

Experiments on rodents and several clinical studies have revealed a strong positive relationship between serum RBP4, obesity, and insulin resistance.

2.4.1   EXPERIMENTAL STUDIES ON RBP4 AND INSULIN RESISTANCE

RBP4 involvement in insulin resistance was first discovered by Yang et al. (2005), by global gene expression analysis of epididymal adipose tissues harboring a primary genetic alteration in GLUT4 expression. These investigators observed that RBP4 expression levels were heightened by 2.3-fold in mice with adipocyte-specific GLUT4 ablation, whereas they were decreased by 54% in GLUT4-overexpressing mice, establishing an inverse relationship between GLUT4 and RBP4 expressions. Yang et al. (2005) also demonstrated that circulating RBP4 levels were increased not only in several mouse models of obesity and insulin resistance (genetically obese, *ob/ob*, and high-fat diet [HFD]-fed), but also in humans with these conditions. Furthermore, an insulin-sensitizing drug (rosiglitazone) lowered elevated RBP4 levels in both adipose tissue and serum of mice. When RBP4 levels were augmented in mice by either overexpressing or injecting RBP4, insulin resistance was induced, whereas knockout of the gene-encoding RBP4 enhanced insulin sensitivity compared to wild type mice.

Fenofibrate, a PPARα activator, greatly diminished RBP4 mRNA levels in adipose tissues but not in the liver, which correlated with decreased circulating RBP4 and improvement of insulin sensitivity in obese rats (Wu et al. 2009). In spontaneously hypertensive rats, elevated plasma RBP4 and increased liver and epididymal RBP4 expression levels were associated with insulin resistance (Ou et al. 2011). Treatment of these rats with fenretinide to augment urinary RBP4 excretion significantly decreased plasma RBP4 levels with improvement of insulin sensitivity. In HFD-fed mice, RBP4 suppression in adipose tissues and liver by anti-RBP4 oligonucleotide resulted in decreased serum RBP4 with improved insulin sensitivity and prevention of metabolic syndrome (Tan et al. 2011). Elevated serum RBP4 levels were lowered in leptin-deficient, insulin-resistant *ob/ob* mice by RA, which resulted in increased insulin sensitivity (Manolescu et al. 2010). It is likely that, in addition to lowering serum RBP4, RA might exert its effects on glucose homeostasis by regulating RA-dependent genes involved in energy metabolism through RA-receptors.

These findings in experimental animals suggest that RBP4 is responsible for obesity-induced insulin resistance and represents a potential target in T2DM.

2.4.2 Clinical Studies on the Relationship between RBP4, Insulin Resistance, and T2DM

In obese nondiabetic and diabetic subjects, serum RBP4 levels are increased and positively correlated with body mass index (BMI) (Yang et al. 2005; Graham et al. 2006). Genetic modifications in RBP4 are linked with adiposity level and predisposition to visceral adipose mass accumulation (Kovacs et al. 2007; Munkhtulga et al. 2010). Several authors have reported an association of increased serum RBP4 and augmented visceral adipose content (Cho et al. 2006; Gavi et al. 2007; Jia et al. 2007; J. W. Lee et al. 2007; Kelly et al. 2010). In addition, higher RBP4 levels and markers of systemic inflammation are coupled with higher waist circumference and waist-to-hip ratio (Hermsdorff et al. 2010).

Decreased weight, achieved by exercise/diet or bariatric surgery, results in reduced serum/adipose RBP4 levels (Graham et al. 2006; Janke et al. 2006; Haider et al. 2007; Vitkova et al. 2007; Lee et al. 2008). Body weight diminution in nondiabetic subjects is accompanied by decreased serum RBP4 levels and improved insulin sensitivity (Lee et al. 2008). These authors also observed a significant correlation between serum RBP4 levels and abdominal visceral fat loss. In obese subjects, serum RBP4 is negatively associated with insulin secretion (Broch et al. 2007).

Linkage between RBP4 and lipid parameters has been noted in obese women after body weight loss due to bariatric surgery (Broch et al. 2010). Circulating RBP4 is positively correlated with liver fat rather than with total adipose fat in nondiabetic subjects (Stefan et al. 2007). Proteomic and metabolomic profiling of serum has revealed an association of RBP4 with obesity and body fat mass changes (Oberbach et al. 2011). In another study, decreased body weight achieved with exercise by obese women resulted in decreased serum RBP4 that predicted improvement in insulin sensitivity with greater specificity than other adipokines (Graham et al. 2006). In fasted, obese subjects, RBP4 was strongly correlated with retinol. These investigators also reported a low retinol-to-RBP ratio in obese subjects owing to elevated apo-RBP4 (Mills et al. 2008). Thus, a strong correlation exists between adiposity and increased serum and adipose tissue RBP4 levels.

Serum RBP4 levels correlated with the magnitude of insulin resistance in subjects with obesity, impaired glucose tolerance, or T2DM, and in nondiabetic subjects with a strong family history of T2DM (Yang et al. 2005; Cho et al. 2006; Graham et al. 2006; Chavez et al. 2009; Kloting et al. 2010). These investigators also observed linkage between serum RBP4 and metabolic syndrome components. Expression levels of serum adipokines in adipose tissues of T2DM patients revealed higher RBP4 expression in subcutaneous fat (Samaras et al. 2010), indicating that serum RBP4 elevation in insulin resistance conditions is the consequence of increased RBP4 synthesis and secretion from subcutaneous fat tissues.

Increased serum RBP4 levels are associated with insulin resistance in postmenopausal and elderly women (An et al. 2009; Suh et al. 2010). In addition, elevated serum RBP4 is coupled with metabolic syndrome components, such as increased BMI, waist-to-hip ratio, serum triglycerides, and systolic blood pressure. Interestingly, exercise training is accompanied by a reduction of serum RBP4 only in subjects in whom insulin resistance improved. Interventions that ameliorated insulin resistance in humans lowered serum RBP4 (Haider et al. 2007; Lim et al. 2008; Ku et al. 2010). Treatment with the insulin sensitizer rosiglitazone reduced serum RBP4 in T2DM patients (Jia et al. 2007). Several RBP4 gene variants are linked with an increased risk of T2DM (Craig et al. 2007; Kovacs et al. 2007; Hu et al. 2008; van Hoek et al. 2008). These findings suggest that serum RBP4 levels are elevated in early stages of insulin resistance and could be a predictor of T2DM development, enabling early interventions. Since factors, such as exercise, weight loss, and insulin sensitizers, lower serum RBP4 levels, it could serve as a serum component in assessing improved insulin sensitivity.

Several studies have shown an association between circulating RBP4 levels and components of metabolic syndrome and other diseases, such as HIV infection, polycystic ovary syndrome, and psoriasis (Graham et al. 2006; von Eynatten et al. 2007; Han et al. 2009; Makino et al. 2009; Wu et al. 2009; Lim et al. 2010; Mohapatra et al. 2011; Mostafaie et al. 2011; Park et al. 2011; Gerdes et al. 2012; Mellati et al. 2012; Sopher et al. 2012). Some authors have postulated that elevated plasma RBP4 is a useful biomarker of atherosclerosis and cardiovascular disease development (Cabre et al. 2007; Ingelsson et al. 2009; Stuck and Kahn 2009; Sasaki et al. 2010; Pala et al. 2012). Higher epicardial RBP4 and lower GLUT-4 levels in epicardial and subcutaneous adipose tissues are associated with coronary artery disease (Salgado-Somoza et al. 2012). Increased plasma RBP4 is seen in patients with dilated inflammatory cardiomyopathy and a high incidence of T2DM (Bobbert et al. 2009). Although all these studies disclosed a relationship between serum RBP4 and coronary heart disease, it is not clear whether the observed increase of RBP4 in these patients is the cause or consequence of coronary heart disease.

2.4.3 RBP4 in Gestational Diabetes Mellitus

One study found that serum RBP4 levels were not elevated in women with gestational diabetes mellitus (GDM). However, RBP4:retinol molar ratio was higher and correlated with fasting glucose in these patients, indicating that RBP4:retinol ratio is a better marker than RBP4 levels when assessing insulin resistance during pregnancy (Krzyzanowska et al. 2008). On the other hand, several authors have reported serum RBP4 elevation in pregnant women with GDM (Klein et al. 2010; Su et al. 2010; Ortega-Senovilla et al. 2011; Saucedo et al. 2011). These studies suggest that serum RBP4 levels could be considered as a valuable marker of insulin resistance and altered lipid metabolism in pregnancy. A positive correlation has been found between cord RBP4 concentrations and fetal birth weight, indicating that cord RBP4 is important in fetal growth (Chan et al. 2011). In another recent study, serum RBP4 levels and their expression in subcutaneous adipose tissues were shown to be elevated in women with GDM, indicating that the higher RBP4 expression in subcutaneous fat tissues may contribute to the increase in serum RBP4 in GDM (Kuzmicki et al. 2011).

2.4.4 RBP4, Obesity, and Insulin Resistance in Children and Adolescents

Linkage between RBP4, obesity, and metabolic syndrome has been investigated in children and adolescents from multiple ethnic backgrounds. Serum RBP4 levels were elevated in both groups (Aeberli et al. 2007; Balagopal et al. 2007; Reinehr et al. 2008; Friebe et al. 2011; Yeste et al. 2010; Kim et al. 2011). In prepubertal and early pubertal children, serum RBP4 and RBP4-to-serum retinol ratio were correlated with obesity and metabolic syndrome components (Aeberli et al. 2007; J. W. Lee et al. 2007) In this study, other parameters, such as dietary vitamin A intake and subclinical inflammation, were also considered. An association between RBP4, insulin resistance, and triglycerides was found in nonobese adolescents and between RBP4 and triglycerides in obese adolescents.

Circulating RBP4 levels are elevated in obese children with glucose intolerance and are directly related to serum insulin, indicating that RBP4 may contribute to the development of muscle insulin resistance (Yeste et al. 2010). Another study found that serum RBP4 levels were correlated with adiposity and insulin resistance in obese children but not with insulin resistance occurring during puberty (Santoro et al. 2009). In a retrospective cohort, increased RBP4 was linked with higher odds of worsening insulin resistance in overweight black adolescents (Goodman et al. 2009). This study also suggested the utility of RBP4 as a biomarker of risk. In obese children, RBP4 has been primarily associated with adipose tissue mass and, secondarily, with metabolic and cardiovascular parameters (Friebe et al. 2011). In early to mid-adolescence, serum RBP4 concentrations are coupled with multiple risk factors for adiposity-related comorbidities (Conroy et al. 2011). Another study discerned that circulating RBP4 levels were independently associated with adiposity and pubertal development but not with insulin resistance (Rhie et al. 2011). In a prospective cohort study, overweight children had significantly higher RBP4 concentrations than normal weight children (Choi et al. 2011).

Lifestyle interventions decreased serum RBP4 levels that correlated with the magnitude of decrease in inflammatory markers (Balagopal et al. 2007). In a longitudinal follow-up study of obese children, weight loss resulted in decreased RBP4 levels, RBP4-to-serum retinol ratio, and improved insulin sensitivity, suggesting linkage between RBP4, obesity, and insulin resistance in children (Reinehr et al. 2008).

2.4.5 RBP4 IN DIABETIC NEPHROPATHY

Nephropathy, a serious microvascular complication in T2DM, is characterized initially by the appearance of low but abnormal levels of albumin in urine—referred to as microalbuminuria (Tomaszewski et al. 2007). Earlier studies showed elevated RBP4 levels in impaired kidney function, attributed to reduced glomerular filtration and catabolism of RBP4 in the kidneys (Scarpioni et al. 1976; Bernard et al. 1988). Furthermore, impaired RBP4 catabolism resulted in accumulation of a truncated variant of RBP4 in plasma of patients with chronic renal failure (Jaconi et al. 1996). In T2DM patients with microalbuminuria, plasma RBP4 levels were found to be higher despite no differences in plasma retinol levels (Raila et al. 2007). These investigators also established no correlation of plasma RBP4 levels and insulin resistance parameters. Renal dysfunction, as measured by serum uric acid levels, albuminuria severity, and glomerular filtration rate in diabetic patients, was associated with elevated RBP4 (Chang et al. 2008; Henze et al. 2008). Heightened plasma RBP4 levels were observed in diabetic patients with renal disease and correlated with creatinine levels as well as 24-hour creatinine clearance (Masaki et al. 2008). An association of RBP4 with impaired glucose regulation and microalbuminuria has been reported in a Chinese population (Xu et al. 2009). In another study, neither retinopathy nor cardiovascular complications affected serum RBP4 levels, but renal function determined RBP4 levels in T2DM (Akbay et al. 2010). In an observational investigation, kidney donors were found to be more susceptible to insulin resistance (Shehab-Elden et al. 2009). It has been reported that unilateral nephrectomy results in increased serum retinol and RBP4 levels (Henze et al. 2011). However, it is not clear whether elevation of serum RBP4 observed after nephrectomy contributes to insulin resistance in kidney donors. In a cross-sectional study, glomerular filtration rate was seen to be the major determinant of serum RBP4 levels rather than the presence of T2DM. All these studies indicate that kidney function but not T2DM contributes to serum RBP4 elevation.

2.5 CLINICAL STUDIES REPORTING A LACK OF CORRELATION BETWEEN RBP4, OBESITY, AND INSULIN RESISTANCE

Not all investigations in humans have shown that circulating RBP4 levels are elevated in obesity or T2DM, and an inverse correlation has been observed between serum RBP4 concentration and insulin resistance. Janke et al. (2006) detected no difference in serum RBP4 levels between lean, overweight, and obese menopausal nondiabetic women. In addition, these investigators found no correlation between RBP4 adipose gene expression and plasma RBP4 concentrations. In male subjects with T2DM or coronary artery disease, serum RBP4 levels did not correlate with insulin resistance (von Eynatten et al. 2007). In another study, plasma RBP4 levels were not altered in healthy, insulin-resistant humans and were not correlated with insulin sensitivity (Promintzer et al. 2007). Unlike plasma RBP4 elevation in T2DM patients, lower plasma RBP4 was reported in one study (Erikstrup et al. 2008). These investigators observed higher RBP4-to-retinol ratios in individuals with T2DM, suggesting that higher levels of RBP4 relative to retinol, expressed as RBP4-to-retinol ratio, are more indicative of T2DM than RBP4 itself.

Calorie restriction in obese, nondiabetic, premenopausal women resulted in reduced adipose RBP4 expression and plasma RBP4 levels. However, no correlation was found between diet-induced changes of RBP4 and insulin sensitivity (Vitkova et al. 2007). In humans with a wide range of BMI and insulin resistance, no significant relationship was observed between either adipose RBP4 expression or plasma RBP4 concentration and insulin resistance (Yao-Borengasser et al. 2007). Serum RBP4 levels are reduced in patients with liver cirrhosis, and are not related to insulin resistance in these patients (Yagmur et al. 2007). In another study, no correlation was apparent between serum RBP4 levels and insulin resistance in women at high or low risk for vascular disease (Silha et al. 2007). Broch et al. (2007) reported no difference in serum RBP4 concentration among lean, overweight, and obese subjects and no association with age, BMI, weight-to-hip ratio, and insulin sensitivity. However, they found a negative association of circulating RBP4 with insulin secretion in obese subjects.

A study of Mexican Americans showed no linkage of RBP4 with obesity, insulin resistance, and impaired insulin secretion. However, elevation of plasma RBP4 and its association with impaired glucose tolerance was observed in this study (Chavez et al. 2009). Data on the study population, comprised of elderly twins, indicated that elevated plasma RBP4 in T2DM is a secondary and predominantly nongenetic phenomenon and may play a minor role in the development of insulin resistance (Ribel-Midsen et al. 2009). In children and adolescents, serum RBP4 levels did not differ between prepubertal and pubertal stages, despite the latter being significantly more insulin resistant, suggesting a lack of causal association between RBP4 and insulin resistance in this study population (Santoro et al. 2009). A recent study disclosed higher serum RBP4 levels in obese men (mean BMI 41.6 kg/m^2) than in obese women. However, no association of RBP4 levels with insulin resistance or metabolic syndrome components was observed (Ulgen et al. 2010). The lack of RBP4 correlation with insulin resistance in this study could have been due to the severity of obesity in patients.

Several alternative explanations have been proposed for the observed lack of positive correlations between serum RBP4 levels, obesity, and insulin resistance in the aforementioned studies. In most investigations in humans, renal function was not taken into consideration. Since T2DM is often associated with renal dysfunction, which leads to serum RBP4 elevation, it is possible that the increased serum RBP4 observed in some studies could have been attributed to impaired renal function rather than insulin resistance per se (Chang et al. 2008; Henze et al. 2008; Masaki et al. 2008). Differences in the study population, including age, gender, ethnicity, sample size, insulin resistance status, degree of obesity (as measured by BMI), and metabolic syndrome parameters, might also have contributed to the divergent results. Age and sex were found to be independent determinants of plasma RBP4 (Cho et al. 2006; Gavi et al. 2007). In several studies, vitamin A status of patients, which may influence serum RBP4 concentrations, was not addressed (Promintzer et al. 2007). Retinol is mainly transported by RBP4, which serves as a precursor for the synthesis of ligands (RAs) for nuclear hormone receptors. Thus, its levels in serum may affect insulin resistance. It is also possible that T2DM in patients receiving glucose-lowering medications or insulin might affect serum RBP4 levels (von Eynatten et al. 2007; Chavez et al. 2009; Ribel-Madsen et al. 2009).

Shortcomings in the methodology of quantifying serum RBP4 levels as well as handling and storage of serum samples have been cited. Most studies adopted competitive enzyme-linked immunosorbent assay (ELISA) to measure serum RBP4. Anticoagulants added during the collection of blood samples may alter absolute RBP4 immunoreactivity values. In addition, high lipid levels present in plasma samples of obese and T2DM patients may interfere with ELISA. Furthermore, experimental design and different types of antibodies employed in immunoassays may impact serum RBP4 levels. Graham et al. (2007) proposed quantitative Western blotting standardized to full-length RBP4 as a reliable method for quantifying serum RBP4. It provides higher and consistent amounts of RBP4 in serum.

2.6 MECHANISMS INVOLVED IN RBP4 EFFECTS ON INSULIN RESISTANCE

The mechanisms by which RBP4 induces insulin resistance are not well understood. Although a correlation exists between adipose RBP4 expression and elevated serum

RBP4 levels in insulin resistance mouse models (Yang et al. 2005; Wu et al. 2009) and in some human subjects with obesity and T2DM (Graham et al. 2006; Kelly et al. 2010; Samaras et al. 2010), the absolute contribution of adipose tissues to increased serum RBP4 values is not known. Since the majority of circulating RBP4 is derived from the liver, it is likely that this organ may also contribute to the serum RBP4 elevation in insulin resistance conditions. It is not clear whether RBP4 synthesized in adipose tissues carries retinol or other ligands, which may be responsible for the development of insulin resistance. However, recent studies have shown that RBP4, synthesized exclusively in the muscles, bound retinol and TTR in the circulation, indicating that extrahepatic RBP4 can transport retinol (Quadro et al. 2002, 2004).

The other possibility is that elevated holoRBP4 (retinol bound to RBP4) rather than apoRBP4 may be involved in the development of insulin resistance. This view is supported by the recent finding that a non-retinoid compound (A1120) binds to RBP4 and lowers serum RBP4 but is unable to enhance insulin sensitivity (Motani et al. 2009), whereas the retinoid fenretinide and RAs not only reduce serum RBP4 but also improve insulin sensitivity (Yang et al. 2005; Manolescu et al. 2010; Preitner et al. 2009). Consistent with a retinoid-dependent mechanism, RBP4 heightened the hepatic expression of a retinoid-regulated gene, the gluconeogenic enzyme phospho-enolpyruvate carboxykinase (Yang et al. 2005). In addition, retinol was shown to be involved in islet development and secretion of insulin from islets (Chertow et al. 1987; Mathews et al. 2004).

In primary human adipocytes, RBP4 inhibited phosphorylation at the site of insulin receptor substrate-1 that may be involved in integrating nutrient sensing with insulin signaling (Ost et al. 2007). Cyclic AMP-induced RBP4 gene and protein expression is modulated by the high mobility group A1 (HMGA1) gene, indicating a novel biochemical pathway comprising the cAMP-HMGA1-RBP4 system in glucose homeostasis (Chiefari et al. 2009). TTR elevation has been found in insulin-resistant *ob/ob* mice, suggesting that increased TTR or alterations in RBP4-TTR binding may contribute to insulin resistance (Mody et al. 2008). Another study implicated iron-associated insulin resistance as being responsible for the impaired insulin action caused by RBP4 (Fernandez-Real et al. 2008). A recent investigation identified RBP-retinol as an activator of a STRA6/JAK2/STAT5 cascade that induces expression of the inhibitor of insulin signaling, SOCS3, providing a molecular basis for insulin resistance caused by elevated serum RBP4 (Berry et al. 2011).

2.7 PERSPECTIVES

Experiments in rodents have clearly demonstrated an association between RBP4 and insulin resistance. In clinical studies, however, the role of RBP4 in the development of insulin resistance remains controversial. It appears that many confounding factors that affect the proposed association should be considered when interpreting the results in humans. The choice of subjects, including ethnic background, age, sex, severity of obesity, and insulin resistance, might contribute to differences in outcomes. Other factors, such as metabolic syndrome parameters and cardiovascular disease, should be carefully evaluated in establishing the relationship between RBP4, obesity, and insulin resistance. Collection, handling of serum samples, and serum RBP4 analysis must be considered carefully and should be standardized with results from insulin-resistant mouse models before application to human samples.

Retinol status, iron values, and kidney function, which all affect serum RBP4, must be taken into account in future studies. More experimental work is needed to understand the synthesis, regulation, and secretion of RBP4 in adipocytes. The role of adipocytes and the liver in the regulation of serum RBP4 levels has to be defined. Limited information is available on the overall metabolism of retinol in obese and insulin resistance conditions. Future studies on retinol storage, transport, and metabolism in obese and insulin-resistant mouse models may provide new insights into the involvement of RBP4-retinol in the development of insulin resistance.

REFERENCES

Abel, E. D., Peroni, O., Kim, J. K. et al. 2001. Adipose-selective targeting of the GLUT4 gene impairs insulin action in muscle and liver. *Nature* 409:729–733.

Aeberli, I., Biebinger, R., Lehmann, R. et al. 2007. Serum retinol-binding 4 concentrations and its ratio to serum retinol are associated with obesity and metabolic syndrome components in children. *J Clin Endocrinol Metab* 92:4359–4365.

Akbay, E., Muslu, N., Nayir, E. et al. 2010. Serum retinol binding protein 4 levels is related with renal functions in type 2 diabetes. *J Endocrinol Invest* 33:725–729.

Aleman, G., Torres, N., and Tovar A, R. 2004. Peroxisome proliferator-activated receptors (PPARs) in obesity, and insulin resistance development. *Rev Invest Clin* 56:351–367.

An, C., Wang, H., Liu, X. et al. 2009. Serum retinol binding protein 4 is elevated and positively associated with insulin resistance in postmenopausal women. *Endocr J* 56:987–996.

Balagopal, P., Graham, T. E., Kahn, B. B. et al. 2007. Reduction of elevated serum retinol binding protein in obese children by lifestyle intervention: Association with subclinical inflammation. *J Clin Endocrinol Metab* 92:1971–1974.

Bernard, A., Vyskocyl, A., Mahieu, P. et al. 1988. Effect of renal insufficiency on concentration of free retinol-binding protein in urine and serum. *Clin Chim Acta* 171:85–93.

Berry, D. C., Jin, H., Majumdar, A. et al. 2011. Signaling by vitamin A and retinol-binding protein regulates gene expression to inhibit insulin responses. *Proc Natl Acad Sci USA* 108:4340–4345.

Bhat, P. V. 2005. Role of retinol dehydrogenase type 1 (RALDH1) in retinoic acid biosynthesis. *Enzymol Mol Biol* 12:66–72.

Bhat, P. V. and Lacroix, A. 1983. Separation and estimation of retinyl fatty acyl esters in tissues of normal rat by high-performance liquid chromatography. *J Chromatogr* 272:269–278.

Blaner, W. S. 1989. Retinol binding protein: The serum transport protein for vitamin A. *Endocr Rev* 10:308–316.

Blaner, W. S. and Goodman, D. S. 1990. Purification and properties of plasma retinol-binding protein. *Methods Enzymol* 189:193–206.

Bobbert, P., Weithauser, A., Andres, J. et al. 2009. Increased plasma retinol binding protein 4 levels in patients with inflammatory cardiomyopathy. *Eur J Heart Fail* 11:1163–1168.

Broch, M., Gomez, J. M., Auguet, M. T. et al. 2010. Association of retinol-binding protein-4 (RBP4) with lipid parameters in obese women. *Obes Surg* 20:1258–1264.

Broch, M., Vendrell, J., Ricart, W. et al. 2007. Circulating retinol-binding protein-4, insulin sensitivity, insulin secretion, and insulin disposition index in obese and nonobese subjects. *Diabetes Care* 30:1802–1806.

Cabre, A., Lazaro, I., Girona, J. et al. 2007. Retinol-binding protein 4 as a plasma biomarker of renal dysfunction and cardiovascular disease in type 2 diabetes. *J Intern Med* 262:496–503.

Chambon, P. 1996. A decade of molecular biology of retinoic acid receptors. *FASEB J* 10:940–954.

Chan, T. F., Tsai, Y. C., Wu, C. H. et al. 2011. The positive correlation between cord serum retinol-binding protein 4 concentrations and fetal growth. *Gynecol Obstet Invest* 72:98–102.

Chang, Y. H., Lin, K. D., Wang, C. L. et al. 2008. Elevated serum retinol-binding protein 4 concentrations are associated with renal dysfunction and uric acid in type 2 diabetic patients. *Diabetes Metab Res Rev* 24:629–634.

Chavez, A. O,. Coletta, D. K., Kamath, S. et al. 2009. Retinol-binding protein 4 is associated with impaired glucose tolerance but not with whole body or hepatic insulin resistance in Mexican Americans. *Am J Physiol Endocrinol Metab* 296:E758–E764.

Chertow, B. S., Blaner, W. S., Baranetsky, N. G. et al. 1987. Effects of vitamin A deficiency and repletion on rat insulin secretion in vivo and in vitro from isolated islets. *J Clin Invest* 79:163–169.

Chiefari, E., Paonessa, F., Liritano, S. et al. 2009. The cAMP-HMGA1-RBP4 system: A novel biochemical pathway for modulating glucose homeostasis. *BMC Biol* 21:7–24.

Cho, Y. M., Youn, B. S., Lee, H. et al. 2006. Plasma retinol-binding protein-4 concentrations are elevated in human subjects with impaired glucose tolerance and type 2 diabetes. *Diabetes Care* 29:2457–2461.

Choi, K. M., Yannakoulia, M., Park, M. S. et al. 2011. Serum adipocyte fatty acid-binding protein, retinol binding protein 4, and adiponectin concentrations in relation to the development of the metabolic syndrome in Korean boys: A 3-y prospective cohort study. *Am J Clin Nutr* 93:19–26.

Conroy, R., Espinal, Y., Fennoy, I. et al. 2011. Retinol binding protein 4 is associated with adiposity-related co-morbidity risk factors in children. *J Pediatr Endocrinol Metab* 24:913–919.

Craig, R. L., Chu, W. S., and Elbein, S. C. 2007. Retinol binding protein 4 as a candidate gene for type 2 diabetes and prediabetic intermediate traits. *Mol Genet Metab* 90:338–344.

DeFronzo, R. A. 1997. Pathogenesis of type 2 diabetes: Metabolic and molecular implications for identifying diabetes gene. *Diabetes Rev* 5:171–269.

Erikstrup, C., Mortensen, O. H., and Nielsen, A. R. 2008. RBP-to-retinol ratio, but not total RBP, is elevated in patients with type 2 diabetes. *Diabetes Obes Metab* 11:204–212.

Fernandez-Real, J. M., Moreno, J. M., and Ricart, W. 2008. Circulating retinol-binding protein-4 concentration might reflect insulin resistance-associated iron overload. *Diabetes* 57:1918–1925.

Friebe, D., Neef, M., Erbs, S. et al. 2011. Retinol binding protein 4 (RBP4) is primarily associated with adipose tissue mass in children. *Int J Pediat Obes* 6:e346–e352.

Frolik, C. A., Roberts, A. B., Tavela, T. E. et al.1979. Isolation and identification of 4-hydroxy- and 4-oxoretinoic acid: In vitro metabolites of *all-trans* retinoic acid in hamster trachea and liver. *Biochemistry* 18:2092–2097.

Gavi, S., Stuart, L. M., Kelly, P. et al. 2007. Retinol-binding protein 4 is associated with insulin resistance and body fat distribution in nonobese subjects without type 2 diabetes. *J Clin Endocrinol. Metab* 92:1886–1890.

Gerdes, S., Osadtschy, S., Rostami-Yazdi, M. et al. 2012. Leptin, adiponectin, visfatin and retinol-binding protein-4-mediators of comorbidities in patients with psoriasis. *Exp Dermatol* 21:43–47.

Goodman, D. S. 1974. Vitamin A transport and RBP metabolism. *Vitam Horm* 32:167–180.

Goodman, D. S. 1982. Retinoid-binding proteins. *J Am Acad Dermatol* 6:583–590.

Goodman, E., Graham, T. E., Dolan, L. M. et al. 2009. The relationship of retinol binding protein 4 to changes in insulin resistance and cardiometabolic risk in overweight black adolescents. *J Pediatr* 154:67–73.

Gottesman, M. E., Quadro, L., and Blaner, W. S. 2001. Studies of vitamin A metabolism in mouse model systems. *BioEssays* 23:409–419.

Graham, T. E., Wason, C. J., Bluher, M. et al. 2007. Shortcomings in methodology complicate measurements of serum retinol binding protein (RBP4) in insulin-resistant human subjects. *Diabetologia* 50:814–823.

Graham, T. E., Yang, Q., Bluher, M. et al. 2006. Retinol-binding protein 4 and insulin resistance in lean, obese, and diabetic subjects. *N Engl J Med* 354:2552–2563.

Haider, D. G., Schindler, K., Prager, G. et al. 2007. Serum retinol-binding protein 4 is reduced after weight loss in morbidly obese subjects. *J Clin Endocrinol Metab* 92:1168–1171.

Han, S. H., Chin, B. S., Lee, H. S. et al. 2009. Serum retinol-binding protein 4 correlates with obesity, insulin resistance, and dyslipidemia in HIV-infected subjects receiving highly active antiretroviral therapy. *Metabolism* 58:1523–1529.

Henze, A., Frey, S. K., Reila, J. et al. 2008. Evidence that kidney function but not type 2 diabetes determines retinol-binding protein 4 serum levels. *Diabetes* 57:3323–3326.

Henze, A., Raila, J., Kempf, C. et al. 2011. Vitamin A metabolism is changed in donors after living kidney transplantation: An observational study. *Lipids Health Dis* 10:231.

Hermsdorff, H. H., Zulet, M. A., Puchau, B. et al. 2010. Central adiposity rather than total adiposity measurements are specifically involved in the inflammatory status from healthy young adults. *Inflammation* 34:161–170.

Hu, C., Jia, W., Zang, R. et al. 2008. Effect of RBP4 gene variants on circulating RBP4 concentration and type 2 diabetes in a Chinese population. *Diabetes Med* 25:11–18.

Ingelsson, E., Sundstrom, J., Melhus, H. et al. 2009. Circulating retinol-binding protein 4, cardiovascular risk factors and prevalent cardiovascular disease in elderly. *Atherosclerosis* 206:239–244.

Jaconi, S., Saurat, J. H., and Siegenthaler, G. 1996. Analysis of normal and truncated holo- and apo-retinol binding protein (RBP) in human serum: Altered ratios in chronic renal failure. *Eur J Encrinol* 134:576–582.

Janke, J., Engeli, S., Boschmann, M. et al. 2006. Retinol-binding protein 4 in human obesity. *Diabetes* 55:2805–2810.

Jia, W., Wu, H., Bao, Y. et al. 2007. Association of serum retinol-binding protein 4 and visceral adiposity in Chinese subjects with and without type 2 diabetes. *J Clin Endocrinol Metab* 92:3224–3229.

Kahn, B. B. and Flier, J. S. 2000. Obesity and insulin resistance. *J Clin Invest* 106:473–481.

Kanai, M., Raz, A., and Goodman, D. S. 1968. Retinol binding protein: The transport protein for vitamin A in human plasma. *J Clin Invest* 47:2025–2044.

Kawaguchi, R., Yu, J., Honda, J. et al. 2007. A membrane receptor for retinol binding protein mediates cellular uptake of vitamin A. *Science* 315:820–825.

Kelly, K. R., Kashyap, S. R., O'Leary, V. B. et al. 2010. Retinol-binding protein 4 (RBP4) protein expression is increased in omental adipose tissue of severely obese patients. *Obesity* 18:663–666.

Kershaw, E. E. and Flier, J. S. 2004. Adipose tissue as an endocrine organ. *J Clin Endocrinol Metab* 89:2548–2556.

Kim, I. K., Lee, H. J., Kang, J. H. et al. 2011. Relationship of serum retinol-binding protein 4 with weight status and lipid profile among Korean children and adults. *Eur J Clin Nutr* 65:226–233.

Klein, K., Bancher-Todesca, D., Leipold, H. et al. 2010. Retinol-binding protein 4 in patients with gestational diabetes mellitus. *J Womens Health* 19:517–521.

Kloting, N., Fasshauer, M., Deitrich, A. et al. 2010. Insulin-sensitive obesity. *Am J Physiol Endocrinol Metab* 299:E506–E515.

Kovacs, P., Geyer, M., Berndt, J. et al. 2007. Effects of genetic variation in the human retinol binding protein 4 gene (RBP4) on insulin resistance and fat depot-specific mRNA expression. *Diabetes* 56:3095–3100.

Krzyzanowska, K., Zemany, L., Krugluger, W. et al. 2008. Serum concentrations of retinol-binding protein 4 in women with and without gestational diabetes. *Diabetologia* 51:1115–1122.

Ku, Y. H., Han, K. A., Ahn, H. et al. 2010. Resistance exercise did not alter intramuscular adipose tissue but reduced retinol-binding protein-4 concentration in individuals with type 2 diabetes mellitus. *J Intl Med Res* 38:782–791.

Kuzmicki, M., Telejko, B., Wawrusiewicz-Kurylonek, N. et al. 2011. Retinol-binding protein 4 in adipose and placental tissue of women with gestational diabetes. *Gynecol Endocrinol* 27:1065–1069.

Labrecque, J., Dumas, F., Lacroix, A. et al. 1995. A novel isozyme of aldehyde dehydrogenase specifically involved in the biosynthesis of 9-*cis* and *all-trans* retinoic acid. *Biochem J* 305:681–684.

Lebovitz, H. E. and Banerji, M. A. 2001. Insulin resistance and its treatment by thiazolodinediones. *Rec Prog Horm Res* 56:265–410.

Lee, D. C., Lee, J. W., and Im J. A. 2007. Association of serum retinol binding protein 4 and insulin resistance in apparently healthy adolescents. *Metabolism* 56:327–331.

Lee, J. W., Im, J. A., Lee, H. R. et al. 2007. Visceral adiposity is associated with serum retinol binding protein-4 levels in healthy women. *Obesity* 15:2225–2232.

Lee, J. W., Lee, H. R., Shim, J. Y. et al. 2008. Abdominal visceral fat reduction is associated with favourable changes of serum retinol-binding protein 4 in nondiabetic subjects. *Endocrine J* 55:811–818.

Lenhard, J. M. 2001. PPAR gamma/RXR as a molecular target for diabetes. *Receptors Channels* 7:248–258.

Lim, S., Choi, S. H., Jeong, I. K. et al. 2008. Insulin-sensitizing effects of exercise on adiponectin and retinol binding-4 concentrations in young and middle aged women. *J Endocrinol Metab* 93:2263–2268.

Lim, S., Yoon, J. W., Choi, S. H. et al. 2010. Combined impact of adiponectin and retinol-binding protein 4 on metabolic syndrome in elderly people: The Korean longitudinal study on health and aging. *Obesity* 18:826–832.

Mader, S., Chen, J. Y., Chen, Z. et al. 1993. The patterns of binding affinities of RAR, RXR and TR homo- and heterodimers to direct repeats are dictated by the binding specificities of the DNA binding domains. *EMBO J* 15:5029–5041.

Makeover, A., Soprano, D. R., Wyatt, M. L. et al. 1989. Localization of retinol-binding protein messenger RNA in the rat kidney and in perinephric fat tissue. *J Lipid Res* 30:171–180.

Makino, S., Fujiwara, M., Suzukawa, K. et al. 2009. Visceral obesity is associated with the metabolic syndrome and elevated plasma retinol-binding protein-4 level in obstructive sleep apnea syndrome. *Horm Metab Res* 41:221–226.

Manolescu, D. C., Sima, A., and Bhat, P. V. 2010. All-*trans* retinoic acid lowers serum retinol-binding protein 4 concentrations and increases insulin sensitivity in diabetic mice. *J Nutr* 140:311–316.

Masaki, T., Anan, F., Tsubone, T. et al. 2008. Retinol binding protein 4 concentrations are influenced by renal function in patients with type 2 diabetes. *Metabolism* 57:1340–1344.

Mathews, K. A., Rhoten, W. B., Driscoll, H. K. et al. 2004. Vitamin A deficiency impairs fetal islet development and causes subsequent glucose intolerance in adult rats. *J Nutr* 134:1958–1963.

Mellati, A. A., Sharifi, F., Sajadinejad, M et al. 2012. The relationship between retinol-binding protein 4 levels, insulin resistance, androgen hormones and polycystic ovary syndrome. *Scand J Clin Lab Invest* 72:39–44.

Metzer, D., Imai T., Jiang, M. et al. 2005. Functional role of RXRs and PPAR gamma in mature adipocytes. *Prostaglandins Leukot Essent Fatty Acids* 73:51–58.

Mills, J. P., Furr, H. C., and Tanumihardjo, S. A. 2008. Retinol to retinol-binding protein (RBP) is low in obese adults due to elevated apo-RBP. *Exp Biol Med* 233:1255–1261.

Mody, N., Graham, T. E., Tsuji, Y et al. 2008. Decreased clearance of serum retinol-binding protein and elevated levels of transthyretin in insulin-resistant *ob/ob* mice. *Am J Physiol Metab* 294:E785–E793.

Mohapatra, J., Sharma, M., Acharya, A. et al. 2011. Retinol-binding protein 4: A possible role in cardiovascular complications. *Br J Pharmacol* 164:1939–1948.

Mostafaie, N., Sebesta, C., Zehetmayer, S. et al. 2011. Circulating retinol-binding protein 4 and metabolic syndrome in the elderly. *Wien Med Wochenschr* 161:505–510.

Motani, A., Zhulun, W., Conn, M. et al. 2009. Identification and characterization of a non-retinoid ligand for retinol-binding 4 levels *in vivo*. *J Biol Chem* 284:7673–7680.

Mukherjee, R., Davies, P. J., Crombie, D. L. et al. 1997. Sensitization of diabetic and obese mice to insulin by retinoid X receptor agonist. *Nature* 386:407–410.

Munkhtulga, L., Nagashima, S., Nakayama, K. et al. 2010. Regulatory SNP in the RBP4 gene modified the expression in adipocytes and associated with BMI. *Obesity* 18:1006–1014.

Napoli, J. L. 2000. Retinoic acid: Its biosynthesis and metabolism. *Prog Nucleic Acid Res* 63:139–188.

Oberbach, A., Bluher, M., Wirth, H. et al. 2011. Combined proteomic and metabolomic profiling of serum reveals association of the complement system with obesity and identifies novel markers of body fat mass changes. *J Proteome Res* 10:4769–4788.

Ortega-Senovilla, H., Schaefer-Graf, U., Meitzner, K. et al. 2011. Gestational diabetes mellitus causes changes in the concentrations of adipocyte fatty acid-binding protein and other adipokines in cord blood. *Diabetes Care* 34:2061–2066.

Ost, A., Danielsson, A., Liden, M. et al. 2007. Retinol-binding protein-4 attenuates insulin-induced phosphorylation of IRS1 and ERK1/2 in primary human adipocytes. *FASEB J* 21:3696–3704.

Ou, H. Y., Wu, H. T., Yang, Y. C. et al. 2011. Elevated retinol binding protein 4 contributes to insulin resistance in spontaneously hypertensive rats. *Horm Metab Res* 43:312–318.

Pala, A., Monami, M., Ciani, S. et al. 2012. Adipokines as possible new predictors of cardiovascular diseases: A case control study. *J Nutr Metab* 2012:253428.

Park, C. S., Ihm, S. H., Park, H. J. et al. 2011. Relationship between plasma adiponectin, retinol binding protein 4 and uric acid in hypertensive patients with metabolic syndrome. *Korean Circ J* 41:198–202.

Pasutto, F., Sticht, H., Hammersen, G. et al. 2007. Mutations in STRA6 cause a broad spectrum of malformations including anophthalmia, congenital heart defects, diaphragmatic hernia, alveolar capillary dysplasia, lung hypoplasia, and mental retardation. *Am J Hum Genet* 80:550–560.

Petersen, K. F. and Shulman, G. I. 2006. Etiology of insulin resistance. *Am J Med* 119:S10–S16.

Peterson, P. A. 1971. Characterization of a vitamin A transport protein complex occurring in human serum. *J Biol Chem* 246:34–43.

Pocai, A., Obici, S., Schwartz, G. J. et al. 2005. A brain-liver circuit regulates glucose homeostasis. *Cell Metab* 1:53–61.

Preitner, F., Mody, N., Graham, T. E. et al. 2009. Long-term Fenretinide treatment prevents high-fat diet-induced obesity, insulin resistance, and hepatic steatosis. *Am J Physiol Endocrinol Metab* 297:E1420–E1429.

Promintzer, M., Krebs, M., Todoric, J. et al. 2007. Insulin resistance is unrelated to circulating retinol binding protein and protein C inhibitor. *J Clin Endocrinol Metab* 92:4306–4312.

Qatanani, M. and Lazar, M. A. 2007. Mechanisms of obesity-associated insulin resistance: Many choices on the menu. *Genes Dev* 21:1443–1455.

Quadro, L., Blaner, W. S., Hamberger, L. et al. 2002. Muscle expression of human retinol-binding protein (RBP). Suppression of the visual defect of RBP knockout mice. *J Biol Chem* 277:30191–30197.

Quadro, L., Blaner, W. S., Hamberger, L. et al. 2004. The role of extrahepatic retinol binding protein in the mobilization of retinoid stores. *J Lipid Res* 45:1975–1982.

Raila, J., Henze, A., Spranger, J. et al. 2007. Microalbuminuria is a major determinant of elevated plasma retinol-binding protein 4 in type 2 diabetic patients. *Kidney Intl* 72:505–511.

Rajala, M. W. and Scherer, P. E. 2003. The adipocyte... at the crossroads of energy homeostasis, inflammation, and atherosclerosis. *Endocrinology* 144:3765–3773.

Reichert, B., Yasmeen, R., Jeyakumar, S. M. et al. 2011. Concerted action of aldehyde dehydrogenases influences depot-specific fat formation. *Mol Endocrinol* 25:799–809.

Reinehr, T., Stoffel-Wagner, B., and Roth, C. L. 2008. Retinol-binding protein 4 and its relation to insulin resistance in obese children before and after weight loss. *J Clin Endocrinol Metab* 93:2287–2293.

Rhie, Y. J., Choi, B. M., Eun, S. H. et al. 2011. Association of serum retinol binding protein 4 with adiposity and pubertal development in Korean children and adolescents. *J Korean Med Sci* 26:797–802.

Ribel-Madsen, R., Friedrichsen, M., Vaag, A. et al. 2009. Retinol-binding protein 4 in twins. Regulatory mechanisms and impact of circulating and tissue expression levels on insulin secretion and action. *Diabetes* 58:54–60.

Rosen, E. D. and Spiegelman, B. M. 2001. PPAR gamma: A nuclear regulator of metabolism, differentiation, and cell growth. *J Biol Chem* 276:37731–37734.

Salgado-Somoza, A., Teijeira-Fernandez, E., Rubio, J. et al. 2012. Coronary artery disease is associated with higher epicardial retinol-binding protein 4 (RBP4) and lower glucose transporter (GLUT) 4 levels in epicardial and subcutaneous adipose tissue. *Clin Endocrinol* 76:51–58.

Saltiel, A. R. and Kahn, C. R. 2001. Insulin signaling and the regulation of glucose and lipid metabolism. *Nature* 414:799–806.

Samaras, K., Botelho, N. K., and Chisholm, D. J. 2010. Subcutaneous and visceral adipose tissue gene expression of serum adipokines that predict type 2 diabetes. *Diabetes* 18:884–889.

Santoro, N., Perrone, L., Cirillo, G. et al. 2009. Variations of retinol binding protein 4 levels are not associated with changes in insulin resistance during puberty. *J Endocrinol Invest* 32:411–414.

Sasaki, M., Otani, T., Kawakami, M. et al. 2010. Elevation of plasma retinol-binding protein 4 and reduction of plasma adiponectin in subjects with cerebral infarction. *Metabolism* 59:527–532.

Saucedo, R., Zarate, A., Basurto, L. et al. 2011. Relationship between circulating adipokines and insulin resistance during pregnancy and postpartum in women with gestational diabetes. *Arch Med Res* 42:318–323.

Scarpioni, L., Dall'aglio, P. P., Poisetti, P. G. et al. 1976. Retinol binding protein in serum and urine of glomerular and tubular nephropathies. *Clin Chim Acta* 68:107–113.

Shehab-Elden, W., Shoeb, S., Khamis, S. et al. 2009. Susceptibility to insulin resistance after kidney donation: A pilot observational study. *Am J Nephrol* 30:371–376.

Shepherd, P. R. and Khan, B. B. 1999. Glucose transporters and insulin action-implications for insulin resistance and diabetic mellitus. *N Engl J Med* 341:248–257.

Silha, J. V., Gregoire-Nyomba, B. L., Leslie, W. D. 2007. Ethinicity, insulin resistance, and inflammatory adipokines in women at high and low risk for vascular disease. *Diabetes Care* 30:286–291.

Sima, A., Manolescu, D-C., and Bhat, P. V. 2011. Retinoids and retinoid-metabolic gene expression in mouse adipose tissues. *Biochem Cell Biol* 89:578–584.

Sopher, A. B., Gerken, A. T., Blaner W. S. et al. 2012. Metabolic manifestations of polycystic ovary syndrome in nonobese adolescents: Retinol-binding protein 4 and ectopic fat deposition. *Fertil Steril* 97:1009–1015.

Soprono, D. R. and Blaner, W. S. 1994. Plasma retinol-binding protein. In: *The Retinoids: Biology, Chemistry and Medicine,* M. B. Sporn, A. B. Roberts, and D. S. Goodman, eds. Raven Press, New York: 257–282.

Stefan, N., Hennige, A. M., and Staiger, H. 2007. High circulating retinol-binding protein 4 is associated with elevated liver fat but not with total, subcutaneous, visceral, or intramyocellular fat in humans. *Diabetes Care* 30:1173–1178.

Stuck, B. J. and Kahn, B. B. 2009. Retinol binding protein 4 (RBP4): A biomarker for subclinical atherosclerosis. *Am J Hypertens* 22:948–949.

Su, Y. X., Hong, J., Yan, Q. et al. 2010. Increased serum retinol-binding protein-4 levels in pregnant women with and without gestational diabetes mellitus. *Diabetes Metab* 36:470–475.

Suh, J. B., Kim, S. M., Cho, G. J. et al. 2010. Elevated serum retinol-binding protein 4 is associated with insulin resistance in older women. *Metabolism* 59:118–122.

Tan, Y., Sun, L. Q., Kamal, M. A. et al. 2011. Suppression of retinol-binding protein 4 with RNA oligonucleotide prevents high-fat diet-induced metabolic syndrome and non-alcoholic fatty liver disease in mice. *Biochim Biophys Acta* 1811:1045–1053.

Tomaszewski, M., Charchar, F. J., Maric, C. et al. 2007. Glomerular hyperfiltration: A new marker of metabolic risk. *Kidney Int* 71:816–821.

Tsutsumi, C., Okuno, M., Tannous, L. et al. 1992. Retinoids and retinoid-binding protein expression in rat adipocytes. *J Biol Chem* 267:1805–1810.

Ulgen, F., Herder, C., Kuhn, M. C. et al. 2010. Association of serum levels of retinol-binding protein 4 with male sex but not with insulin resistance in obese patients. *Arch Physiol Biochem* 116:57–62.

van Hoek, M., Dehghan, A., Zillikens, M. C. et al. 2008. An RBP4 promoter polymorphism increases risk of type 2 diabetes. *Diabetologia* 51:1423–1428.

Villarroya, F., Giralt, M., and Iglesias, R. 1999. Retinoids and adipose tissue: Metabolism, cell differentiation and gene expression. *Int J Obes Relat Metab Disord* 23:1–6.

Vitkova, M., Klimcakova, E., Kovacikova, M. et al. 2007. Plasma levels and adipose tissue messenger ribonucleic acid expression of retinol-binding protein 4 are reduced during calorie restriction in obese subjects but are not related to diet-induced changes in insulin sensitivity. *J Clin Endocrinol Metab* 92:2330–2335.

Vogel, S., Piantedosi, R., O'Bryme, S. M. et al. 2002. Retinol-binding protein-deficient mice: Biochemical basis for impaired vision. *Biochemistry* 41:15360–15368.

von Eynatten, M., Lepper, P. M., Liu, D. et al. 2007. Retinol-binding protein 4 is associated with components of metabolic syndrome, but not with insulin resistance, in men with type 2 diabetes or coronary heart disease. *Diabetologia* 50:1930–1937.

Weisberg, S. P. McCann, D. Desai, M. et al. 2003. Obesity is associated with macrophage accumulation in adipose tissue. *J Clin Invest* 112:1789–1808.

Wu, H., Wei, L., Bao, Y. et al. 2009. Fenofibrate reduces serum retinol-binding protein-4 by suppressing its expression in adipose tissue. *Am J Physiol Endocrinol Metab* 296:E628–E634.

Xu, M., Li, X. Y., Wang, J. G. et al. 2009. Retinol binding protein 4 is associated with impaired glucose regulation and microalbuminuria in a Chinese population. *Diabetologia* 52:1511–1519.

Yagmur, E., Weiskirchen, R., Gressner, A. M. et al. 2007. Insulin resistance in liver cirrhosis is not associated with circulating retinol-binding protein 4. *Diabetes Care* 30:1168–1172.

Yang, Q., Graham, T. E., Mody, N. et al. 2005. Serum retinol binding protein 4 contributes to insulin resistance in obesity and type 2 diabetes. *Nature* 436:356–362.

Yao-Borengasser, A., Varma, V., Bodles, A. M. et al. 2007. Retinol binding protein 4 expression in humans: Relationship to insulin resistance, inflammation, and response to pioglitazone. *J Clin Endocrinol Metab* 92:2590–2597.

Yeste, D., Vendrell, J., Tomasini, R. et al. 2010. Retinol-binding protein 4 levels in obese children and adolescents with glucose intolerence. *Horm Res Pediat* 73:335–340.

Zang, X. K., Lehman, J., Hoffmann, B. et al. 1992. Homodimer formation of retinoid X receptor induced by 9-*cis* retinoic acid. *Nature* 358:587–591.

3 Retinoic Acid and Immunity

Yoshishige Miyabe, Chie Miyabe, and Toshihiro Nanki

CONTENTS

3.1 INTRODUCTION

Vitamin A (retinol) is a fat-soluble vitamin important for the maintenance of skin, bone, and blood vessels, as well as for the promotion of vision (Theodosiou et al. 2010). It is obtained from the diet either as all-trans-retinol, retinyl esters, or β-carotene (Blomhoff and Blomhoff 2006) and is stored in the liver (Moise et al. 2007). Vitamin A is converted to retinoic acid (RA), which is formed mainly through intracellular oxidative metabolism by retinal dehydrogenases (RALDHs) (Lampen et al. 2000). RA plays important roles in embryonic development, organogenesis, tissue homeostasis, cell proliferation, differentiation, and apoptosis (Theodosiou et al. 2010). In adult mammals, RALDH is found in intestinal epithelial cells (IECs) and gut associated-dendritic cells (DCs) from Peyer's patches and mesenteric lymph nodes (Iwata 2004, Coombes et al. 2007). Gut-associated DCs and IECs can metabolize vitamin A to RA *in vitro* (Lampen 2000), which indicates they may be a source of RA in gut mucosa. RA binds to two families of nuclear receptors, RA receptor (RAR) isotypes (α, β, and γ) and retinoic X receptor (RXR) isotypes (α, β, and γ). RAR and RXR form heterodimers and interact with retinoic acid response elements (RAREs) within the promoters of retinoic acid responsive genes (Blomhoff and Blomhoff 2006). RAR is ubiquitously expressed and up-regulated by RA. RXR also forms the RXR-RXR homodimer. In addition, RXR can also bind to other nuclear receptors, including nuclear vitamin D receptor, thyroid hormone receptor, peroxisome proliferator activated receptor, and liver X receptor. Recently, RA was shown to play an important role in immune function through RAR and RXR (Blomhoff and Blomhoff 2006). In this chapter, the immunological function of RA and effects of retinoids on autoimmune diseases are reviewed.

3.2 IMMUNOLOGICAL FUNCTION OF RETINOIC ACID

3.2.1 IMMUNOMODULATORY ROLE OF RETINOIC ACID

3.2.1.1 T Cells

We know that diarrhea with infectious disease is a major cause of infant mortality in developing countries and that vitamin A supplementation significantly reduces this mortality (Sommer et al. 1983, Sommer 1984). It has been suggested that vitamin A plays an important role in gut immunity.

Stimulation of T cells with all-trans retinoic acid (ATRA) induces cell proliferation. ATRA does not alter the expression of IL-2 receptor (IL-2R) but strongly enhances the level of IL-2 secreted by T cells. Secreted IL-2 affects the cell cycle and DNA synthesis of T cells. Blocking of IL-2R markedly inhibits ATRA-induced cell proliferation. Therefore, RA has an effect on T cell proliferation by inducing IL-2 secretion, which stimulates T cells in an autocrine manner (Ertesvag et al. 2002). Vitamin A–deficient mice exhibit sharply diminished IgG responses to protein Ag. T cells derived from vitamin A–deficient mice provide inadequate stimulatory signals to B cells, whereas the function of B cells and macrophages from vitamin A–deficient mice is normal. The T cell functional block is reversible by supplementation with RA, which restores the capacity of T cells to stimulate B cells. Therefore, T cell activity and immunological functions may partially depend on RA (Carman et al. 1989).

In addition, it was reported that vitamin A can modulate the function of the immune response, such as the T-helper 1 (Th1)–T-helper 2 (Th2) cell balance and differentiation of T-helper 17 (Th17) cells and forkhead box protein 3 (FOXP3)-positive regulatory T (Treg) cells. Vitamin A deficiency results in decreased Th2 cells and vitamin A supplementation blocks the production of IFN-γ, a Th1 cell–specific cytokine *in vitro* and *in vivo*. These effects of vitamin A on Th1 and Th2 differentiation are dependent on RA. Actually, RA enhances Th2 differentiation by inducing IL-4 gene expression (Lovett-Racke and Racke 2002). In addition, RA inhibits differentiation of Th1 cells by down-regulating Th1-specific T box transcription factor (T-bet) expression, promotes Th2 cell differentiation by inducing GATA-binding protein 3 (GATA3) and macrophage-activating factor expression, and activates signal transducer and activator of transcription 6 (Figure 3.1) (Iwata et al. 2003).

Recently, it was reported that RA could also regulate differentiation of Th17 and Treg cells. Gut-associated DCs enhanced Treg cell differentiation in an RA dose–dependent manner and RA could also induce the expression of gut-homing receptors α4β7 integrin and CC-chemokine receptor 9 (CCR9) on T cells *in vitro*. In up-regulating the expression of gut-homing receptors, RA could block Th17 cell differentiation and induce Treg cells by down-regulating receptor-related orphan receptor-γt (RORγt) and induce FOXP3 expression in T cells, depending on the concentration of RA (Figure 3.1) (Iwata et al. 2004). Thus, the vitamin A metabolite, RA, can regulate differentiation of Th1, Th2, Th17, and Treg cells.

3.2.1.2 B Cells

RA also stimulates B cells with p38 MAPK phosphorylation, which induces IL-10 and IgG secretion and cyclin D3 expression. IL-10 is in turn essential for both cyclin D3 expression and IgG secretion. Cyclin D3 plays an important role in B cell proliferation. Thus, as a result of the p38 MAPK phosphorylation induced by RA, RA enhanced the proliferation of memory B cells, and this proliferative response is accompanied by increased immunoglobulin (Ig) secretion indicative of plasma-cell formation (Ertesvag et al. 2007).

Another report showed that RA regulates the dynamics of B cell antigen receptor/ CD38-stimulated B cell activation/differentiation in both early and later stages of cell activation. In the early stages of B cell activation, RA restrains cell proliferation and reduces the frequency of cell division, germ-line transcript levels, and expression of the *pax-5* gene, which is a member of the paired box family of transcription factors and encodes the B cell lineage–specific activator protein that was expressed at early, but not late stages of B cell differentiation. In the intermediate and later stages of B cell differentiation, RA increases expression of activation-induced cytidine deaminase, which is currently thought to be the master regulator of secondary antibody diversification and enriches the population of B cells that express IgG1. Overall, RA could promote B cell differentiation, antibody production, and isotype switching (Chen and Ross 2005). Thus, RA has also an effect on B cell activation and proliferation through RAR receptors.

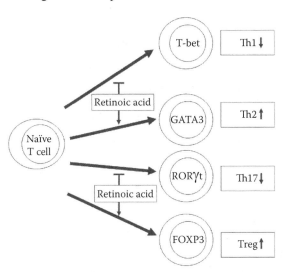

FIGURE 3.1 The effect of retinoic acid (RA) on Th differentiation. RA induces differentiation of Th2 and Treg, and suppresses Th1 and Th17 differentiation.

3.2.1.3 DCs

RA increases expression of matrix metalloproteinases, which have the potential to boost tumor-specific T cell responses by increasing migration of tumor-infiltrating DCs to draining lymph nodes. In the presence of inflammation, RA induces DC maturation and capacity of antigen present through RXR receptors (Darmanin et al. 2007). Gut-associated DCs enhance Treg cell differentiation and production of IgA in an RA dependent manner *in vitro* (Sun et al. 2007, Saurer et al. 2007). It was reported that vitamin A–depleted mice showed decreased IgA in the small bowel lamina propria, and oral administration of an RAR agonist significantly increased serum IgA levels (Kuwabara et al. 1996). These reports suggested that RA could stimulate gut-associated DCs to induce IgA production by B cells. Recently, nitric oxide synthase (iNOS) and nitric oxide (NO) are known to induce the production of IgA. In addition, RA enhances iNOS expression in several organs and plasma concentrations of nitrates and nitrites (Seguin et al. 2002). Therefore, RA may also indirectly generate IgA secretion through the generation of iNOS and nitric oxide.

3.2.2 RETINOIC ACID AND LYMPHOCYTE HOMING

Effector and memory lymphocytes acquire trafficking-related molecules that endow them with the capacity to migrate to selected extralymphoid tissues and sites of inflammation. Gastrointestinal mucosa and the skin are two main body surfaces exposed to environmental antigens, for which tissue-specific adhesion and chemoat-tractant receptors have been characterized in detail. Migration of effector and memory lymphocytes to the small bowel requires expression of $\alpha4\beta7$-integrin and CCR9, whereas migration to the skin relies on the expression of ligands for E- and P-selectin and CCR4 or CCR10 (Mora et al. 2005). In the lymphoid microenvironment, DCs are essential for efficient T cell activation, sufficient enough to induce the expression of $\alpha4\beta7$-integrin and CCR9 and, therefore, imprint gut-homing capacity on activated T cells. It has been reported that mice depleted of vitamin A show decreased numbers of effector and memory T cells in the gut mucosa, gut-associated DCs that express RALDH enzymes, essential for retinoic acid biosynthesis. Inhibition of RAR signaling significantly decreases induction of $\alpha4\beta7$-integrin expression on T cells, and RA could induce the expression of $\alpha4\beta7$-integrin and CCR9 on activated T cells, even if DCs are absent (Figure 3.2) (McDermott et al. 1982, Mora et al. 2003, Iwata et al. 2004). Therefore, RA is necessary to induce colon-homing of lymphocytes.

RA could regulate activation of T cells, B cells, and DCs; differentiation of T cells and B cells; and production of immunoglobulin and T cell homing to gut. These immunological functions of RA might contribute to human immunity, including preventing gut infection.

3.3 EFFECTS OF RETINOIDS ON AUTOIMMUNE DISEASES

3.3.1 RETINOIDS

Retinoid, a derivative of vitamin A, is a general term for compounds that bind to and activate RARs and/or RXRs, members of the nuclear receptor superfamily. ATRA is the

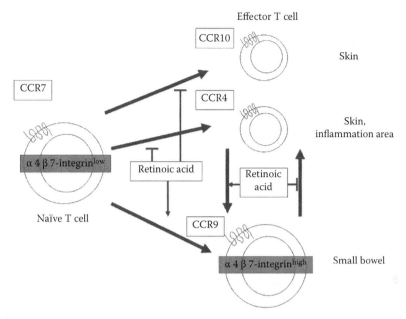

FIGURE 3.2 Roles of retinoic acid (RA) in T cell homing. RA induces the expression of α4β7-integrin and CC-chemokine receptor 9 (CCR9), and suppresses the expression of CCR10 or CCR4 on an effector T cell.

most notable endogenous retinoid, which is a ligand for RARα, β, and γ. The synthetic retinoid, Am80, is a specific ligand for RARα and β, but not for RARγ, and is characterized by higher stability, fewer side effects, and superior bioavailability than those with ATRA. Synthetic retinoids were also reported to promote differentiation of Th2 and Treg and suppress Th1 and Th17 differentiation. In addition, retinoids inhibit TNF-α and NO production by murine peritoneal macrophages and human keratinocytes. Clinically, retinoids are often used for the treatment of cutaneous inflammatory disorders such as psoriasis and acne. In addition, ATRA and Am80 are used for the treatment of acute promyelocytic leukemia.

3.3.2   RETINOIDS AND AUTOIMMUNE DISEASES

Several studies have reported that retinoids are effective for animal models of autoimmune diseases. Lupus nephritis is a major cause of mortality among systemic lupus erythematosus patients. ATRA inhibits IFN-γ cytokine production from Th1 and production and deposition to the kidneys of anti-DNA antibody IgG2a, and suppresses proteinuria and renal involvement in NZB/WF1 mice, which are used as a lupus nephritis model (Nozaki et al. 2005). In an open clinical trial, seven patients with active lupus nephritis were treated with ATRA. As a result, four patients showed improvements in clinical symptoms and laboratory findings, including proteinuria and anti-dsDNA antibody levels. There were no adverse effects of ATRA therapy in any patient (Kinoshita et al. 2009).

In addition, Am80 has been successfully treated in some models, including experimental autoimmune encephalomyelitis (Wang et al. 2000), adjuvant arthritis (Brinckerhoff et al. 1983), collagen-induced arthritis (CIA) (Sato et al. 2010), experimental autoimmune myositis (Ohyanagi et al. 2009), and experimental nephritis (Escribese et al. 2007). In these models, the effects of retinoid were, at least partially, due to the inhibition of Th17 cell differentiation and induction of Th2 cell responses. In addition, suppression of pro-inflammatory cytokine production and chemokines, decreased expression of $\alpha4\beta1$-integrin on effector T cells, and suppression of antibody production from B cells also contributed to the attenuation of these diseases. Conversely, RA did not enhance the differentiation of Tregs in murine CIA (Sato et al. 2010). On the other hand, retinoid has been used for the treatment of psoriasis by many dermatological doctors since the 1950s, and it is also effective for contact dermatitis in mice and humans without causing severe adverse effects, including infection, compared to prednisolone and immunosuppressive therapies. Therefore, RA and RAR agonists could be promising therapies for autoimmune disease.

3.4 CONCLUSIONS

RA contributes to the immune response for T cells, B cells, and DCs. Moreover, retinoids may be therapeutic agents for autoimmune diseases.

REFERENCES

Blomhoff R, Blomhoff HK. 2006. Overview of retinoid metabolism and function. *J. Neurobiol.* 66: 606–630.

Brinckerhoff CE, Coffey JW, Sullivan AC. 1983. Inflammation and collagenase production in rats with adjuvant arthritis reduced with 13-cis-retinoic acid. *Science* 221: 756–758.

Campbell DJ, Butcher EC. 2002. Rapid acquisition of tissue-specific homing phenotypes by CD4+ T cells activated in cutaneous or mucosal lymphoid tissues. *J. Exp. Med.* 195: 135–141.

Carman JA, Smith SM, Hayes CE. 1989. Characterization of a helper T lymphocyte defect in vitamin A-deficient mice. *J Immunol.* 142: 388–393.

Chen Q, Ross AC. 2005. Vitamin A and immune function: Retinoic acid modulates population dynamics in antigen receptor and CD38-stimulated splenic B cells. *Proc Natl Acad Sci U S A.* 102: 14142–14149.

Coombes JL, Siddiqui KR, Arancibia-Carcamo CV et al. 2007. A functionally specialized population of mucosal CD103+ DCs induces Foxp3+regulatory T cells via a TGF-β- and retinoic acid dependent mechanism. *J. Exp. Med.* 204: 1757–1764.

Darmanin S, Chen J, Zhao S et al. 2007. All-trans retinoic acid enhances murine dendritic cell migration to draining lymph nodes via the balance of matrix metalloproteinases and their inhibitors. *J. Immunol.* 179: 4616–4625.

Ertesvag A, Aasheim HC, Naderi S et al. 2007. Vitamin A potentiates CpG-mediated memory B-cell proliferation and differentiation: Involvement of early activation of p38MAPK. *Blood* 109: 3865–3872.

Ertesvag A, Engedal N, Naderi S et al. 2002. Retinoic acid stimulates the cell cycle machinery in normal T cells: Involvement of retinoic acid receptor mediated IL-2 secretion. *J. Immunol.* 169: 5555–5563.

Escribese MM, Conde E, Martin A et al. 2007. Therapeutic effect of *all-trans*-retinoic acid (at-RA) on an autoimmune nephritis experimental model: Role of the VLA-4 integrin. *BMC Nephrol.* 8: 3.

Iwata M, Eshima Y, Kagechika H. 2003. Retinoic acids exert direct effects on T cells to suppress Th1 development and enhance Th2 development via retinoic acid receptors. *Int. Immunol.* 15: 1017–1025.

Iwata M, Hirakiyama A, Eshima Y et al. 2004. Retinoic acid imprints gut-homing specificity on T cells. *Immunity* 21: 527–538.

Kinoshita K, Kishimoto K, Funauchi M et al. 2009. Successful treatment with retinoids in patients with lupus nephritis. *Am J Kidney Dis.* 55: 344–347.

Kuwabara K, Shudo K, Hori Y. 1996. Novel synthetic retinoic acid inhibits rat collagen arthritis and differentially affects serum immunoglobulin subclass levels. *FEBS Lett.* 378: 153–156.

Lampen A, Meyer S, Arnhold T et al. 2000. Metabolism of vitamin A and its active metabolite *all-trans*-retinoic acid in small intestinal enterocytes. *J. Pharmacol. Exp. Ther.* 295: 979–985.

Lovett-Racke AE, Racke MK. 2002. Retinoic acid promotes the development of Th2-like human myelin basic protein-reactive T cells. *Cell. Immunol.* 215: 54–60.

McDermott MR, Mark DA, Befus AD et al. 1982. Impaired intestinal localization of mesenteric lymphoblasts associated with vitamin A deficiency and protein-calorie malnutrition. *Immunol.* 45: 1–5.

Moise AR, Noy N, Palczewski K et al. 2007. Delivery of retinoid-based therapies to target tissues. *Biochemistry* 46: 4449–4458.

Mora JR. 2008. Homing imprinting and immunomodulation in the gut: Role of dendritic cells and retinoids. *Inflamm. Bowel Dis.* 14: 275–289.

Mora JR, Bono MR, Manjunath N et al. 2003. Selective imprinting of gut-homing T cells by Peyer's patch dendritic cells. *Nature* 424: 88–93.

Mora JR, Cheng G, Picarella D et al. 2005. Reciprocal and dynamic control of CD8 T cell homing by dendritic cells from skin- and gut-associated lymphoid tissues. *J. Exp. Med.* 201: 303–316.

Mora JR, Iwata M, Eksteen B et al. 2006. Generation of gut-homing IgA secreting B cells by intestinal dendritic cells. *Science* 314: 1157–1160.

Nozaki Y, Yamagata T, Kanamaru A et al. 2005. The beneficial effects of treatment with all-trans-retinoic acid plus corticosteroid on autoimmune nephritis in NZB/WF mice. *Clin Exp Immunol.* 139: 74–83.

Ohyanagi N, Miyasaka N, Nanki T et al. 2009. Retinoid ameliorates experimental autoimmune myositis, with modulation of Th cell differentiation and antibody production *in vivo*. *Arthritis Rheum.* 60: 3118–27.

Sato A, Miyasaka N, Nanki T et al. 2010. The effect of synthetic retinoid, Am80, on T helper cell development and antibody production in murine collagen-induced arthritis. *Mod Rheumatol.* 20: 244–51.

Saurer L, McCullough KC, Summerfield A. 2007. In vitro induction of mucosa-type dendritic cells by all-trans retinoic acid. *J. Immunol.* 179: 3504–3514.

Seguin-Devaux C, Devaux Y, Latger-Cannard V et al. 2002. Enhancement of the inducible NO synthase activation by retinoic acid is mimicked by RARα agonist *in vivo*. *Am. J. Physiol. Endocrinol. Metab.* 283: E525–E535.

Semba RD, Ndugwa C, Perry RT et al. 2005. Effect of periodic vitamin A supplementation on mortality and morbidity of human immunodeficiency virus-infected children in Uganda: A controlled clinical trial. *Nutrition* 21: 25–31.

Sommer A. 1984. Vitamin A deficiency and mortality risk. *Lancet* 1: 347–348.

Sommer A, Tarwotjo I, Hussaini G et al. 1983. Increased mortality in children with mild vitamin A deficiency. *Lancet* 2: 585–588.

Sun CM, Hall JA, Blank RB et al. 2007. Small intestine lamina propria dendritic cells promote *de novo* generation of Foxp3⁺ T reg cells via retinoic acid. *J. Exp. Med.* 204: 1775–1785.

Theodosiou M, Laudet V, Schubert M. 2010. From carrot to clinic: An overview of the retinoic acid signaling pathway. *Cell Mol Life Sci.* 67: 1423–1445.

Wang T, Niwa S, Nagai H et al. 2000. The effect of Am-80, one of retinoids derivatives on experimental allergic encephalomyelitis in rats. *Life Sci.* 67: 1869–1879.

4 Vitamin D₃ Up-Regulated Protein 1 (VDUP1) and the Immune System

Hyun Woo Suh, Haiyoung Jung,
Young Jun Park, and Inpyo Choi

CONTENTS

4.1 1,25 DIHYDROXY-VITAMIN D AND VDUP1

The active form of vitamin D, 1,25 dihydroxy-vitamin D ($1,25(OH)_2D$), binds to the vitamin D receptor (VDR), which is expressed in most cells. $1,25(OH)_2D$ functions as a steroid hormone, usually by heterodimerization with the retinoid X receptor, binding to vitamin-D response elements (VDRE) located in promoter regions. It induces the expression of target genes involved in many cellular responses, including the immune responses.

$1,25(OH)_2D$ decreases dendritic cell proliferation (Griffin et al. 2001) and maturation by inhibiting CD80 and CD86 expression (Almerighi et al. 2009, Gauzzi et al. 2005). It decreases IL-12 production and T-cell stimulation, but increases IL-10 production (Barrat et al. 2002). $1,25(OH)_2D$ increases the Th2 response and IL-15 expression in T cells (Boonstra et al. 2001, Daniel et al. 2008) and decreases plasma differentiation and Ig production in B cells (Chen S. et al. 2007). Additionally, it regulates MHC class II expression and antigen presentation in macrophages (Helming et al. 2005), and it increases macrophage phagocytosis (Liu et al. 2006). Overall, a lot of evidence shows that $1,25(OH)_2D$ regulates adaptive and innate immunity.

$1,25(OH)_2D$ is also involved in many immune diseases. It prevents collagen-induced arthritis, and its deficiency was observed in patients with rheumatoid arthritis (Cantorna et al. 1998, Hein and Oelzner 2000, Merlino et al. 2004). In a type I diabetes model, it delayed the progression of diabetes (Gregori et al. 2002, Hypponen et al. 2001). 1,25(OH)$_2$D also has immune modulating effects on many types of immune cells, as well as many inflammatory and autoimmune diseases.

In an investigation of the cellular target genes of 1,25(OH)$_2$D, the vitamin D$_3$ up-regulated protein 1 (VDUP1) was identified in HL-60 cells treated with 1,25(OH)$_2$D (Chen and DeLuca 1994). Later, two independent studies (Junn et al. 2000, Nishiyama et al. 1999) determined that VDUP1 interacts with thioredoxin (TRX), which is a key antioxidant protein in cells, and that this interaction inhibits the antioxidant function of TRX. VDUP1, also known as TRX binding protein-2 (TBP-2) or thio-redoxin interacting protein (TXNIP), recognizes the catalytically active center of TRX via the cysteine 247 residue. In addition to this biochemical property, VDUP1 is involved in many cellular activities. It inhibits tumor cell growth by regulating cell cycle progress (Han et al. 2003). It regulates the protein stability of p27 by interact-ing with jab1 (Jeon et al. 2005). VDUP1 also regulates inflammation by forming the inflammasome complex with NALP3 to regulate IL-1β processing (Zhou R. et al. 2010). In addition, it is also involved in glucose metabolism and diabetes (Chen J. et al. 2010, Chutkow et al. 2010, Yu et al. 2010).

1,25(OH)$_2$D and VDUP1 share common properties in the regulation of many cellular responses, including immune regulation (Figure 4.1). They regulate immune cell function and inflammation; 1,25(OH)$_2$D and VDUP1 regulate dendritic cell function and cytokine production. As a target molecule of 1,25(OH)$_2$D, the roles of VDUP1 in immune regulation have been disclosed.

4.2   VDUP1 AND IMMUNE CELL FUNCTION

It has been reported that VDUP1 is involved in immune cell regulation (Chung et al. 2006). VDUP1 plays an important role in regulating dendritic cells (DCs) to promote the T cell response. DCs derived from VDUP1$^{-/-}$ mice were defective in inducing T cell activation and proliferation (Son et al. 2008). The levels of IL-12p40,

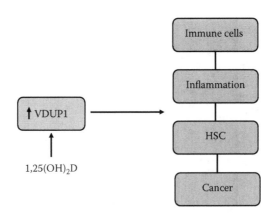

FIGURE 4.1   The common properties of 1,25(OH)$_2$D and VDUP1. VDUP1 and 1,25(OH)$_2$D have many common effects on cellular activities such as immune cell regulation, inflamma-tion, and functions of HSC and cancer cells.

IL-12p70, and IL-6 were attenuated in VDUP1⁻/⁻ DC. However, VDUP1⁻/⁻ DCs and WT DCs expressed comparable levels of MHC class II and co-stimulatory molecules such as CD40, CD80, and CD86 (Son et al. 2008).

VDUP1 is involved in regulating immune cell apoptosis. The glucocorticoid (GC) class of steroid hormone has many diverse effects, including harmful side effects. GC hormones may induce apoptosis in lymphoid cells. VDUP1 induced by GC is responsible for GC-induced apoptosis in adult T cell leukemia (ATL) cells (Chen Z. et al. 2011). Knockdown of VDUP1 consistently reduced the incidence of GC-induced apoptosis (Chen Z. et al. 2010). Human T cell leukemia virus type I (HTLV1) is the causative agent of ATL. HTLV1-infected T cells transit from the IL-2-dependent to IL-2-independent growth during long-term culture *in vitro*. The expression of VDUP1 was shown to be lost during the transition of HTLV1-infected T cell lines (Ahsan et al. 2006, Nishinaka et al. 2004b). Meanwhile, VDUP1 was significantly induced in these T cells in response to GC treatment, with growth arrest and apoptosis in the IL-2-dependent growth stage (Chen Z. et al. 2010). Thus, VDUP1 plays a crucial role in regulating T cell growth, and the loss of VDUP1 expression in HTLV1-infected T cells is one of the key events involved in the multistep progression of ATL leukemogenesis. Ceramide, a tumor-suppressor lipid, is generated by sphingomyelin hydrolysis when cells are activated by various stress stimuli. Ceramide significantly up-regulates VDUP1 expression in mouse T-hybridoma and human Jurkat T cells, resulting in ASK1 activation, ER stress, and p38 and JNK phosphorylation, all leading to cell apoptosis (Chen C. L. et al. 2008).

VDUP1⁻/⁻ mice displayed immune system defects, including perturbed development and function of natural killer (NK) cells. There was a profound reduction in the numbers of NK cells and in NK activity in VDUP1⁻/⁻ mice, which showed minimal changes in T and B cell development (Lee et al. 2005). The NK cell abnormality in the VDUP1⁻/⁻ mice caused severe lymphoid hyperplasia in the small intestine and reduced tumor rejection (Lee et al. 2005). Moreover, the percentage of hepatic natural killer T (NKT) cells in VDUP1⁻/⁻ mice was decreased, and the percentage

•Increased growth retardation/aging syndrome

•Increased hepatic NKT cells

•Increased apoptosis in ATL/HTLV1-infected T cells in the presence of GC

•Decreased NK/ hepatic NKT cells

•Decreased functional DC cells for T cell activation

•Decreased apoptosis in ATL/HTLV1-infected T cells in the presence of GC

FIGURE 4.2 Functions of VDUP1 observed in mice. VDUP1 TG mice exhibited a syndrome resembling aging human phenotypes and increased the percentage of hepatic NKT cells. Defection of NK and hepatic NKT cells were observed in VDUP1-deficient mice.

of hepatic NKT cells in VDUP1 TG mice was increased (Okuyama et al. 2009). In addition, VDUP1 expression was increased significantly in mice deficient in klotho (a gene involved in aging), which exhibit a syndrome resembling the human phenotypes of aging (Kuro-O et al. 1997, Okuyama et al. 2009, Yoshida et al. 2005). Overall, VDUP1 is one of the key molecules regulating innate immunity and the aging process (Figure 4.2).

4.3   VDUP1 AND INFLAMMATION

Inflammation is a part of the complex biological response of the body to burns, chemical irritants, infection by pathogens, and immune reactions due to hypersensitivity. Inflammatory disorders are associated with many diseases, such as asthma, autoimmune diseases, glomerulonephritis, and inflammatory bowel diseases. Toll-like receptors (TLRs) are a type of pattern recognition receptor (PRR) that have a central role in host cell recognition and the response to microbial pathogens. They recognize molecular structures that are broadly shared by pathogens but distinguishable from host molecules, such as pathogen-associated molecular patterns (PAMPs) (Akira et al. 2006, Hoffmann 2003). There are an estimated ten to fifteen types of TLRs in humans and mice. TLRs are highly specific, each TLR detecting distinct PAMPs derived from viruses, bacteria, mycobacteria, fungi, and parasites (Kawai and Akira 2011). Liu et al. (2006) suggested that vitamin D3 is a TLR trigger. Induction of the antimicrobial peptide cathelicidin by vitamin D3 improves the host immune defense against mycobacterial infection. Expression of the vitamin D receptor and vitamin D-1-hydroxylase genes is induced by activation of TLRs, and it stimulates the induction of the antimicrobial peptide cathelicidin (Liu et al. 2006).

VDUP1 is a member of the arrestin protein family, containing two characteristic arrestin-like domains with a PXXP sequence, a known binding motif for SH3-domain containing proteins, and a PPXY sequence, a known binding motif for the WW domain (Patwari et al. 2006). Subsequently, it was identified independently by several investigators as a thioredoxin-binding protein that negatively regulates the expression and activity of thioredoxin; thus, VDUP1 is involved in redox regulation (Junn et al. 2000). There are many reports demonstrating that VDUP1 regulates the production of reactive oxygen species (ROS) in cells. Recently, it was reported that VDUP1 interacts with the NLRP3 inflammasome, which is vital for the production of mature IL-1β and is associated with several inflammatory diseases (Zhou R. et al. 2010). Three distinct signaling pathways for inflammasome activation have been proposed. First, activators enter the cytoplasm via pannexin 1 (Marina-Garcia et al. 2008). Secondly, cathepsin B is released from ruptured lysosomes following phagocytosis of large particles (Hornung et al. 2008). Finally, ROS activate the inflammasome (Martinon et al. 2009). The NLRP3 inflammasome activation is mediated by its association with VDUP1, which dissociates from TRX in a ROS-sensitive manner (Zhou R. et al. 2010). In the absence of VDUP1, inflammasome function is defective and secretion of mature IL-1β is reduced. Based on these observations, it seems that VDUP1 is one of the key regulatory molecules in inflammasome and inflammation.

4.4 VDUP1 AND HSCS

Hematopoietic stem cells (HSCs) are distinct cells that are predominantly located in the bone marrow (BM) microenvironment. They maintain their self-renewal capability and differentiate into various cell types within the blood lineages (Morrison and Weissman 1994, Zhang J. et al. 2003). Under normal conditions, the most primitive HSCs are in a quiescent state and continue to produce all types of blood cells throughout a prolonged life span without depleting the regenerative cell pool. However, various stresses, such as oxidative stress, cell cycling, and aging, can induce disruption of HSC quiescence and hematopoiesis, causing hematological failure (Cheng et al. 2000, Hock et al. 2004). HSCs can move from the osteoblastic niche to the vascular niche to undergo cell division and migrate into secondary organs, such as the spleen and liver. Following this, the mobilized HSCs may return to their BM niche and regain their quiescent state. Recent studies have elucidated the regulatory mechanisms in HSCs. Transcriptional regulators, cell cycle regulatory proteins, tumor suppressors, and proto-oncogenes intrinsically regulate the balance between HSC quiescence and activation (Jones and Wagers 2008, Wilson and Trumpp 2006). A number of intrinsic factors, including c-Myc, p21, Mef/Elf4, Foxo3a, Gfi1, Pbx1, and Fbw7, have been implicated in the regulation of HSC self-renewal (Calvi et al. 2003, Ficara et al. 2008, Lacorazza et al. 2006, Miyamoto et al. 2007, Perry and Li 2008, Wilson et al. 2004, Zeng et al. 2004). In addition, there are extrinsic specialized microenvironment factors, such as supportive cells, that express adhesion molecules and secrete soluble factors (e.g., N-cadherin, Tie2/Ang-1, osteopontin (Opn)/integrins, Notch/Jagged1, and Dkk-1/Lrp5) that regulate HSC maintenance and migration (Arai et al. 2004, Calvi et al. 2003, DiMascio et al. 2007, Fleming et al. 2008, Stier et al. 2005).

VDUP1 acts as a mediator of oxidative stress, either by inhibiting TRX activity or by limiting its bioavailability. VDUP1 is also a transcriptional repressor that may act as a bridge molecule between transcription factors and co-repressor complexes; its overexpression induces G_0/G_1 cell cycle arrest. VDUP1 functions as a tumor and metastasis suppressor. Reduced VDUP1 expression has been demonstrated in many types of tumors, and VDUP1 overexpression inhibits tumor growth by blocking cell cycle progression (Han et al. 2003). VDUP1 suppressed cell invasiveness and tumor metastasis, which were also recovered by blocking the nuclear export of HIF1α by the pVHL/VDUP1 complex (Shin D. et al. 2008a).

Recently, it was reported that VDUP1 plays a pivotal role in controlling the HSC fate decision and the interaction between HSCs and the BM niche. Although VDUP1 was detected in all HSCs, VDUP1 expression was substantially reduced as the HSCs differentiated into ST-HSC and MPP, a process that is accompanied by the loss of potential for self-renewal and an increased rate of proliferation (Adolfsson et al. 2001, Jeong et al. 2009, Min et al. 2008). To understand the role of VDUP1 in HSC maintenance and regulation, VDUP1$^{-/-}$ mice were generated by disrupting exons one to eight. In this study, the immunophenotype and HSC pool size in 2- and 12-month-old VDUP1$^{-/-}$ mice were analyzed by flow cytometry. Various lineages of the HSC population were not significantly altered in young animals under steady state, but aged mice displayed a decreased number of LT-HSCs (LKS CD34$^-$Flk2$^-$ and LKS

CD150$^+$CD48$^-$ cells). BM transplantation revealed that VDUP1$^{-/-}$ mice have a defective capacity for HSC self-renewal and reduced quiescence under stress conditions. These mice also displayed increased mobilization of HSCs caused by reduced niche interactions and cell cycling by activation of Wnt signaling. This study demonstrated that VDUP1 is a regulator of HSC quiescence and migration under stress conditions. These results can be applied to the development of strategies to effectively stimulate the *ex vivo* expansion of HSC for clinical use and to control the over-proliferation of HSCs that causes leukemia. The role of VDUP1 in controlling HSC fate and quiescence is closely related to cell cycle regulation of HSCs by VDUP1.

4.5 VDUP1 AND CANCER

Recent studies suggest tumorigenesis is controlled by immune responses and inflammation (Bui and Schreiber 2007, Chow et al. 2012). Based on the known functions of VDUP1, it is not surprising that it plays a crucial role in the regulation of various cancer cells. Upon stimulation by growth-inhibitory signals such as TGF-beta 1 and 1,25(OH)$_2$D$_3$, VDUP1 expression is strongly up-regulated. Ectopic overexpression of VDUP1 in tumor cells reduces cell growth (Han et al. 2003). Ectopically expressed VDUP1 in the cancer cell line was localized predominantly in the nucleus, and the nuclear localization of VDUP1 with importin alpha1 (Rch1) exhibited growth-suppressing activity (Nishinaka et al. 2004a). VDUP1 forms a transcriptional repressor complex with promyelocytic leukemia zinc-finger, Franconi anemia zinc-finger, and histone deacetylase 1, which are known to be transcriptional co-repressors (Han et al. 2003). The ubiquitin ligase Itch mediates polyubiquitination of VDUP1, leading to its proteosomal degradation both *in vitro* and *in vivo* (Zhang P. et al. 2010).

VDUP1 plays a dominant role in the process of apoptosis, and its role is dependent on cell type. The overexpression of VDUP1 alone is not sufficient to induce apoptosis; additional stimuli, such as H$_2$O$_2$ and UV, are required (Junn et al. 2000). A synthetic retinoid, CD437, has an antitumor effect in human osteosarcoma cells via apoptosis. CD437 treatment induced JNK1 activation and apoptosis through the up-regulation of VDUP1 mRNA (Hashiguchi et al. 2010, Matsuoka et al. 2008). 3-Deazaneplanocin A (DZNep), a histone methyltransferase inhibitor, also up-regulated VDUP1 expression and induced robust apoptosis in AML cell lines (Zhou J. et al. 2011). Similarly, intracellular VDUP1 expression was markedly enhanced in MOLT-4F (T-cell lymphoblastic leukemia) cells and HuH-7 hepatocellular carcinoma cells by $_D$-allose treatment, and in human head and neck cancer cells by radiation combined with $_D$-allose (Hirata et al. 2009, Hoshikawa et al. 2010, 2011, Yamaguchi et al. 2008). Elevated VDUP1 expression under various stress stimuli primarily results in intracellular ROS production and cell apoptosis.

Hyperglycemia regulates TRX-ROS activity through the induction of VDUP1 during apoptosis of pancreatic beta cells and various cancer-derived cells. Elevated glucose levels (1-15 mM) up-regulated VDUP1 gene expression two- to four-fold in human prostate carcinoma cells (LNCaP) and hepatocellular carcinoma cells HepG2 (Pang et al. 2009). The promoter activity of the VDUP1 gene is up-regulated by glucose, 3-O-methylglucose, and maltos, but not by mannitol (Minn et al. 2006, Pang et al. 2009). In addition, a mitotic inhibitor, paclitaxel, aggravated cytotoxicity via hyperglycemia-mediated VDUP1 up-regulation in metastatic breast cancer-derived MDA-MB-231 cells (Turturro et al. 2007a, Turturro et al. 2007b).

Heterogeneous nuclear ribonucleoproteins (hnRNPs) are nucleic acid-binding proteins that have critical roles in DNA repair and transcriptional gene regulation. hnRNP G, a member of the hnRNP family, has tumor-suppressive activity in human oral squamous cell carcinoma cells. hnRNP G binds the VDUP1 promoter, increasing the expression of VDUP1 and mediating its tumor-suppressive effect (Shin K. H. et al. 2008b, Shin K. H. et al. 2006). In addition, VDUP1 is induced in response to hypoxia in an HIF-1α-dependent manner in pancreatic cancer cells, resulting in increased apoptosis in response to platinum anticancer therapy (Baker et al. 2008, Le Jan et al. 2006). Suberoylanilide hydroxamic acid (SAHA), an inhibitor of histone deacetylases (HDACs), strongly induces the expression of VDUP1, which causes growth arrest and/or apoptosis of many tumor types (Butler et al. 2002). VDUP1 is also highly up-regulated in response to 5-fluorouracil cytotoxicity in SW620 colon carcinoma cells (Takahashi et al. 2002).

VDUP1 expression is reduced in many tumor cells. Renal cell carcinoma cells mostly showed undetectable levels of VDUP1 (Dutta et al. 2005). VDUP1 is also down-regulated in human bladder cancer according to grade and stage (Nishizawa et al. 2011). Similarly, VDUP1 expression is decreased in colorectal and gastric cancers compared with their adjacent normal tissues (Ikarashi et al. 2002). VDUP1 expression is also reduced during human hepatic carcinogenesis (HCC) and mice lacking VDUP1 displayed aggravated diethylnitrosamine-induced HCC (Kwon et al. 2012, Sheth et al. 2006, Yamamoto et al. 2001). VDUP1 deficiency also promoted

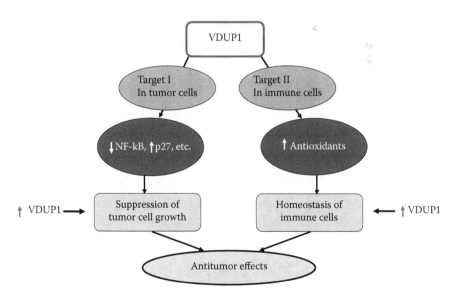

FIGURE 4.3 Antitumor effects of VDUP1 in two ways. VDUP1 inhibits tumor cell growth by modulating cell growth-related genes. In an alternative way, it maintains homeostasis of immune cells to induce the antitumor immunity.

N-methyl-N-nitrosourea or *helicobacter pylori*-induced gastric carcinogenesis in mice (Kwon et al. 2012). In addition, knockdown of VDUP1 accelerated liver regeneration after hepatectomy in mice (Kwon et al. 2011).

Angiogenesis and metastasis are key events in tumor progression. Induced VDUP1 expression suppresses cell invasiveness and tumor metastasis (Goldberg et al. 2003, Shin D. et al. 2008a). The down-regulation of metastasis suppressor genes in malignant pheochromocytoma may play a role in malignancy. VDUP1 was down-regulated significantly in malignant compared to benign pheochromocytoma (Ohta et al. 2005). Down-regulation of metastasis-related genes is highly indicative of malignant pheochromocytoma. The cyclin-dependent kinase inhibitor p21(WAF1) induces TRX secretion and angiogenesis in cancer cells by repressing the transcription of VDUP1 (Kuljaca et al. 2009). Overall, VDUP1 has diverse functions in regulating cellular phenotypes and functions. VDUP1 inhibits tumor cell proliferation by modulating tumor-related genes, and it also regulates homeostasis of immune cells to maintain antitumor immunity (Figure 4.3).

4.6 CONCLUSIONS

In this chapter, we described the effects of VDUP1 on immune regulation, HSC expansion, and tumor progression. VDUP1 deficiency results in significantly reduced populations of NK cells and hepatic NKT cells, and VDUP1 can also control DC functions and T cell growth. It is also involved in inflammatory cytokine production and inflammation *in vivo*. During these processes, VDUP1 can either positively or negatively regulate gene transcription in response to immune stimulation. VDUP1 has diverse effects on cell proliferation, apoptosis, and tumorigenesis due to its control of and interaction with many cellular components. Gaining insights into the role of VDUP1 in tumorigenesis and immune regulation may serve as a novel strategy for antiproliferative cancer therapy and immunotherapy for cancer.

ACKNOWLEDGMENTS

This work was supported in part by grants from the GRL project and the New Drug Target Discovery Project (M10848000352-08N4800-35210), the Ministry of Education, Science & Technology, Republic of Korea.

REFERENCES

Adolfsson J, Borge OJ, Bryder D, Theilgaard-Monch K, Astrand-Grundstrom I, Sitnicka E, Sasaki Y, Jacobsen SE. 2001. Upregulation of Flt3 expression within the bone marrow Lin(-)Sca1(+)c-kit(+) stem cell compartment is accompanied by loss of self-renewal capacity. *Immunity* 15: 659–669.

Ahsan MK, Masutani H, Yamaguchi Y, Kim YC, Nosaka K, Matsuoka M, Nishinaka Y, Maeda M, Yodoi J. 2006. Loss of interleukin-2-dependency in HTLV-I-infected T cells on gene silencing of thioredoxin-binding protein-2. *Oncogene* 25: 2181–2191.

Akira S, Uematsu S, Takeuchi O. 2006. Pathogen recognition and innate immunity. *Cell* 124: 783–801.

Almerighi C, Sinistro A, Cavazza A, Ciaprini C, Rocchi G, Bergamini A. 2009. 1Alpha,25-dihydroxyvitamin D3 inhibits CD40L-induced pro-inflammatory and immunomodulatory activity in human monocytes. *Cytokine* 45: 190–197.

Arai F, Hirao A, Ohmura M, Sato H, Matsuoka S, Takubo K, Ito K, Koh GY, Suda T. 2004. Tie2/angiopoietin-1 signaling regulates hematopoietic stem cell quiescence in the bone marrow niche. *Cell* 118: 149–161.

Baker AF, Koh MY, Williams RR, James B, Wang H, Tate WR, Gallegos A, Von Hoff DD, Han H, Powis G. 2008. Identification of thioredoxin-interacting protein 1 as a hypoxia-inducible factor 1alpha-induced gene in pancreatic cancer. *Pancreas* 36: 178–186.

Barrat FJ, Cua DJ, Boonstra A, Richards DF, Crain C, Savelkoul HF, de Waal-Malefyt R, Coffman RL, Hawrylowicz CM, O'Garra A. 2002. *In vitro* generation of interleukin 10-producing regulatory CD4(+) T cells is induced by immunosuppressive drugs and inhibited by T helper type 1 (Th1)- and Th2-inducing cytokines. *J Exp Med* 195: 603–616.

Boonstra A, Barrat FJ, Crain C, Heath VL, Savelkoul HF, O'Garra A. 2001. 1alpha,25-Dihydroxyvitamin d3 has a direct effect on naive CD4(+) T cells to enhance the development of Th2 cells. *J Immunol* 167: 4974–4980.

Bui JD, Schreiber RD. 2007. Cancer immunosurveillance, immunoediting and inflammation: Independent or interdependent processes? *Curr Opin Immunol* 19: 203–208.

Butler LM, Zhou X, Xu WS, Scher HI, Rifkind RA, Marks PA, Richon VM. 2002. The histone deacetylase inhibitor SAHA arrests cancer cell growth, up-regulates thioredoxin-binding protein-2, and down-regulates thioredoxin. *Proc Natl Acad Sci U S A* 99: 11700–11705.

Calvi LM, Adams GB, Weibrecht KW, Weber JM, Olson DP, Knight MC, et al. 2003. Osteoblastic cells regulate the haematopoietic stem cell niche. *Nature* 425: 841–846.

Cantorna MT, Hayes CE, DeLuca HF. 1998. 1,25-Dihydroxycholecalciferol inhibits the progression of arthritis in murine models of human arthritis. *J Nutr* 128: 68–72.

Chen CL, Lin CF, Chang WT, Huang WC, Teng CF, Lin YS. 2008. Ceramide induces p38 MAPK and JNK activation through a mechanism involving a thioredoxin-interacting protein-mediated pathway. *Blood* 111: 4365–4374.

Chen J, Fontes G, Saxena G, Poitout V, Shalev A. 2010. Lack of TXNIP protects against mitochondria-mediated apoptosis but not against fatty acid-induced ER stress-mediated beta-cell death. *Diabetes* 59: 440–447.

Chen KS, DeLuca HF. 1994. Isolation and characterization of a novel cDNA from HL-60 cells treated with 1,25-dihydroxyvitamin D-3. *Biochim Biophys Acta* 1219: 26–32.

Chen S, Sims GP, Chen XX, Gu YY, Lipsky PE. 2007. Modulatory effects of 1,25-dihydroxyvitamin D3 on human B cell differentiation. *J Immunol* 179: 1634–1647.

Chen Z, Lopez-Ramos DA, Yoshihara E, Maeda Y, Masutani H, Sugie K, Maeda M, Yodoi J. 2011. Thioredoxin-binding protein-2 (TBP-2/VDUP1/TXNIP) regulates T-cell sensitivity to glucocorticoid during HTLV-I-induced transformation. *Leukemia* 25: 440–448.

Chen Z, Yoshihara E, Son A, Matsuo Y, Masutani H, Sugie K, Maeda M, Yodoi J. 2010. Differential roles of Annexin A1 (ANXA1/lipocortin-1/lipomodulin) and thioredoxin binding protein-2 (TBP-2/VDUP1/TXNIP) in glucocorticoid signaling of HTLV-I-transformed T cells. *Immunol Lett* 131: 11–18.

Cheng T, Rodrigues N, Shen H, Yang Y, Dombkowski D, Sykes M, Scadden DT. 2000. Hematopoietic stem cell quiescence maintained by p21cip1/waf1. *Science* 287: 1804–1808.

Chow MT, Moller A, Smyth MJ. 2012. Inflammation and immune surveillance in cancer. *Semin Cancer Biol* 22: 23–32.

Chung JW, Jeon JH, Yoon SR, Choi I. 2006. Vitamin D3 upregulated protein 1 (VDUP1) is a regulator for redox signaling and stress-mediated diseases. *J Dermatol* 33: 662–669.

Chutkow WA, Birkenfeld AL, Brown JD, Lee HY, Frederick DW, Yoshioka J, et al. 2010. Deletion of the alpha-arrestin protein Txnip in mice promotes adiposity and adipogenesis while preserving insulin sensitivity. *Diabetes* 59: 1424–1434.

Daniel C, Sartory NA, Zahn N, Radeke HH, Stein JM. 2008. Immune modulatory treatment of trinitrobenzene sulfonic acid colitis with calcitriol is associated with a change of a T helper (Th) 1/Th17 to a Th2 and regulatory T cell profile. *J Pharmacol Exp Ther* 324: 23–33.

DiMascio L, Voermans C, Uqoezwa M, Duncan A, Lu D, Wu J, Sankar U, Reya T. 2007. Identification of adiponectin as a novel hemopoietic stem cell growth factor. *J Immunol* 178: 3511–3520.

Dutta KK, Nishinaka Y, Masutani H, Akatsuka S, Aung TT, Shirase T, et al. 2005. Two distinct mechanisms for loss of thioredoxin-binding protein-2 in oxidative stress-induced renal carcinogenesis. *Lab Invest* 85: 798–807.

Ficara F, Murphy MJ, Lin M, Cleary ML. 2008. Pbx1 regulates self-renewal of long-term hematopoietic stem cells by maintaining their quiescence. *Cell Stem Cell* 2: 484–496.

Fleming HE, Janzen V, Lo Celso C, Guo J, Leahy KM, Kronenberg HM, Scadden DT. 2008. Wnt signaling in the niche enforces hematopoietic stem cell quiescence and is necessary to preserve self-renewal in vivo. *Cell Stem Cell* 2: 274–283.

Gauzzi MC, Purificato C, Donato K, Jin Y, Wang L, Daniel KC, Maghazachi AA, Belardelli F, Adorini L, Gessani S. 2005. Suppressive effect of 1alpha,25-dihydroxyvitamin D3 on type I IFN-mediated monocyte differentiation into dendritic cells: Impairment of functional activities and chemotaxis. *J Immunol* 174: 270–276.

Goldberg SF, Miele ME, Hatta N, Takata M, Paquette-Straub C, Freedman LP, Welch DR. 2003. Melanoma metastasis suppression by chromosome 6: Evidence for a pathway regulated by CRSP3 and TXNIP. *Cancer Res* 63: 432–440.

Gregori S, Giarratana N, Smiroldo S, Uskokovic M, Adorini L. 2002. A 1alpha,25-dihydroxyvitamin D(3) analog enhances regulatory T-cells and arrests autoimmune diabetes in NOD mice. *Diabetes* 51: 1367–1374.

Griffin MD, Lutz W, Phan VA, Bachman LA, McKean DJ, Kumar R. 2001. Dendritic cell modulation by 1alpha,25 dihydroxyvitamin D3 and its analogs: A vitamin D receptor-dependent pathway that promotes a persistent state of immaturity *in vitro* and *in vivo*. *Proc Natl Acad Sci U S A* 98: 6800–6805.

Han SH, Jeon JH, Ju HR, Jung U, Kim KY, Yoo HS, et al. 2003. VDUP1 upregulated by TGF-beta1 and 1,25-dihydroxyvitamin D3 inhibits tumor cell growth by blocking cell-cycle progression. *Oncogene* 22: 4035–4046.

Hashiguchi K, Tsuchiya H, Tomita A, Ueda C, Akechi Y, Sakabe T, et al. 2010. Involvement of ETS1 in thioredoxin-binding protein 2 transcription induced by a synthetic retinoid CD437 in human osteosarcoma cells. *Biochem Biophys Res Commun* 391: 621–626.

Hein G, Oelzner P. 2000. [Vitamin D metabolites in rheumatoid arthritis: findings—hypotheses—consequences]. *Z Rheumatol* 59 Suppl 1: 28–32.

Helming L, Bose J, Ehrchen J, Schiebe S, Frahm T, Geffers R, Probst-Kepper M, Balling R, Lengeling A. 2005. 1alpha,25-Dihydroxyvitamin D3 is a potent suppressor of interferon gamma-mediated macrophage activation. *Blood* 106: 4351–4358.

Hirata Y, Saito M, Tsukamoto I, Yamaguchi F, Sui L, Kamitori K, et al. 2009. Analysis of the inhibitory mechanism of D-allose on MOLT-4F leukemia cell proliferation. *J Biosci Bioeng* 107: 562–568.

Hock H, Hamblen MJ, Rooke HM, Schindler JW, Saleque S, Fujiwara Y, Orkin SH. 2004. Gfi-1 restricts proliferation and preserves functional integrity of haematopoietic stem cells. *Nature* 431: 1002–1007.

Hoffmann JA. 2003. The immune response of Drosophila. *Nature* 426: 33–38.

Hornung V, Bauernfeind F, Halle A, Samstad EO, Kono H, Rock KL, Fitzgerald KA, Latz E. 2008. Silica crystals and aluminum salts activate the NALP3 inflammasome through phagosomal destabilization. *Nat Immunol* 9: 847–856.

Hoshikawa H, Indo K, Mori T, Mori N. 2011. Enhancement of the radiation effects by D-allose in head and neck cancer cells. *Cancer Lett* 306: 60–66.

Hoshikawa H, Mori T, Mori N. 2010. *In vitro* and *in vivo* effects of D-allose: Up-regulation of thioredoxin-interacting protein in head and neck cancer cells. *Ann Otol Rhinol Laryngol* 119: 567–571.

Hypponen E, Laara E, Reunanen A, Jarvelin MR, Virtanen SM. 2001. Intake of vitamin D and risk of type 1 diabetes: A birth-cohort study. *Lancet* 358: 1500–1503.

Ikarashi M, Takahashi Y, Ishii Y, Nagata T, Asai S, Ishikawa K. 2002. Vitamin D3 up-regulated protein 1 (VDUP1) expression in gastrointestinal cancer and its relation to stage of disease. *Anticancer Res* 22: 4045–4048.

Jeon JH, Lee KN, Hwang CY, Kwon KS, You KH, Choi I. 2005. Tumor suppressor VDUP1 increases p27(kip1) stability by inhibiting JAB1. *Cancer Res* 65: 4485–4489.

Jeong M, Piao ZH, Kim MS, Lee SH, Yun S, Sun HN, et al. 2009. Thioredoxin-interacting protein regulates hematopoietic stem cell quiescence and mobilization under stress conditions. *J Immunol* 183: 2495–2505.

Jones DL, Wagers AJ. 2008. No place like home: Anatomy and function of the stem cell niche. *Nat Rev Mol Cell Biol* 9: 11–21.

Junn E, Han SH, Im JY, Yang Y, Cho EW, Um HD, et al. 2000. Vitamin D3 up-regulated protein 1 mediates oxidative stress via suppressing the thioredoxin function. *J Immunol* 164: 6287–6295.

Kawai T, Akira S. 2011. Toll-like receptors and their crosstalk with other innate receptors in infection and immunity. *Immunity* 34: 637–650.

Kuljaca S, Liu T, Dwarte T, Kavallaris M, Haber M, Norris MD, Martin-Caballero J, Marshall GM. 2009. The cyclin-dependent kinase inhibitor, p21(WAF1), promotes angiogenesis by repressing gene transcription of thioredoxin-binding protein 2 in cancer cells. *Carcinogenesis* 30: 1865–1871.

Kuro-O M, Matsumura Y, Aizawa H, Kawaguchi H, Suga T, Utsugi T, et al. 1997. Mutation of the mouse klotho gene leads to a syndrome resembling ageing. *Nature* 390: 45–51.

Kwon HJ, Won YS, Yoon YD, Yoon WK, Nam KH, Choi IP, Kim DY, Kim HC. 2011. Vitamin D3 up-regulated protein 1 deficiency accelerates liver regeneration after partial hepatectomy in mice. *J Hepatol* 54: 1168–1176.

Kwon HJ, Won YS, Nam KT, Yoon YD, Jee H, Yoon WK, et al. 2012. Vitamin D upregulated protein 1 deficiency promotes N-methyl-N-nitrosourea and *Helicobacter pylori*-induced gastric carcinogenesis in mice. *Gut* 61: 53–63.

Kwon HJ, Won YS, Suh HW, Jeon JH, Shao Y, Yoon SR, et al. 2010. Vitamin D3 upregulated protein 1 suppresses TNF-alpha-induced NF-kappaB activation in hepatocarcinogenesis. *J Immunol* 185: 3980–3989.

Lacorazza HD, Yamada T, Liu Y, Miyata Y, Sivina M, Nunes J, Nimer SD. 2006. The transcription factor MEF/ELF4 regulates the quiescence of primitive hematopoietic cells. *Cancer Cell* 9: 175–187.

Lee KN, Kang HS, Jeon JH, Kim EM, Yoon SR, Song H, et al. 2005. VDUP1 is required for the development of natural killer cells. *Immunity* 22: 195-208.

Le Jan S, Le Meur N, Cazes A, Philippe J, Le Cunff M, Leger J, Corvol P, Germain S. 2006. Characterization of the expression of the hypoxia-induced genes neuritin, TXNIP and IGFBP3 in cancer. *FEBS Lett* 580: 3395–3400.

Liu PT, Stenger S, Li H, Wenzel L, Tan BH, Krutzik SR, et al. 2006. Toll-like receptor triggering of a vitamin D-mediated human antimicrobial response. *Science* 311: 1770–1773.

Marina-Garcia N, Franchi L, Kim YG, Miller D, McDonald C, Boons GJ, Nunez G. 2008. Pannexin-1-mediated intracellular delivery of muramyl dipeptide induces caspase-1 activation via cryopyrin/NLRP3 independently of Nod2. *J Immunol* 180: 4050–4057.

Martinon F, Mayor A, Tschopp J. 2009. The inflammasomes: Guardians of the body. *Annu Rev Immunol* 27: 229–265.

Matsuoka S, Tsuchiya H, Sakabe T, Watanabe Y, Hoshikawa Y, Kurimasa A, et al. 2008. Involvement of thioredoxin-binding protein 2 in the antitumor activity of CD437. *Cancer Sci* 99: 2485–2490.

Merlino LA, Curtis J, Mikuls TR, Cerhan JR, Criswell LA, Saag KG. 2004. Vitamin D intake is inversely associated with rheumatoid arthritis: Results from the Iowa Women's Health Study. *Arthritis Rheum* 50: 72–77.

Min IM, Pietramaggiori G, Kim FS, Passegue E, Stevenson KE, Wagers AJ. 2008. The transcription factor EGR1 controls both the proliferation and localization of hematopoietic stem cells. *Cell Stem Cell* 2: 380–391.

Minn AH, Couto FM, Shalev A. 2006. Metabolism-independent sugar effects on gene transcription: The role of 3-O-methylglucose. *Biochemistry* 45: 11047–11051.

Miyamoto K, Araki KY, Naka K, Arai F, Takubo K, Yamazaki S, et al. 2007. Foxo3a is essential for maintenance of the hematopoietic stem cell pool. *Cell Stem Cell* 1: 101–112.

Morrison SJ, Weissman IL. 1994. The long-term repopulating subset of hematopoietic stem cells is deterministic and isolatable by phenotype. *Immunity* 1: 661–673.

Nishinaka Y, Masutani H, Oka S, Matsuo Y, Yamaguchi Y, Nishio K, Ishii Y, Yodoi J. 2004a. Importin alpha1 (Rch1) mediates nuclear translocation of thioredoxin-binding protein-2/vitamin D(3)-up-regulated protein 1. *J Biol Chem* 279: 37559–37565.

Nishinaka Y, Nishiyama A, Masutani H, Oka S, Ahsan KM, Nakayama Y, Ishii Y, Nakamura H, Maeda M, Yodoi J. 2004b. Loss of thioredoxin-binding protein-2/vitamin D3 up-regulated protein 1 in human T-cell leukemia virus type I-dependent T-cell transformation: Implications for adult T-cell leukemia leukemogenesis. *Cancer Res* 64: 1287–1292.

Nishiyama A, Matsui M, Iwata S, Hirota K, Masutani H, Nakamura H, Takagi Y, Sono H, Gon Y, Yodoi J. 1999. Identification of thioredoxin-binding protein-2/vitamin D(3) up-regulated protein 1 as a negative regulator of thioredoxin function and expression. *J Biol Chem* 274: 21645–21650.

Nishizawa K, Nishiyama H, Matsui Y, Kobayashi T, Saito R, Kotani H, et al. 2011 Thioredoxin-interacting protein suppresses bladder carcinogenesis. *Carcinogenesis* 32: 1459–1466.

Ohta S, Lai EW, Pang AL, Brouwers FM, Chan WY, Eisenhofer G, et al. 2005. Downregulation of metastasis suppressor genes in malignant pheochromocytoma. *Int J Cancer* 114: 139–143.

Okuyama H, Yoshida T, Son A, Oka S, Wang D, Nakayama R, Masutani H, Nakamura H, Nabeshima Y, Yodoi J. 2009. Thioredoxin binding protein 2 modulates natural killer T cell-dependent innate immunity in the liver: Possible link to lipid metabolism. *Antioxid Redox Signal* 11: 2585–2593.

Pang ST, Hsieh WC, Chuang CK, Chao CH, Weng WH, Juang HH. 2009. Thioredoxin-interacting protein: An oxidative stress-related gene is upregulated by glucose in human prostate carcinoma cells. *J Mol Endocrinol* 42: 205–214.

Patwari P, Higgins LJ, Chutkow WA, Yoshioka J, Lee RT. 2006. The interaction of thioredoxin with Txnip: Evidence for formation of a mixed disulfide by disulfide exchange. *J Biol Chem* 281: 21884–21891.

Perry JM, Li L. 2008. Self-renewal versus transformation: Fbxw7 deletion leads to stem cell activation and leukemogenesis. *Genes Dev* 22: 1107–1109.

Sheth SS, Bodnar JS, Ghazalpour A, Thipphavong CK, Tsutsumi S, Tward AD, Demant P, Kodama T, Aburatani H, Lusis AJ. 2006. Hepatocellular carcinoma in Txnip-deficient mice. *Oncogene* 25: 3528–3536.

Shin D, Jeon JH, Jeong M, Suh HW, Kim S, Kim HC, et al. 2008a. VDUP1 mediates nuclear export of HIF1alpha via CRM1-dependent pathway. *Biochim Biophys Acta* 1783: 838–848.

Shin KH, Kang MK, Kim RH, Christensen R, Park NH. 2006. Heterogeneous nuclear ribonucleoprotein G shows tumor suppressive effect against oral squamous cell carcinoma cells. *Clin Cancer Res* 12: 3222–3228.

Shin KH, Kim RH, Kang MK, Park NH. 2008b. hnRNP G elicits tumor-suppressive activity in part by upregulating the expression of Txnip. *Biochem Biophys Res Commun* 372: 880–885.

Son A, Nakamura H, Okuyama H, Oka S, Yoshihara E, Liu W, et al. 2008. Dendritic cells derived from TBP-2-deficient mice are defective in inducing T cell responses. *Eur J Immunol* 38: 1358–1367.

Stier S, et al. 2005. Osteopontin is a hematopoietic stem cell niche component that negatively regulates stem cell pool size. *J Exp Med* 201: 1781–1791.

Takahashi Y, Nagata T, Ishii Y, Ikarashi M, Ishikawa K, Asai S. 2002. Up-regulation of vitamin D3 up-regulated protein 1 gene in response to 5-fluorouracil in colon carcinoma SW620. *Oncol Rep* 9: 75–79.

Turturro F, Friday E, Welbourne T. 2007a. Hyperglycemia regulates thioredoxin-ROS activity through induction of thioredoxin-interacting protein (TXNIP) in metastatic breast cancer-derived cells MDA-MB-231. *BMC Cancer* 7: 96.

Turturro F, Von Burton G, Friday E. 2007b. Hyperglycemia-induced thioredoxin-interacting protein expression differs in breast cancer-derived cells and regulates paclitaxel IC50. *Clin Cancer Res* 13: 3724–3730.

Wilson A, Murphy MJ, Oskarsson T, Kaloulis K, Bettess MD, Oser GM, Pasche AC, Knabenhans C, Macdonald HR, Trumpp A. 2004. c-Myc controls the balance between hematopoietic stem cell self-renewal and differentiation. *Genes Dev* 18: 2747–2763.

Wilson A, Trumpp A. 2006. Bone-marrow haematopoietic-stem-cell niches. *Nat Rev Immunol* 6: 93–106.

Yamaguchi F, Takata M, Kamitori K, Nonaka M, Dong Y, Sui L, Tokuda M. 2008. Rare sugar D-allose induces specific up-regulation of TXNIP and subsequent G1 cell cycle arrest in hepatocellular carcinoma cells by stabilization of p27kip1. *Int J Oncol* 32: 377–385.

Yamamoto Y, Sakamoto M, Fujii G, Kanetaka K, Asaka M, Hirohashi S. 2001. Cloning and characterization of a novel gene, DRH1, down-regulated in advanced human hepatocellular carcinoma. *Clin Cancer Res* 7: 297–303.

Yoshida T, Nakamura H, Masutani H, Yodoi J. 2005. The involvement of thioredoxin and thioredoxin binding protein-2 on cellular proliferation and aging process. *Ann N Y Acad Sci* 1055: 1–12.

Yu FX, Chai TF, He H, Hagen T, Luo Y. 2010. Thioredoxin-interacting protein (Txnip) gene expression: Sensing oxidative phosphorylation status and glycolytic rate. *J Biol Chem* 285: 25822–25830.

Zeng H, Yucel R, Kosan C, Klein-Hitpass L, Moroy T. 2004. Transcription factor Gfi1 regulates self-renewal and engraftment of hematopoietic stem cells. *EMBO J* 23: 4116–4125.

Zhang J, Niu C, Ye L, Huang H, He X, Tong WG, et al. 2003. Identification of the haematopoietic stem cell niche and control of the niche size. *Nature* 425: 836–841.

Zhang P, Wang C, Gao K, Wang D, Mao J, An J, et al. 2010. The ubiquitin ligase itch regulates apoptosis by targeting thioredoxin-interacting protein for ubiquitin-dependent degradation. *J Biol Chem* 285: 8869-8879.

Zhou J, Bi C, Cheong LL, Mahara S, Liu SC, Tay KG, Koh TL, Yu Q, Chng WJ. 2011. The histone methyltransferase inhibitor, DZNep, up-regulates TXNIP, increases ROS production, and targets leukemia cells in AML. *Blood* 118: 2830–2839.

Zhou R, Tardivel A, Thorens B, Choi I, Tschopp J. 2010. Thioredoxin-interacting protein links oxidative stress to inflammasome activation. *Nat Immunol* 11: 136–140.

5 Rapid Pre-Genomic Responses of Vitamin D

Tremaine M. Sterling, Ramesh C. Khanal,
Yu Meng, Yang Zhang, and Ilka Nemere

CONTENTS

5.1   INTRODUCTION

We have come to realize that vitamin D is actually a prehormone, serving as the parent compound for the production of three active metabolites: 25-hydroxyvitamin D_3, 1,25-dihydroxyvitamin D_3, and 24,25-dihydroxyvitamin D_3. This chapter summarizes evidence that all three metabolites produce physiological effects in vitro and in vivo, and where possible, discusses what is known about their receptors.

5.2   25-HYDROXYVITAMIN D_3

25-Hydroxyvitamin D_3, 25(OH)D_3, is generally considered a prehormone that is activated through the action of 25-hydroxyvitamin D_3 1 α-hydroxylase. However, there exists some evidence in the literature that it indeed can also be a functional metabolite (Blunt and DeLuca 1969). Early indicators of its significance as a functional metabolite were shown in experiments with chick hatchability. For instance, 24-hydroxylation of 25(OH)D_3 was not required for the embryonic development of chicks (Ameenuddin et al. 1982), whereas 25(OH)D_3 was suggested to be essential for chick hatchability. Similarly, normal embryonic development was found in eggs from hens given 25(OH)D_3 or 24,24-difluoro-25-hydroxyvitamin D_3, in contrast to embryos from hens given 1,25(OH)$_2D_3$, 24,25(OH)$_2D_3$, or in combination: the latter were abnormal and failed to hatch (Hart and DeLuca 1985). Similarly, maximum embryonic survival was obtained when 0.20 µg/egg of 1,25(OH)$_2D_3$ or 0.60 µg/egg of 25(OH)D_3 were injected (Ameenuddin et al. 1983). On the other hand, embryos from hens fed 1,25(OH)$_2D_3$ and/or 24,25(OH)$_2D_3$ had vitamin D

deficiency that resulted in low bone ash, low plasma calcium, low total body calcium, and extremely high plasma phosphorus (Hart and DeLuca 1985).

More recently, treatment of chicken intestinal cells with 25(OH)D$_3$ resulted in
stimulation of [45]Ca uptake 5–10 min after steroid, which was inhibited by preincubation with RpcAMP, a PKA antagonist (Sterling and Nemere 2007). Microscopically,
uniform diffuse staining, which would indicate cytoplasmic calcium transport,
was observed only in cells devoid of steroid treatment (Sterling and Nemere 2007).
Rapid calcium transport was observed (within 2 min) after perfusion of isolated
duodenal loops of normal chickens with 100 nM 25(OH)D$_3$ relative to the vehicle
controls (Nemere et al. 1984; Yoshimoto and Norman 1986; Phadnis and Nemere
2003). Although physiologically relevant, 100 nM 25(OH)D$_3$ did not stimulate [33]P
movement from the lumen to venous effluent but, 20 nM steroid did from 6–40 min
(Nemere 1996). In fetal rat calvaria cultures, significant increases in alkaline phosphatase and tartrate-resistant acid phosphatase activities were observed when treated
with the metabolite compared with the untreated controls at 10 and 30 min (Municio
and Traba 2004). Furthermore, 25(OH)D$_3$ had no effect on PKC activity, but PKA
activity in isolated enterocytes was stimulated by 200%, relative to vehicle controls
(Phadnis and Nemere 2003).

In another experiment, membranes isolated from chick intestinal epithelial
cells containing both a Ca-dependent alkaline phosphatase and the more abundant
Mg-dependent alkaline phosphatase, which are critical for phosphate transport, that
were treated with the steroid have been shown to stimulate activities of both the enzymes
(Birge and Avioli 1981). Furthermore, the stimulation of phosphate uptake paralleled
the increase in Ca-dependent alkaline phosphatase activity (Birge and Avioli 1981). In
humans, low levels of circulating 25(OH)D$_3$ in pregnancy are associated with reduced
bone mass in the offspring (review by Earl et al. 2010), suggesting that 25(OH)D$_3$ may
possibly be involved in placental calcium (and phosphate) transport.

A series of five trials were conducted in chickens to determine the effect of
25(OH)D$_3$ on their performance and bone development under adequate calcium and
phosphate supplementation, and under moderate dietary restriction of calcium and
phosphate (Bar et al. 2003). Five to 10 µg of vitamin D$_3$ or 25(OH)D$_3$/kg diet were
sufficient to ensure normal body weight and bone ash in chickens under continuous
lighting. The two materials had similar effects on body weight and bone ash. In one
out of the three experiments, 25(OH)D$_3$ increased body weight and body weight
gain, while in the others it had a similar effect to that of vitamin D$_3$, or even a slight
negative effect in a trial conducted on the floor, in which the diets were supplemented
with the D sources at 75 µg/kg. The effects of both D sources on bone ash and on
the severity or frequency of tibial dischondroplasia were similar. Finally, 25(OH)D$_3$
restrained the effect of moderate dietary phosphate restriction, but not of calcium
restriction, on BW gain and bone ash in 22-day-old chickens. This effect could not
be explained by a higher phosphate bioavailability in the 25(OH)D$_3$-fed chickens.
These results suggested that the role of 25(OH)$_2$D$_3$ may be more important than currently perceived.

25(OH)D$_3$ has been shown to suppress parathyroid hormone synthesis by parathyroid cells (Ritter et al. 2006; Ritter and Brown 2011). Lou et al. (2004) suggested that
the metabolite at a high but physiological concentration acts as an active hormone

with respect to vitamin D_3 responsive gene regulation and suppression of cell proliferation. Quantitative real-time RT-PCR analysis revealed that $25(OH)D_3$ induced 25-hydroxyvitamin D_3 24-hydroxylase mRNA in a dose- and time-dependent manner (Lou et al. 2004). In prostate cell cancer studies, $25(OH)$-D_3 has been shown to inhibit the proliferation of primary prostatic epithelial cells (Barreto et al. 2000). Moreover, Lambert et al. (2007) demonstrated that the antiproliferative effect of $25(OH)D_3$-3-bromoacetate is predominantly mediated by the classical vitamin D receptor (VDR) in ALVA-31 prostate cancer cells. In another study related to oxidative stress, Peng et al. (2010) used multiple in vitro stress models including serum starvation, hypoxia, oxidative stress, and apoptosis induction in breast epithelial cells using an established breast epithelial cell line MCF12F. $25(OH)D_3$ (250 nmol/L) significantly protected cells against cell death under all situations, demonstrating the prehormone was involved in the stress process. Microarray analysis further demonstrated that stress induced by serum starvation caused significant alteration in the expression of multiple miRNAs, but the presence of $25(OH)D_3$ effectively reversed this alteration. These data suggested that there is a significant protective role for $25(OH)D_3$ against cellular stress in the breast epithelial cells, and these effects may be mediated by altered miRNA expression.

Overall, the preceding discussion not only suggests the physiological significance of the metabolite $25(OH)D_3$ but also the possible genomic and pre-genomic pathways for eliciting the effects. Further investigation is needed in at least a few primary areas: 1. to provide conclusive evidence on its physiological roles, 2. to elucidate the complete signal transduction pathways originating from actions at the level of plasma membrane, 3. to identify and characterize any cell surface binding proteins, and 4. to fully characterize its binding to the VDR if it indeed is the receptor for $25(OH)D_3$.

5.3 1,25-DIHYDROXYVITAMIN D_3

As the most studied vitamin D metabolite, 1,25-dihydroxyvitamin D_3 ($1,25(OH)_2D_3$) was first shown to regulate calcium (Schachtner and Rosen 1959, Wasserman et al. 1966, Taylor and Wasserman 1969) and phosphate (Wasserman and Taylor 1973) transport in the intestine. Since then, $1,25(OH)_2D_3$ has been identified as a key regulator of both calcium and phosphate homeostasis acting on bone, the intestine, the kidneys, and the parathyroid glands as its primary target tissues. These classical actions of $1,25(OH)_2D_3$ are shown to elicit cellular responses through a vitamin D receptor (nVDR; Haussler et al. 1998). $1,25(OH)_2D_3$ binding to this nuclear receptor in these target tissues elicits very specific genomic cellular responses to maintain calcium-phosphate homeostasis by regulating genes involved in a plethora of transcription events ranging from the synthesis of hydroxylases, calcium binding proteins and transporters, phosphate binding proteins and transporters, transcription factors, and receptor synthesis (see Table 5.1).

Although the regulation of cellular changes through a nuclear receptor are known to facilitate cellular effects that are observed within the range of hours to days, the VDR has also been identified as one receptor involved in $1,25(OH)_2D_3$-mediated pre-genomic effects that are observed within seconds to minutes. For example,

TABLE 5.1
Genes Regulated to Elicit Classical VDR Genomic Effects

Gene(s) Regulated	Type of Regulation	Target Tissue	Physiological Effect	References
CYP24A1 (1α-hydroxylase)	–	Kidneys	↓ 1,25(OH)$_2$D$_3$ synthesis	Canaff et al. 2002; Murayama et al. 1998; Meyer et al. 2010
CYP24A1/CYP-24 (24-hydroxylase)	+	Kidneys, bone	1,25(OH)$_2$D$_3$ catabolism	Zierold et al. 1994; Kim et al. 2005
Calbindin D$_{9K}$/D$_{28K}$	+	Kidney, intestine, bone	Calcium transport	Gagnon et al. 1994; Huang et al. 1989; Perret et al. 1985; Varghese et al. 1989, Li and Christakos 1991; Huang et al. 1989; Breheir and Thomasset 1990; Dupret et al. 1987; Ferrari et al. 1992; Song et al. 2003a; Gill and Christakos 1993
CASR	+	Kidneys, parathyroids	↓ 1,25(OH)$_2$D$_3$ and PTH synthesis	Ximnim et al. 2003; Canaff and Hendy 2002
FGF-23	+	Bone	↓ 1,25(OH)$_2$D$_3$ synthesis and phosphate transport	Kolek et al. 2005; Masuyama et al. 2006; Kolek et al. 2005
NCX1	+	Kidney	Calcium extrusion	Lytton et al. 1996
NPT2	+	Bone, intestine	Phosphate transport	Masuyama et al. 2006; Xu et al. 2002
nVDR	+	Intestine, kidneys	↑nVDR levels	Strom et al. 1989; Xinmin et al. 2003
Osteocalcin	+	Bone	Bone mineralization	Kerner et al. 1989; Demay et al. 1990; Markose et al. 1990
Osteopontin (sialoprotein)	+	Bone	Bone mineralization	Kim et al. 2005; Noda et al. 1990; Li and Sodek 1993
PTH	–	Parathyroids	↓ PTH and 1,25(OH)$_2$D$_3$ synthesis	Kim et al. 2007
RANKL	+	Bone	Bone resorption	Kitazawa et al. 2003; Pike et al. 2007
TRPV5 (CaT2, CaC1)	+	Kidneys	Calcium transport	Hoenderop et al. 2001
TRPV6 (CaT1, CaC2)	+	Intestine	Calcium transport	Pike et al. 2007; Song et al. 2003b; Meyer et al. 2006

+, Up-regulated; –, down-regulated; ↓, decreases; ↑, increases.

suppression or knockout of the VDR in rat osteosarcoma-derived cells (Bravo et al. 2006) or mouse osteoblasts (Zanello and Norman 2004a,b) abolishes $1,25(OH)_2D_3$-mediated increases in intracellular calcium that are observed within 1–5 min; similarly, decreases in calcium intestinal absorption are observed in VDR knockout mice, although these effects were most likely genomic (Song et al. 2003a). However, there is also evidence that targeted knockout or suppression of the VDR does not completely abolish rapid (pre-genomic) $1,25(OH)_2D_3$-mediated calcium uptake or transport in similar species or cell types. For instance, different analogs of $1,25(OH)_2D_3$ that selectively bind to the VDR in rat osteosarcoma-derived cells (Farach-Carson et al. 1991) or chick intestinal cells (Zhou et al. 1992) demonstrate minimal ability to open calcium channels, while analogs that poorly bind VDR facilitate calcium influx or transport. In addition, knockout of the VDR in mouse osteoblasts does not alter $1,25(OH)_2D_3$-induced intracellular calcium (Wali et al. 2003), and inhibition of nuclear transcription in chick intestinal cells does not inhibit calcium transport (Norman et al. 1992). These results, taken along with the finding of Lieberherr et al. (1989) that a membrane receptor is necessary for $1,25(OH)_2D_3$-induced phosphoinositide metabolism in rat enterocytes, suggest a mechanism that does not require gene regulation to elicit the rapid, pre-genomic effects of $1,25(OH)_2D_3$.

This led researchers to investigate the role of a plasma membrane receptor to explain these rapid, nonnuclear effects (Nemere and Szego 1981a,b; Lieberherr et al. 1989; Nemere et al. 1994; Pietras et al. 2001; Farach-Carson and Nemere 2003) and subsequent characterization of a receptor protein distinct from the VDR (Nemere et al. 1998, 2000; Nemere and Campbell 2000; Larsson and Nemere 2003a,b; Nemere et al. 2004a,b). The role of this protein, termed the $1,25(OH)_2D_3$ membrane associated, rapid response steroid-binding ($1,25D_3$-MARRS) receptor, facilitates pre-genomic $1,25(OH)_2D_3$-mediated effects in intestinal cells, chondrocytes, and osteoblasts. In intestinal cells, the $1,25D_3$-MARRS receptor facilitates hormone-mediated calcium (Larsson and Nemere 2003a,b; Khanal et al. 2008; Nemere et al. 2010) and phosphate (Zhao and Nemere 2002; Nemere et al. 2004a; Sterling and Nemere 2005) absorption and transport within minutes via the protein kinase A (PKA) and protein kinase C (PKC) pathways, respectively; also, the $1,25D_3$-MARRS receptor is involved in the mineralization of growth plate chondrocytes (Boyan et al. 1999) and osteoblast maturation (Chen et al. 2010) that is mediated by PKC. As observed in additional cell types, $1,25(OH)_2D_3$ binding to the $1,25D_3$-MARRS receptor also activates signal transduction pathways that include protein kinase C (PKC), protein kinase A (PKA), extracellular signal-regulated kinases (ERK ½), phospholipase A (PLA), and phospholipase C (PLC) activation and other signaling molecules (see Table 5.2). Further investigations have also identified the $1,25D_3$-MARRS receptor as identical to ERp57/ERp60/GRp58/PDIA3, thereby extending its functions to include those of chaperone protein, DNA-binding protein, a contributor in immune function, as well as an enzyme (for review, see Khanal and Nemere 2007).

While binding of $1,25(OH)_2D_3$ to the $1,25D_3$-MARRS receptor/ERp57/ERp60/GRp58/PDIA3 stimulates rapid pre-genomic effects, there is evidence that the $1,25D_3$-MARRS receptor translocates to the nucleus in promyelocytic leukemia cells (Wu et al. 2010), chick enterocytes (Nemere et al. 2000), and rat intestinal cells (Rohe et al. 2005). In addition, Chen et al. (2010) demonstrate that the

TABLE 5.2

Pre-Genomic Effects Facilitated by 1,25D₃MARRS/ERp 57/ERp60/PDIA3/

Target Tissue	Cell Type: Species	Physiological Effect	References
Bone	Osteoblasts: human, rat, mouse	PKC activation	Boyan et al. 2002; Nemere et al. 1998; Schwartz et al. 2002; Chen et al. 2010; Boyan and Schwartz 2009; Nemere et al. 2010; Boyan et al. 1999
	Chondrocytes: human, rat	PKC alpha activation	Boyan and Schwartz 2009
	Ameloblasts/ odontoblasts: human, mouse	cell differentiation	Mesbah et al. 2002; Berdal et al. 2003
		biomineralization	Mesbah et al. 2002; Teillaud et al. 2005; Berdal et al. 2003
		↑ERK 1/2	Chen et al. 2010
		PLA₂ activation	Chen et al. 2010; Boyan and Schwartz 2009; Boyan et al. 1999
		↑PGE₂ PLC activation	Chen et al. 2010; Boyan et al. 1999
		PLC gamma activation	Boyan and Schwartz 2009
Intestine	Chick, mouse	PKA activation	Nemere et al. 2004b; Nemere and Campbell 2000; Nemere et al. 2010, 2012
		PKC activation	Nemere and Campbell 2000; Zhao and Nemere 2002; Sterling and Nemere 2005; Larsson and Nemere 2003a,b
		PKC alpha activation	Boyan and Schwartz 2009; Tunsophon and Nemere 2010
		PKC beta activation	Tunsophon and Nemere 2010, 2011
		PO₄⁻ uptake	Nemere et al. 2004a; Zhao and Nemere 2002; Sterling and Nemere 2005; Tunsophon and Nemere 2010; Nemere et al. 2012
		Ca²⁺ extrusion/efflux	Nemere and Campbell 2000; Tunsophon and Nemere 2011
		Ca²⁺ uptake/Caᵢ	Nemere et al. 2000; Khanal et al. 2008; Nemere et al. 2010
		Ca²⁺ transport	Larsson and Nemere 2003a,b
Kidneys	Chick	PKC activation	Khanal et al. 2007
			Khanal et al. 2007

PKC, protein kinase C; ↑, increases; ERK, extra-cellular-signal related kinase; PLA, phospholipase A; PGE, prostaglandin E; PLC, phospholipase C; PKA, protein kinase A.

1,25D$_3$-MARRS receptor increases genetic expression of alkaline phosphatase, collagenase 3, and osteopontin while decreasing genetic expression of osteocalcin, osteoprotegrin, and SMAD2 to facilitate bone mineralization in murine osteoblasts. These findings suggest that the 1,25D$_3$-MARRS receptor may stimulate non-genomic effects in some cell types that act in concert with the VDR to stimulate or enhance 1,25(OH)$_2$D$_3$-mediated genomic effects.

5.4 24,25-DIHYDROXYVITAMIN D$_3$

To optimize nutrient utilization by humans and animals during growth, it is important to understand both stimulatory and inhibitory pathways mediated by vitamin D metabolites that regulate phosphate and calcium homeostasis; 24,25-dihydroxyvitamin D$_3$ [24,25(OH)$_2$D$_3$] is a vitamin D metabolite that contributes to this process by acting as an endogenous inhibitor of 1,25(OH)$_2$D$_3$-mediated signaling. It is synthesized when serum levels of calcium and phosphate are high, so that 25(OH)$_2$D$_3$ is then hydroxylated on carbon-24 in the kidneys. Even though 24,25(OH)$_2$D$_3$ was initially considered a "scrap" metabolite, there is increasing evidence its endocrine actions are distinct from 1,25(OH)$_2$D$_3$.

Lam et al. (1973) first introduced the biological activity and metabolism of 24,25(OH)$_2$D$_3$ in studies indicating that supraphysiological concentrations of 24,25(OH)$_2$D$_3$ contributed to marked increases of both intestinal calcium transport and serum calcium levels in vitamin D-deficient rats fed on a diet containing adequate calcium and phosphorus. However, other studies investigating the physiological concentrations of this metabolite have reported that 24,25(OH)$_2$D$_3$ not only inhibits 1,25(OH)$_2$D$_3$-mediated stimulated intestinal calcium transport (Nemere 1996; Larsson et al. 1995, 2002) but phosphate transport as well (Nemere 1999; Peery and Nemere 2007). 24,25(OH)$_2$D$_3$ is also shown to stimulate bone formation and mineralization (Ornoy et al. 1978; Corvol et al. 1978; Endo et al. 1980; Wu et al. 2006), cartilage growth (Sömjen et al. 1982a,b; Binderman and Sömjen 1984; Kato et al. 1998; Pedrozo et al. 1999; Boyan et al. 2002), egg hatchability (Henry and Norman 1978), and suppress parathyroid hormone secretion (Carpenter et al. 1996; Nemere 1999). The inhibitory actions of 24,25(OH)$_2$D$_3$ on calcium and phosphate absorption are observed across species. In addition to the work performed in chicks, it has been reported that 24,25(OH)$_2$D$_3$ causes hypocalcemia in rats (Maeda et al. 1985), and that 24,25(OH)$_2$D$_3$ decreases calcium absorption in dogs (Tryfonidou et al. 2002). Currently, 24R,25(OH)$_2$D$_3$ and 24S,25(OH)$_2$D$_3$ are two stereoisomers of 24,25(OH)$_2$D$_3$ that are shown to elicit cellular effects linked to several signaling events that are dependent upon species, cell type, and/or dosage. For example, Sömjen and colleagues found that 24R,25(OH)$_2$D$_3$ increases DNA synthesis and ornithine decarboxylase activity as well as creatine kinase, to regulate epiphyseal growth to induce endochondral bone formation in rats (Sömjen et al. 1983) and chicks (Sömjen et al. 1983); however, 24S,25(OH)$_2$D$_3$ had no effect. In addition, Boyan et al. (2002) found that 24R,25(OH)$_2$D$_3$ stimulated PKC activity in undifferentiated or moderately differentiated rat osteoblast cell lines, but 24S,25(OH)$_2$D$_3$ did not. However, other investigators have demonstrated that both stereoisomers are active. Yamamoto et al. (1998) demonstrated that physiological concentrations (10^{-9}–10^{-8}

M) of 24R,25(OH)$_2$D$_3$ significantly increased the cyclic guanosine 5'-monophosphate (cGMP) content in the human osteoblastic cells by approximately 200% in 5 to 15 min; in contrast, 24S,25-dihydroxyvitamin D$_3$ had a weak effect on the cGMP content. Larsson et al. (2002) have demonstrated that 24R,25(OH)$_2$D$_3$ and 24S,25(OH)$_2$D$_3$ are equally effective in suppressing Ca uptake in fish enterocytes. Meng and Nemere (2013) recently found that 24R,25(OH)$_2$D$_3$ and 24S,25(OH)$_2$D$_3$ have significant but different effects on phosphate absorption in chicken intestine. While 24R,25(OH)$_2$D$_3$ inhibited phosphate absorption as early as 1 hr over a sustained period of 18 hr, 24S,25(OH)$_2$D$_3$ stimulated intestinal phosphate absorption at 5 hr but had no other effects at the other time points tested (Meng and Nemere 2013). They also found in dose response studies that low levels (100 µg per chick) of 24R,25(OH)$_2$D$_3$ stimulated phosphate absorption while higher levels (200–300 µg per bird) of 24R,25(OH)$_2$D$_3$ decreased phosphate absorption.

On a cellular level, 24,25(OH)$_2$D$_3$ has been found to inhibit the rapid (pre-genomic) 1,25(OH)$_2$D$_3$-mediated effects that facilitate the opening of calcium channels in osteoblasts and osteosarcoma cells (Yukohori et al. 1996; Khoury et al. 1995; Takeuchi and Guggino 1996) and PKC activation in chick enterocytes (Nemere 1999; Nemere 2002).

The putative receptors for 24,25(OH)$_2$D$_3$ show allosteric binding to catalase (most binding activity was found intracellularly) even though the 24,25(OH)$_2$D$_3$ binding protein on the cell surface is evident. The cell surface protein mediates rapid membrane-initiated responses regulating calcium homeostasis, cell differentiation, and maturation, and triggers second messenger systems in chondrocytes, osteoblasts, and intestinal cells. For example, Li et al. (1996) have also demonstrated that 24,25(OH)$_2$D$_3$–mediated changes in L-type Ca channels is not only biphasic in rat osteoblasts but is also pre-genomic: low concentrations of 24,25(OH)$_2$D$_3$ stimulated channel activation through the PKA pathway, and high concentrations of 24,25(OH)$_2$D$_3$ inhibited channel activation through the PKC pathway within minutes of hormone exposure. Schwartz et al. (2002) have shown that 24,25(OH)$_2$D$_3$ activated the MAP kinase pathway implicating signaling molecules that include ERK 1 and 2, phospholipase D (PLD)-dependent PKC, and H$_2$O$_2$. This experimental evidence involving signaling molecules that facilitate membrane receptor initiated events, in addition to the presence of a 24,25(OH)$_2$D$_3$ binding protein found in the plasma membranes and matrix vesicles of chondrocytes (Pedrozo et al. 1999), lysosomal and basal lateral membrane fractions of chick duodenal loops (Nemere et al. 2002), and the basal lateral membranes of chick and fish enterocytes (Larsson et al. 2001) suggested a putative receptor for 24,25(OH)$_2$D$_3$.

While other investigators have partially characterized this putative receptor for 24,25(OH)$_2$D$_3$ (Sömjen 1982a,b; Pedrozo et al. 1999; Larsson et al. 2001), Larsson et al. (2006) have successfully identified this putative receptor for 24,25(OH)$_2$D$_3$ as catalase. Sequence analysis of a 24,25(OH)$_2$D$_3$ binding protein was performed, and five regions of homology with the enzyme catalase were found (Larsson et al. 2006). These findings not only identified catalase as the 24,25(OH)$_2$D$_3$ binding protein but also led Nemere et al. (2006) to test catalase activity following 24,25(OH)$_2$D$_3$ treatment. Addition of physiological levels of 24,25(OH)$_2$D$_3$ to isolated intestinal cells resulted in a time-dependent decrease in catalase activity, whereas 130 pM

1,25(OH)$_2$D$_3$ did not. The modest 30% decrease in catalase-specific activity was paralleled by an equivalent increase in H$_2$O$_2$ following 24,25(OH)$_2$D$_3$ treatment.

Earlier work in the Nemere lab has indicated that a cellular binding protein for 24,25(OH)$_2$D$_3$ is the enzyme catalase (Larsson et al. 2006). A functional consequence of 24,25(OH)$_2$D$_3$ binding to catalase was found to be a decrease in enzyme activity, with concomitant increase in H$_2$O$_2$ production (Nemere et al. 2006). One mechanism of inhibition was then found to be oxidation of the 1,25D$_3$-MARRS receptor (which contains two thioredoxin domains [Nemere et al. 2004a]) and inhibition of 1,25(OH)$_2$D$_3$ binding (Nemere et al. 2006). Antioxidant diets nearly doubled phosphate absorption in vivo (Nemere et al. 2006). Peery and Nemere (2007) then proved the hypothesis that antioxidant conditions, such as an anti-catalase antibody or excess exogenous catalase protein, could block the inhibitory action of 24,25(OH)$_2$D$_3$, whereas pro-oxidant conditions, such as a triazole inhibitor of catalase activity, could mimic the inhibitory action of the secosteroid. It was further recognized that H$_2$O$_2$ could also influence the signal transduction pathway beyond ligand binding to the 1,25(OH)$_2$D$_3$-MARRS receptor (Peery and Nemere 2007).

The 1,25D$_3$-MARRS receptor, which contains 2 thioredoxin homology domains, is critical for stimulating PKC, which in turn mediates phosphate transport. Catalase is a binding protein for 24,25(OH)$_2$D$_3$; the complex results in decreased enzyme activity and increased H$_2$O$_2$. H$_2$O$_2$ in turn inhibits phosphate transport by two routes: oxidizing thiols in the 1,25D$_3$-MARRS receptor, which inhibits hormone binding; and inhibiting PKC activity. These observations demonstrate that 24,25(OH)$_2$D$_3$ provides an important feedback loop to protect against excess 1,25(OH)$_2$D$_3$ (Larsson et al. 1995). The model for the interaction between the 1,25D$_3$-MARRS receptor and catalase is described in Figure 5.1. Upon binding 1,25(OH)$_2$D$_3$, the 1,25D$_3$-MARRS receptor dimerizes and increases signal transduction through PKA for enhanced

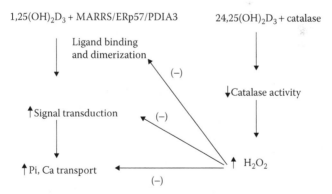

FIGURE 5.1   Schematic presentation of the stimulatory effects of 1,25(OH)$_2$D$_3$ and feedback inhibition by 24,25(OH)$_2$D$_3$. Binding of 1,25(OH)$_2$D$_3$ to the 1,25D$_3$-MARRS receptor leads to dimerization, stimulation of the PK A and PK C pathways, and ultimately an increase in calcium and phosphate uptake. Binding of 24,25(OH)$_2$D$_3$ to catalase results in decreased enzymatic activity with an increase in hydrogen peroxide. This, in turn, decreases the stimulatory activity of 1,25(OH)$_2$D$_3$ by decreasing dimerization of the 1,25D$_3$-MARRS receptor, as well as a direct decrease on the PK C pathway, ultimately resulting in decreased calcium and phosphate uptake.

calcium uptake and PKC for enhanced phosphate uptake. Feedback inhibition of this stimulatory pathway is initiated when $24,25(OH)_2D_3$ binds to catalase, thereby decreasing enzymatic activity. The increased H_2O_2 levels are known to inhibit stimulated ion uptake, and in this way the rapid stimulation by $1,25(OH)_2D_3$ is inhibited by $24,25(OH)_2D_3$.

5.5   CONCLUSIONS

It is evident from considering the wealth of evidence presented here that $1,25(OH)_2D_3$ is not the only hormonally active metabolite. Each of the metabolites discussed herein has pre-genomic effects, and for two of these, novel receptors have been identified. The broader implications are that other steroid hormones may have additional active metabolites whose effects have yet to be elucidated, as well as novel receptors for steroid hormones that are already recognized to have biological activity.

REFERENCES

Ameenuddin, S., Sunde, M., DeLuca, H.F., et al. 1982. 24-Hydroxylation of 25 hydroxyvitamin D_3: Is it required for embryonic development in chicks? *Science* 217: 451–2.

Ameenuddin, S., Sunde, M.L., DeLuca, H.F., et al. 1983. Support of embryonic chick survival by vitamin D metabolites. *Arch. Biochem. Biophys.* 226: 666–70.

Bar, A., Razaphkovsky,V., Vax, E., Plavnik, I. 2003. Performance and bone development in broiler chickens given 25-hydroxycholecalciferol. *Br. Poult. Sci.* 44: 224–33.

Barreto, A.M., Schwartz, G.G., Woodruff, R., Cramer, S.D. 2000. 25-Hydroxyvitamin D_3, the prohormone of 1,25-dihydroxyvitamin D_3, inhibits the proliferation of primary prostatic epithelial cells. *Cancer Epidemiol. Biomarkers Prev.* 9: 265–70.

Berdal, A., Mesbah, M., Papagerakis, P., Nemere, I. 2003. Putative membrane receptor for 1,25(OH)2 vitamin D3 in human mineralized tissues during prenatal development. *Connect. Tissue Res.* 44 (1): 136–40.

Binderman, I. and Sömjen, D. 1984. 24,25-Dihydroxycholecalciferol induces the growth of chick cartilage in vitro. *Endocrinology* 115(1): 430–2.

Birge, S.J. and Avioli, R.C. 1981. Intestinal phosphate transport and alkaline phosphatase activity in the chick. *Am. J. Physiol.* 240: E384–90.

Blunt, J.W. and Deluca, H.F. 1969. 25-Hydroxycholecalciferol. A biologically active metabolite of vitamin D_3. *Biochemistry* 7: 10.

Boyan, B.D., Bonewald, L.F., Sylvia,V.L., et al. 2002. Evidence for distinct membrane receptors for 1 alpha,25-(OH)(2)D(3) and 24R,25-(OH)(2)D(3) in osteoblasts. *Steroids* 67(3-4): 235–46.

Boyan, B.D. and Schwartz, Z. 2009. 1,25-Dihydroxy vitamin D_3 is an autocrine regulator of extracellular matrix turnover and growth factor release via ERp60-activated matrix vesicle matrix metalloproteinases. *Cells Tissues Organs* 189(1-4): 70–4.

Boyan, B.D., Sylvia,V.L,. Dean, D.D., et al. 1999. 1,25-$(OH)_2D_3$ modulates growth plate chondrocytes via membrane receptor-mediated protein kinase C by a mechanism that involves changes in phospholipid metabolism and the action of arachidonic acid and PGE2. *Steroids* 64(1-2): 129–36.

Bravo, S., Paredes, R., Izaurieta, P., et al. 2006. The classic receptor for 1alpha,25-dihydroxy vitamin D3 is required for non-genomic actions of 1a,25-dihydroxy vitamin d3 in osteosarcoma cells. *J. Cell Biochem.* 99: 995–1000.

Brehier, A. and Thomasset, M. 1990. Stimulation of calbindin-D9K (CaBP9K) gene expression by calcium and 1,25-dihydroxycholecalciferol in fetal rat duodenal organ culture. *Endocrinology* 127: 580–585.

Canaff, L., Hendy, G.N. 2002. Vitamin D response elements in promoters P1 and P2 confer transcriptional responsiveness to 1,25-dihydroxyvitamin D. *J. Biol. Chem.* 277(33): 30337–50.

Carpenter, T.O., Keller, M., Scwartz, D., et al. 1996. 24,25-Dihydroxy vitamin D supplementation corrects hyperparathyroidism and improves skeletal abnormalities in X-linked hypophosphatemic rickets—a clinical research center study. *J. Clin. Endocrinol. Metab.* 81(6): 2381–8.

Chen, J., Olivares-Navarrete, R., Wang, Y. 2010. Protein-disulfide isomerase-associated 3 (PDIA3) mediates the membrane response to 1,25-dihydroxyvitamin D3 in osteoblasts. *J. Biol. Chem.* 285(47): 37041–50.

Corvol, M.T., Garabedian, M., Du Bois, M.B., et al. 1978. Vitamin D and cartilage. II. Biological activity of 25-hydroxycholecalciferol and 24,25- and 1,25-dihydroxycholecalciferols on cultured growth plate chondrocytes. *Endocrinology* 102: 1269–74.

Demay, M.B., Gerardi, J.M., DeLuca, H.F. and Kronenberg, H.M. 1990. DNA sequences in the rat osteocalcin gene that bind the 1,25-dihydroxyvitamin D_3 receptor and confer responsiveness to 1,25-dihydroxyvitamin D3. *Proc. Natl. Acad. Sci USA* 87: 369–373.

Dupret, J.M., Brun, P., Perret, C., et al. 1987. Transcriptional and post-transcriptional regulation of vitamin D–dependent calcium-binding protein gene expression in the rat duodenum by 1,25-dihydroxycholecalciferol. *J. Biol. Chem.* 262: 16553–55.

Earl, S., Cole, Z.A., Holroyd, C., Cooper, C. and Harvey, N.C. 2010. Session 2: Other diseases: Dietary management of osteoporosis throughout the life course. *Proc. Nutr. Soc.* 69: 25–33.

Endo, H., Kiyoki, M., Kawashima, K., et al. 1980. Vitamin D3 metabolites and PTH synergistically stimulate bone formation of chick embryonic femur in vitro. *Nature* 286: 262–4.

Farach-Carson, M.C. and Nemere, I. 2003. Membrane receptors for vitamin D steroid hormones: Potential new drug targets. *Curr. Drug Targets* 4(1): 67–76.

Farach-Carson, M.C., Sergeev, I., and Norman, A.W. 1991. Nongenomic actions of 1,25-dihydroxyvitamin D3 in rat osteosarcoma cells: Structure-function studies using ligand analogs. *Endocrinology* 129(4): 1876–84.

Ferrari, S., Molinari, S., Battini, R., et al. 1992. Induction of calbindin-D28k by 1,25-dihydroxyvitamin D3 in cultured chicken intestinal cells. *Exp. Cell Res.* 200: 528–31.

Gagnon, A., Simboli-Campbell, M., Welsh, J. 1994. Induction of calbindin D-28k in Madin-Darby bovine kidney cells by 1,25(OH)2D3. *Kidney Int.* 45: 95–102.

Gill,R.K. and Christakos, S. 1993. Identification of sequence elements in mouse calbindin-D28k gene that confer 1,25-dihydroxyvitamin D3- and butyrate-inducible responses. *Proc. Natl. Acad. Sci. USA* 90(7): 2984–8.

Hart, L.E. and DeLuca, H.F. 1985. Effect of vitamin D_3 metabolites on calcium and phosphorus metabolism in chick embryos. *Am. J. Physiol.* 248: E281–5.

Haussler, M.R., Whitfield, G.K., Haussler, C.A., et al. 1998. The nuclear vitamin D receptor: Biological and molecular regulatory properties revealed. *J. Bone Mineral Res.* 13(3): 325–49.

Henry, H.L. and Norman, A.W. 1978. Vitamin D: Two hydroxylated metabolites are required for normal chick egg hatchability. *Science* 201: 835–5.

Hoenderop, J.G., Muller, D., Van Der Kemp, A.W., et al. 2001. Calcitriol controls the epithelial calcium channel in kidney. *J. Am. Soc. Nephrol* 12(7): 1342–9.

Huang, Y., Lee, S., Stolz, R., et al. 1989. Effect of hormones and development on the expression of the rat 1,25-dyhydroxyvitamin D_3 receptor gene: Comparison with calbindin gene expression. *J. Biol. Chem.* 264: 17454–61.

Kato, A., Seo, E.G., Einhorn, T.A., et al. 1998. Studies on 24R,25-Dihydroxyvitamin D_3: Evidence for a nonnuclear membrane receptor in the chick tibial fracture-healing callus. *Bone* 23: 141–6.

Kerner, S.A., Scott, R.A., Pike, J.W. 1989. Sequence elements in the human osteocalcin gene confer basal activation and inducible response to hormonal vitamin D3. *Proc. Natl. Acad. Sci. USA* 86(12): 4455–9.

Khanal, R.C. and Nemere, I. 2005. The ERp57/GRp58/1,25D_3-MARRS receptor: Multiple functional roles in diverse cell systems. *Curr. Med. Chem.* 14(10): 1087–93.

Khanal, R. and Nemere I. 2007. The ERp57/GRp58/1,25D_3-MARRS receptor: Multiple functional roles in diverse cell systems. *Curr. Medic. Chem.* 14: 1087–93.

Khanal, R.C., Peters, T.M., Smith, N.M., Nemere, I. 2008. Membrane receptor-initiated signaling in 1,25$(OH)_2D_3$-stimulated calcium uptake in intestinal epithelial cells. *J. Cell. Biochem.* 105: 1109–1116.

Khanal, R.C., Smith, N.M., Nemere, I. 2007. Phosphate uptake in chick kidney cells: Effects of 1,25(OH)2D3 and 24,25(OH)2D3. *Steroids* 72(2): 158–64.

Khoury, R.S., Weber, J., and Farach-Carson, M.C. 1995. Vitamin D metabolites modulate osteoblast activity by Ca+2 influx-independent genomic and Ca+2 influx-dependent nongenomic pathways. *J. Nutr.* 125(6 Suppl): 1699S–1703S.

Kim, M.S., Fujiki, R., Murayama, A., et al. 2007. 1Alpha,25$(OH)_2D_3$-induced transrepression by vitamin D receptor through E-box-type elements in the human parathyroid hormone gene promoter. *Mol. Endocrinol.* 21(2): 334–42.

Kim, S., Shevde, N.K., and Pike, J.W. 2005. 1,25-Dihydroxyvitamin D3 stimulates cyclic vitamin D receptor/retinoid X receptor DNA-binding, co-activator recruitment, and histone acetylation in intact osteoblasts. *J. Bone Miner. Res.* 20(2): 305–15.

Kitazawa, S., Kajimoto, K., Kondo, T., Kitazawa, R. 2003. Vitamin D_3 supports osteoclastogenesis via functional vitamin D response element of human RANKL gene promoter. *J. Cell. Biochem.* 89: 771–5.

Kolek, O.I., Hines E.R., Jones M.D., et al. 2005. 1-alpha, 25-Dihydroxyvitamin D_3 upregulates FGF23 gene expression in bone: the final link in a renal-gastrointestinal-skeletal axis that controls phosphate transport. *Am. J. Physiol. Gastrointest. Liver Physiol.* 289: G1036–42.

Lam, H.Y., Schnoes, H.K., DeLuca, H.F., et al. 1973. 24,25-Dihydroxyvitamin D: Synthesis and biological activity. *Biochemistry* 12(24): 4851–55.

Lambert, J.R., Young, C.D., Persons, K.S. and Ray, R. 2007. Mechanistic and pharmacodynamic studies of a 25-hydroxyvitamin D_3 derivative in prostate cancer cells. *Biochem. Biophys. Res. Commun.* 361: 189–95.

Larsson, B. and Nemere, I. 2003a. Effect of growth and maturation on membrane-initiated actions of 1,25-dihydroxyvitamin D3. I. Calcium transport, receptor kinetics, and signal transduction in intestine of male chickens. *Endocrinology* 144: 1726–35.

Larsson, B. and Nemere, I. 2003b. Effect of growth and maturation on membrane-initiated actions of 1,25-dihydroxyvitamin D3. II. Calcium transport, receptor kinetics, and signal transduction in intestine of female chickens. *J. Cellular Biochem.* 90 (5): 901–13.

Larsson, D., Anderson, D., Smith, N. and Nemere, I. 2006. 24,25-Dihydroxyvitamin D_3 binds to catalase. *J. Cellular Biochem* 97: 1259-66.

Larsson D., Aksnes L., Th Björnsson B., Larsson B., Lundgren T. and Sundell K. 2002. Antagonistic effects of 24R,25-dihydroxyvitamin D3 and 25-hydroxyvitamin D3 on L-type Ca2+ channels and Na+/Ca2+ exchange in enterocytes from Atlantic cod (Gadus morhua). *J. Mol. Endocrinol.* 28(1): 53–68.

Larsson D., Björnsson B.T. and Sundell K. 1995. Physiological concentrations of 24,25-dihydroxyvitamin D3 rapidly decrease the in vitro intestinal calcium uptake in the Atlantic cod, Gadus morhua. *Gen. Comp. Endocrinol.* 100(2): 211–5.

Larsson, D., Larsson, D., Aksnes, L., et al. 2002. Antagonistic effects of 24R,25-dihydroxyvitamin D3 and 25-hydroxyvitamin D3 on L-type Ca2+ channels and Na+/Ca2+ exchange in enterocytes from Atlantic cod (Gadus morhua). *J. Mol. Endocrinol.* 28(1): 53–68.

Larsson, D., Nemere, I., Sundell, K. 2001. Putative basal lateral membrane receptors for 24,25-dihydroxyvitamin D(3) in carp and Atlantic cod enterocytes: Characterization of binding and effects on intracellular calcium regulation. *J. Cell Biochem.* 83(2): 171–86.

Li, B., Chik, C.L., Taniguchi, N., Ho, A.K., Karpinski, E. 1996. 24,25(OH)2 vitamin D3 modulates the L-type Ca2+ channel current in UMR 106 cells: Involvement of protein kinase A and protein kinase C. *Cell Calcium* 19(3): 193–200.

Li, H. and Christakos, S. 1991. Differential regulation by 1,25-dihydroxyvitamin D3 of calbindin-D9 and calbindin-D28k gene expression in mouse kidney. *Endocrinology* 128: 2844–52.

Li, J.J. and Sodek, J. 1993. Cloning and characterization of the rat bone sialoprotein gene promoter. *Biochem. J.* 289(3): 625–9.

Li, X., Zheng, W. and Li, Y.C. 2003. Altered gene expression profile in the kidney of vitamin D receptor knockout mice. *J. Cell. Biochem.* 89: 709–9.

Lieberherr, M., Grosse, B., Duchambon, P., Drueke, T.A. 1989. Functional cell surface type receptor is required for the early action of 1,25-dihydroxyvitamin D_3 on the phosphoinositide metabolism in rat enterocytes. *J. Biol. Chem.* 264(34): 20403–6.

Lou, Y.R., Laaksi, I., Syvälä, H., et al. 2004. 25-hydroxyvitamin D_3 is an active hormone in human primary prostatic stromal cells. *FASEB J.* 18: 332–4.

Lytton, J., Lee, S., Lee,W., et al. 1996. The kidney sodium-calcium exchanger. *Ann. N. Y. Acad. Sci.* 779: 58–72.

Maeda, Y., Yamato, H., Katoh, T., et al. 1985. Hypocalcemic effect of 24R,25-dihydroxyvitamin D_3 in rats. *In vivo* 1(6): 347–50.

Markose, E.R., Stein, J.L., Stein, G.S. and Lian, J.B. 1990. Vitamin D-mediated modifications in protein-DNA interactions at two promoter elements of the osteocalcin gene. *Proc. Natl. Acad. Sci. USA* 87: 1701–5.

Masuyama, R., Stockmans, I., Torrekens S, et al. 2006. Vitamin D receptor in chondrocytes promotes osteoclastogenesis and regulates FGF23 production in osteoblasts. *J. Clin. Invest.* 116(12): 3150–9.

Meng, Y. and Nemere, I. 2013. Effect of 24,25-dihydroxyvitamin D_3 on phosphate absorption in vivo. *Immunol. Endocrinol., and Met. Agents in Medic. Chem.* 13: 60–67.

Mesbah, M., Nemere, I., Papagerakis, P., et al. 2002. Expression of a 1,25-dihydroxyvitamin D3 membrane-associated rapid-response steroid binding protein during human tooth and bone development and biomineralization. *J. Bone Mineral Res.* 17(9): 1588–96.

Meyer, M.B., Goetsch, P.D., Pike, J.W. 2010. A downstream intergenic cluster of regulatory enhancers contributes to the induction of CYP24A1 expression by 1alpha,25-dihydroxyvitamin D3. *J. Biol. Chem.* 285(20): 15599–610.

Meyer, M.B., Watanuki, M., Kim, S., et al. 2006. The human transient receptor potential vanilloid type 6 distal promoter contains multiple vitamin D receptor binding sites that mediate activation by 1,25-dihydroxyvitamin D_3 in intestinal cells. *Mol. Endocrinol.* 20: 1447–61.

Municio, M.J., Traba, M.L. 2004. Effects of 24,25(OH)$_2$D$_3$, 1,25(OH)$_2$D$_3$ and 25(OH)D$_3$ on alkaline and tartrate-resistant acid phosphatase activities in fetal rat calvaria. *J. Physiol. Biochem.* 60: 219–224.

Murayama, A., Takeyama, K., Kitanaka,S., Kodera, Y., Hosoya, T., Kato, S. 1998. The promoter of the human 25-hydroxyvitamin D_3 1 alpha-hydroxylase gene confers positive and negative responsiveness to PTH, calcitonin, and 1 alpha,25(OH)$_2$D$_3$. *Biochem. Biophys. Res.* 249(1): 11–6.

Nemere, I. 1996. Apparent nonnuclear regulation of intestinal phosphate transport: Effects of 1,25-dihydroxyvitamin D_3, 24,25-dihydroxyvitamin D_3, and 25-hydroxyvitamin D_3. *Endocrinology* 137: 2254–61.

Nemere I. 1999. 24,25-dihydroxyvitamin D3 suppresses the rapid actions of 1, 25-dihydroxyvitamin D3 and parathyroid hormone on calcium transport in chick intestine. *J. Bone Miner. Res.* 14(9): 1543–9.

Nemere, I. and Campbell, K. 2000. Immunochemical studies on the putative plasmalemmal receptor for 1, 25-dihydroxyvitamin D(3). III. Vitamin D status. *Steroids* 65(8): 451–5.

Nemere, I., Dormanen, M.C., Hammond, M.W., et al. 1994. Identification of a specific binding protein for 1 alpha,25-dihydroxyvitamin D_3 in basal-lateral membranes of chick intestinal epithelium and relationship to transcaltachia. *J. Biol. Chem.* 269(38): 23750–6.

Nemere, I., Farach-Carson, M.C., Rohe, B., et al. 2004a. Ribozyme knockdown functionally links a 1,25(OH)$_2D_3$ membrane binding protein (1,25D_3-MARRS) and phosphate uptake in intestinal cells. *Proc. Natl. Acad. Sci. USA* 101(19): 7392–5.

Nemere, I., Garbi, N., Hämmerling, G.J., and Khanal, R.C. 2010. Intestinal cell calcium uptake and the targeted knockout of the 1,25D3-MARRS (membrane-associated, rapid response steroid-binding) receptor/PDIA3/Erp55. *J. Biol. Chem.* 285(41): 31859–66.

Nemere, I., Garcia-Garbi, N., Hämmerling, G.J., Winger, Q. 2012. Intestinal cell phosphate uptake and the targeted knockout of the 1,25D3-MARRS Receptor/PDIA3/ERp55. *Endocrinology* http://endo.endojournals.org/content/early/2012/02/08/en.2011-1850. long.

Nemere, I., Ray, R., and McManus, W. 2000. Immunochemical studies on the putative plasmalemmal receptor for 1, 25(OH)(2)D(3). I. Chick intestine. *Am. J. Physiol. Endocrinol. Metab.* 278(6): E1104–14.

Nemere, I., Safford, S.E., Rohe, B., et al. 2004b. Identification and characterization of 1,25D_3-membrane-associated rapid response, steroid (1,25D_3-MARRS) binding protein. *J. Steroid Biochem. Mol. Biol.* 89-90(1-5): 281–5.

Nemere, I., Schwartz, Z., Pedrozo, H, et al. 1998. Identification of a membrane receptor for 1,25-dihydroxyvitamin D_3 which mediates rapid activation of protein kinase C. *J. Bone Mineral Res.* 13(9): 1353–9.

Nemere, I. and Szego, C.M. 1981a. Early actions of parathyroid hormone and 1,25-dihydroxycholecalciferol on isolated epithelial cells from rat intestine: I. Limited lysosomal enzyme release and calcium uptake. *Endocrinology* 108(4): 1450–62.

Nemere, I. and Szego, C.M. 1981b. Early actions of parathyroid hormone and 1,25-dihydroxycholecalciferol on isolated epithelial cells from rat intestine: II. Analyses of additivity, contribution of calcium, and modulatory influence of indomethacin. *Endocrinology* 109(6): 2180–5.

Nemere, I. Wilson, C., Jensen, W., et al. 2006. Mechanism of 24,25-dhydroxyvitamin D_3-mediated inhibition of rapid, 1,25-dihydroxyvitamin D_3-induced responses: Role of reactive oxygen species *J. Cell. Biochem.* 99: 1572–81.

Nemere, I., Yazzie-Atkinson, D., Johns, D.O., Larsson, D. 2002. Biochemical characterization and purification of a binding protein for 24,25-dihydroxyvitamin D3 from chick intestine. *J. Endocrinol.* 172(1): 211–9.

Nemere, I., Yoshimoto, Y., Norman, A.W. 1984. Calcium transport in perfused duodena from normal chicks: Enhancement within fourteen minutes of exposure to 1,25-dihydroxyvitamin D_3. *Endocrinology* 115: 1476–83.

Noda, M., Vogel, R.L., Craig, A.M., Prahl, J., DeLuca, H.F. and Denhardt, D.T. 1990. Identification of a DNA sequence responsible for binding of the 1-alpha,25-dihydroxyvitamin D3 receptor and 1,25-dihydroxyvitamin D_3 enhancement of mouse secreted phosphoprotein 1 (SPP-1 or osteopontin) gene expression. *Proc. Natl. Acad. Sci. USA* 87(24): 9995–9.

Norman, A.W., Nemere, I., Zhou, L., et al. 1992. 1,25(OH)2-vitamin D3, a steroid hormone that produces biologic effects via both genomic and nongenomic pathways. *J. Steroid Biohem.* 41(3-8): 231–40.

Ornoy, A., Goodwin, D., Noff, D., et al. 1978. 24,25-dihydroxyvitamin D is a metabolite of vitamin D essential for bone formation. *Nature* 276: 517–19.

Pedrozo, H.A., Schwartz, Z., Rimes, S., et al. 1999. Physiological importance of the 1,25(OH)2D3 membrane receptor and evidence for a membrane receptor specific for 24,25(OH)2D3. *J. Bone Miner. Res.* 14(6): 856–65.

Peery, S.L. and Nemere I. 2007. Contributions of pro-oxidant and anti-oxidant conditions to the actions of 24,25-dihydroxyvitamin D3 and 1,25-dihydroxyvitamin D3 on phosphate uptake in intestinal cells. *J. Cell Biochem.* 101(5): 1176–84.

Peng, X., Vaishnav, A., Murillo, G., et al. 2010. Protection against cellular stress by 25-hydroxyvitamin D3 in breast epithelial cells. *J. Cell. Biochem.* 110: 1324–33.

Perret, C., Desplan, C., Thomasset, M. 1985. Cholecalcin (a 9-kDa cholecalciferol-induced calcium-binding protein) messenger RNA. Distribution and induction by calcitriol in the rat digestive tract. *Eur. J. Biochem.* 150: 211–15.

Phadnis R. and Nemere, I. 2003. Direct, rapid effects of 25-hydroxyvitamin D$_3$ on isolated intestinal cells. *J. Cell. Biochem.* 90: 287–93.

Pietras, R.J., Nemere, I., Szego, C.M. 2001. Steroid hormone receptors in target cell membranes. *Endocrine* 14(3): 417–25.

Pike, J., Zella, L., Meyer, M., et al. 2007. Molecular actions of 1,25-dihydroxyvitamin D$_3$ on genes involved in calcium homeostasis. *J Bone Miner Res* 22(2): V16–9.

Ritter, C.S., Armbrecht, H.J., Slatopolsky, E., Brown, A.J. 2006. 25-Hydroxyvitamin D$_3$ suppresses PTH synthesis and secretion by bovine parathyroid cells. *Kidney Int.* 70: 654–9.

Ritter, C.S. and Brown, A.J. 2011. Direct suppression of PTH gene expression by the vitamin D prohormones doxercalciferol and calcidiol requires the vitamin D receptor. *J. Mol. Endocrinol.* 46: 63–6.

Rohe, B., Safford, S.E., Nemere, I. and Farach-Carson, M.C. 2005. Identification and characterization of 1,25D3-membrane-associated rapid response, steroid (1,25D3-MARRS)-binding protein in rat IEC-6 cells. *Steroids* 70(5-7): 458–63.

Schachter, D. and Rosen, S.M. 1959. Active transport of Ca45 by the small intestine and its dependence on vitamin D. *Am. J. Physiol.* 196(2): 357–62.

Schwartz, Z., Ehland, H., Sylvia, V.L., et al. 2002. 1alpha,25-dihydroxyvitamin D(3) and 24R,25-dihydroxyvitamin D(3) modulate growth plate chondrocyte physiology via protein kinase C-dependent phosphorylation of extracellular signal-regulated kinase 1/2 mitogen-activated protein kinase. *Endocrinology* 143(7): 2775–86.

Sömjen, D., Binderman, I., Weisman, Y. 1983. The effects of 24R,25-dihydroxycholecalciferol and of 1 alpha,25-dihydroxycholecalciferol on ornithine decarboxylase activity and on DNA synthesis in the epiphysis and diaphysis of rat bone and in the duodenum. *Biochem J.* 214(2): 293–8.

Sömjen, D., Kaye, A.M., Binderman, I. 1984a. 24R,25-dihydroxyvitamin D stimulates creatine kinase BB activity in chick cartilage cells in culture. *FEBS Lett.* 167(2): 281–4.

Sömjen, D., Sömjen, G.J., Harell, A., Mechanic, G.L., Binderman, I. 1982a. Partial characterization of a specific high affinity binding macromolecule for 24R,25 dihydroxyvitamin D3 in differentiating skeletal mesenchyme. *Biochem. Biophys. Res. Commun.* 106(2): 644–51.

Sömjen, D., Sömjen, G.J., Weisman, Y., Binderman, I. 1982b. Evidence for 24,25-dihydroxycholecalciferol receptors in long bones of newborn rats. *Biochem. J.* 204(1): 31–6.

Song, Y., Kato, S., Fleet, J.C. 2003a. Vitamin D receptor (VDR) knockout mice reveal VDR-independent regulation of intestinal calcium absorption and ECaC2 and calbindin D-9k mRNA. *J. Nutr.* 133: 374–80.

Song,Y., Peng, X., Porta,A., et al. 2003b. Calcium transporter 1 and epithelial calcium channel messenger ribonucleic acid are differentially regulated by 1,25 dihydroxyvitamin D3 in the intestine and kidney of mice. *Endocrinology* 144(9): 3885–94.

Sterling, T.M. and Nemere, I. 2005a. 1,25-dihydroxyvitamin D3 stimulates vesicular transport within 5s in polarized intestinal epithelial cells. *J. Endocrinol.* 185(1): 81-91.

Sterling T.M., and Nemere I. 2007. Calcium uptake and membrane trafficking in response to PTH or 25(OH)D$_3$ in polarized intestinal epithelial cells. *Steroids* 72: 151–5.

Strom M., Sandgren, M.E., Brown, T.A., DeLuca, H.F. 1989. 1,25-Dihydroxyvitamin D$_3$ up-regulates the 1,25-dihydroxyvitaminD3 receptor in vivo. *Proc. Natl. Acad. Sci. USA* 86: 9770–73.

Takeuchi, K., and Guggino S.E. 1996. 24R,25-(OH)2 Vitamin D3 inhibits 1alpha,25-(OH)2 vitamin D3 and testosterone potentiation of calcium channels in osteosarcoma cells. *J. Biol. Chem.* 271: 4529–37.

Taylor, A.N. and Wasserman, R.H. 1969. Correlations between the vitamin D-induced calcium binding protein and intestinal absorption of calcium. *Fed. Proc.* 28(6): 1834–8.

Teillaud, C., Nemere, I., Boukhobza, F., et al. 2005. Modulation of 1alpha,25-dihydroxyvitamin D3-membrane associated, rapid response steroid binding protein expression in mouse odontoblasts by 1alpha,25-(OH)2D3. *J. Cell. Biochem.* 94(1): 139–52.

Tryfonidou, M.A., Stevenhagen, J.J., van den Bemd, G.J., et al 2002. Moderate cholecalciferol supplementation depresses intestinal calcium absorption in growing dogs. *J. Nutr.* 132: 3363–8.

Tunsophon, S., and Nemere, I. 2010. Protein kinase C isotypes in signal transduction for the 1,25D3-MARRS receptor (ERp57/PDIA3) in steroid hormone-stimulated phosphate uptake. *Steroids* 75(4-5): 307–13.

Tunsophon, S. and Nemere, I. 2011. Role of PKC-beta in signal transduction for the 1,25D3MARRS receptor (ERP57/PDIA3) in steroid hormone-stimulated calcium extrusion. *Anat. Physiol.* http://dx.doi.org/10.4172/2162-0940.S1-002.

Varghese, S., Deaven, L.L., Huang, Y.C., et al. 1989. Transcriptional regulation and chromosomal assignment of the mammalian calbindin-D28k gene. *Mol. Endocrinol.* 3: 495–502.

Wali, R.K., Kong, J., Sitrin, M.D., et al. 2003. Vitamin D receptor is not required for the rapid actions of 1,25-dihydroxyvitamin D3 to increase intracellular calcium and activate protein kinase C in mouse osteoblasts. *J. Cell. Biochem.* 88(4): 794–801.

Wang, Y., Chen, J., Lee, C.S., et al. 2010. Disruption of PDIA3 gene results in bone abnormality and affects 1alpha,25-dihydroxy-vitamin D3-induced rapid activation of PKC. *J. Steroid Biochem. Mol. Biol.* 121(1-2): 257–60.

Wasserman, R.H., Taylor, A.N. and Kallfelz, F.A. 1966. Vitamin D and transfer of plasma calcium to intestinal lumen in chicks and rats. *Am. J. Physiol.* 211(2): 419–23.

Wasserman, R.H., Taylor, A.N. 1973. Intestinal absorption of phosphate in the chick: Effect of vitamin D and other parameters *J. Nutr.* 103(4): 586–99.

Wu, L.N., Genge, B.R., Ishikawa, Y., Ishikawa, T. and Wuthier, R.E. 2006. Effects of 24R,25- and 1alpha,25-dihydroxyvitamin D3 on mineralizing growth plate chondrocytes. *J. Cell Biochem.* 98(2): 309–34.

Wu, W., Beilhartz, G., Roy Y., et al. 2010. Nuclear translocation of the 1,25D$_3$-MARRS (membrane associated rapid response to steroids) receptor protein and NFkappaB in differentiating NB4 leukemia cells. *Exp. Cell Res.* 316(7): 1101–8.

Ximnin, L., Wei, Z., and Yan, C.L. 2003. Altered gene expression profile in the kidney of vitamin D receptor knockout mice. *J. Cell. Biochem.* 89: 709–19.

Xu, H., Bai, L., Collins, J.F., Ghishan, F.K. 2002. Age-dependent regulation of rat intestinal type IIb sodium-phosphate cotransporter by 1,25-(OH)(2) vitamin D(3). *Am. J. Physiol. Cell Physiol.* 282(3): C487–93.

Yamamoto T., Ozono K., Shima M., Yamaoka K. and Okada S. 1998. 24R,25-dihydroxyvitamin D3 increases cyclic GMP contents, leading to an enhancement of osteocalcin synthesis by 1,25-dihydroxyvitamin D3 in cultured human osteoblastic cells. *Exp. Cell Res.* 244: 71–6.

Yoshimoto, Y. and Norman, A.W. 1986. Biological activity of vitamin D metabolites and analogs: Dose-response study of ^{45}Ca transport in an isolated chick duodenum perfusion system. *J. Steroid Biochem.* 25: 905–9.

Yukohiri, S., Posner, G.H., and Guggino S. 1996. Vitamin D3 analogs stimulate calcium currents in rat osteosarcoma cells. *J. Biol. Chem.* 269: 23889–93.

Zanello, L.N and Norman, A.W. 2004a. Electrical responses to $1\alpha,25(OH)_2$ vitamin D_3 and their physiological significance in osteoblasts. *Steroids* 69: 561–5.

Zanello, L. and Norman, A.W. 2004b. Rapid modulation of osteoblast ion channel responses by 1a,25(OH)2-vitamin D3 requires the presence of a functional vitamin D nuclear receptor. *Proc. Natl. Acad. Sci. USA* 101(6): 1589–94.

Zhao, B. and Nemere, I. 2002. $1,25(OH)_2D_3$-mediated phosphate uptake in isolated chick intestinal cells: Effect of $24,25(OH)_2D_3$, signal transduction activators, and age. *J. Cell. Biochem.* 86(3): 497–508.

Zhou, L.X., Nemere I., Norman, A.W. 1992. $1,25$-Dihydroxyvitamin D_3 analog structure-function assessment of the rapid stimulation of intestinal calcium absorption (transcal-tachia). *J. Bone Mineral Res.* 7(4): 457–63.

Zierold, C., Darwish, H.M. and DeLuca, H.F. 1994. Identification of a vitamin D-response element in the rat calcidiol (25-hydroxyvitamin D3) 24-hydroxylase gene. *Proc. Natl. Acad. Sci. USA* 91(3): 900–5.

6 The Role of Vitamin D in Infectious Processes

Russell W. Chesney

CONTENTS

6.1 INTRODUCTION

6.1.1 NOT A NEW CONCEPT

Before vitamin D was discovered, a widely held concept was that rickets and infection were related. Children, with rickets as the manifestation of their vitamin D deficiency, suffered from the common infectious maladies of the late eighteenth, nineteenth, and early twentieth centuries, namely, tuberculosis, pneumonia, dysentery, and severe measles (Park 1923). The rachitic lung was listed as one of the clinical features of rickets (Khajavi and Amirhakimi 1977). In a severe form of rickets leading to inanition and death, children often succumbed to serious infections (Chesney 2010). Indeed, rickets was felt by many child health professionals to be the result of infection (Jenner 1895).

6.1.2 DID AN INFECTIOUS PROCESS CAUSE RICKETS?

Before the late 1880s, the concepts of infectious agents causing disease, and of nutrients that could prevent deficiency diseases, were held by only a few prescient physicians (Chesney 2012a). More commonly, these maladies were felt to be due to a miasma. Once the concept of infection due to microbes emerged, several scholars postulated an infectious origin of rickets (Iovane and Forte 1907; Moussu 1903). However, it was the studies of Morpurgo (1902) and Koch (1914) who propagated this concept (Edlefsen 1901, 1902). They both injected organisms, including *Streptococcus longus*, into animals and noted rickets in rats and dogs, respectively. They also speculated that enlargement of the spleen in rickets was due to infection. This theory was quickly rejected as a nutritional basis for rickets became established (Mellanby 1919; Park 1923).

6.1.3 ASSOCIATION OF RICKETS WITH TUBERCULOSIS AND PNEUMONIA

A link between rickets and respiratory tract infections in general, and tuberculosis (TB) in particular, has been recognized since the nineteenth century (Chesney 2010). Sir William Jenner, a leading pediatrician and surgeon in London in the 1890s, wrote about the association of TB and rickets (Jenner 1895). Holt and Howland observed that "rachitic lung" was often interstitial pneumonitis (Holt and Howland 1911). Recent themes of the past decade, including the seasonality of respiratory tract infections and the association of vitamin D deficiency with influenza outbreaks, active TB, and pneumonia (Modlin and Cheng 2004), were noted by scholars of rickets over one hundred years ago (Park 1923).

6.1.4 STUDIES FROM THE DEVELOPING WORLD

Among children in Africa, the Middle East, and Asia, vitamin D status appears to influence the risk of developing both upper and lower respiratory tract infections. A thirteen-fold higher risk of pneumonia was found in vitamin D–deficient Ethiopian children less than five years of age compared to vitamin D–sufficient age-matched controls (Muhe et al. 1997). In Yemen, 50% of children admitted for pneumonia had rickets (Banajeh et al. 1997), and 43% of children in a Kuwait series also were rachitic (Lubani et al. 1989). Subclinical vitamin D deficiency in conjunction with nonexclusive breastfeeding during the first 120 days of life emerged as a risk factor for lower respiratory tract infection in Indian children (Wayse et al. 2004). These reports, dating from 1989 to 2004, are eerily reminiscent of the high rates of lower respiratory infection in Glasgow (Arneil 1975; Findlay 1915) and in Baltimore (Park 1923) in the early decades of the twentieth century.

6.1.5 NEONATAL INFECTIONS

Because vitamin D–deficient and rachitic infants frequently are born of vitamin D–deficient mothers, there are reports of rickets and infections developing early in life (Park 1923; Wayse et al. 2004). However, a more relevant issue is whether infants born of vitamin D–deficient mothers are more prone to develop neonatal infections. To answer this would require prospective trials in which the incidence of infection was compared in the offspring of women with and without vitamin D supplementation during pregnancy.

6.2 IS INFECTION MORE COMMON IN PEOPLE WITH VITAMIN D DEFICIENCY?

6.2.1 ROLE IN IMMUNITY

Vitamin D plays a role in the immune system (Table 6.1). A cataloging of the non-osseous activities of the most active endogenous metabolite, $1,25(OH)_2D_3$, includes T cell–associated effects in Th1/Th17-mediated immunity as well as in regulatory T cell populations (Adams and Hewison 2008, 2011; Bruce et al. 2010). Th2 immune responses and the innate immune system are also influenced by $1,25(OH)_2D_3$ (Bruce et al. 2010). In essence, vitamin D appears to suppress the generation of a Th1 response both in vitro and in vivo (Lemire 1992; Mathieu and Adorini 2002; Mathieu et al. 2004). Another potential role of vitamin D analogs is to suppress the production of IL-17, which occurs through the Th17 pathway that is important in

---

TABLE 6.1
The Anti-Infectious Properties of Vitamin D: Types of Immunity Involved

Th1 T cell function
Th2 T cell function
Th17 T cell function
Regulatory T cells
Dendritic cells
Innate immunity
Synthesis of antimicrobial peptides
Dependent upon vitamin D status of individual

---

autoimmunity (Tang et al. 2009). Vitamin D and $1,25(OH)_2D_3$ contribute to T cell regulatory function, which acts to suppress inflammation (Yu and Cantorna 2008). The effect of $1,25(OH)_2D_3$ on Th2 function is complex and may act to increase IL-4 and IL-10 production by CD4+ T cells (Imazeki et al. 2006) or even to suppress Th-2 function (Pichler et al. 2002). The impact of $1,25(OH)_2D_3$ on the innate immune system is paramount and will be examined more fully later in this review (Walker and Modlin 2009).

6.2.2   RESPIRATORY TRACT INFECTIONS

In humans, the circulating values of 25(OH)D appear to influence the frequency of respiratory tract infections, with an inverse relationship between vitamin metabolite concentrations and rates of infection (Ginde et al. 2009a). It has been suggested that enhanced production of the antimicrobial peptide cathelicidin (LL-37) is protective against respiratory tract pathogens (Ginde et al. 2009b). Among the organisms more prevalent in humans with vitamin D deficiency are *Streptococcus pneumonia*, *Mycobacterium tuberculosis*, influenza virus, and respiratory viruses (Bruce et al. 2010; Sabetta et al. 2010). Key among these viral pathogens that might be influenced by vitamin D status is the influenza virus (Ginde et al. 2009a). The role of vitamin D status in respiratory infections has emerged as one of the "hot topics" in understanding the biologic roles of the vitamin.

6.2.3   URINARY TRACT INFECTIONS

The role of vitamin D in infections of the urinary tract is much less clear than it is for respiratory infections. Because of the proximity of the bladder to the perineum, a barrier function of the bladder wall to prevent infection has been postulated, as infection would otherwise be common. The human antimicrobial peptide cathelicidin (LL-37) appears to be up-regulated after *Escherichia coli* infection, which may play a role in maintaining the integrity of the bladder epithelium against infection (Chromek and Brauner 2008; Zasloff 2007). Cathelicidin has been postulated to protect the urinary tract against bacterial invasion (Chromek et al. 2006). A recent study examined the impact of supplementation of $25(OH)D_3$ on bladder tissue obtained and cultured pre-and post-vitamin D supplementation (Hertting et al. 2010). Not only was cathelicidin production enhanced, but vitamin D supplementation was associated with more killing of uropathic *E. coli* by cultured bladder epithelial cells. This phenomenon is an in vitro observation that might indicate that clinical trials are worthwhile in patients prone to urinary tract infections.

6.2.4   EPIDEMIOLOGY: ASSOCIATION OR CAUSATION?

The increased rate of infections, particularly of the respiratory system in vitamin D–deficient subjects, must be viewed as an association for which there is epidemiologic evidence. Although mechanisms exist in vitro that can explain how locally produced $1,25(OH)_2D$ could enhance innate immunity, no causation can be assigned, as yet, to an anti-infective role for vitamin D (Ginde et al. 2009b; Hansdottir and

Monick 2011). While epidemiologic studies suggest that vitamin D deficiency can be important in the onset of viral and mycobacterial respiratory tract infections, prospective controlled trials are lacking. The mechanisms by which lung epithelial cells can be affected (Hansdottir et al. 2008) will be discussed later.

Recent studies have found interesting associations relative to infection status. An association of serum 25(OH)D values of less than 40 nmol/L with acute respiratory tract infections was evident in 800 young Finnish military recruits. Those men with the lowest quartile for 25(OH)D values had more frequent absences from duty because of infections (Laaksi et al. 2007). Vitamin D deficiency was also associated with greater latent tuberculosis among Africans who had immigrated to Australia. Higher vitamin D concentrations were associated with lower probability of active or past tuberculosis (Gibney et al. 2008). In a study of 6789 participants from the 1958 Nationwide British birth cohort who had a measurement of 25(OH)D in serum, lung function (forced vital capacity at age 45 years) and respiratory infection were higher when 25(OH)D was lower (to wit, vitamin D status has a linear relationship with respiratory infection rates and lung function; Berry et al. 2011).

An alternative point of view has been proposed by which vitamin D supplementation could actually suppress the immune system and interfere with recovery of patients with infections or autoimmune disorders (Marshall 2008). Also at issue is that in fully developed disease, such as in patients with active asthma or tuberculosis, supplementation with vitamin D may not be effective in changing the clinical course of established disease (Wejse et al. 2009; Yamshchikov et al. 2009). As stated clearly by Yamshchikov, more rigorously designed clinical trials are needed for more complete evaluation of the relationship between vitamin D status and the immune response to infection (Yamshchikov et al. 2009).

6.3   IMMUNE MECHANISMS AND VITAMIN D

6.3.1   INNATE IMMUNITY

The innate immune system forms the epithelial barrier between bacteria prominent in the outside environment and the sterile host. As bacterial products gain entry into the host, they are recognized by a class of receptors in macrophage and monocyte plasma membranes termed Toll-like receptors (TLRs). These TLRs recognize a wide array of antigens, including nucleic acids, lipids, and peptides (Bouillon et al. 2008; Modlin and Cheng 2004). As antigens are bound to TLRs, various adapter proteins (e.g., MyD88, or myeloid differentiation factor 88) are recruited and turn on signaling pathways, many of which terminate in the transactivation of NFκB. In terms of local vitamin D action, the response of the innate immune system is to up-regulate the vitamin D receptor (VDR) with ensuing activation stimulated by the most active vitamin metabolite, $1,25(OH)_2D$ (Prehn et al. 1992). Up-regulation of the VDR response promotes antigen processing, phagocytosis, and production of IL-1β and TNFα (Fagan et al. 1991) and leads to removal of foreign antigens. Relative to innate immunity, the macrophage is able to take up 25(OH)D and locally synthesize $1,25(OH)_2D$. Adams and his colleagues (Adams 2006) have speculated that this inter-macrophage production of $1,25(OH)_2D$ may actually form "a primitive non-endocrine biological system designed to control immune responsiveness to invading antigens." They further speculate that this primitive immune system may predate the vitamin D–PTH–feedback axis, since local macrophage $1,25(OH)_2D$ synthesis is not regulated by extracellular divalent mineral concentrations or by PTH.

6.3.2   LEUKOCYTE FUNCTION AND PHAGOCYTOSIS

While the major immune effects of vitamin D are on T cells and macrophages, some studies have shown an influence on leukocytes and their function. Vitamin D status influences leukocyte telomere length such that women with higher 25(OH)D values have longer telomere length (Richards et al. 2007). Phagocytosis by macrophages is also enhanced by higher circulating 25(OH)D values. In contrast to other immune cells, the leukocyte lacks a vitamin D receptor; in fact, in the VDR null mouse and in humans with significant mutations in VDR, myelopoiesis is quite normal (Bouillon et al. 2008).

6.3.3   AUTOIMMUNE FUNCTION AND INFECTION

The Th1 immune response, which is pathological, is greatly attenuated by the action of $1,25(OH)_2D$ after binding to the VDR (Bouillon et al. 2008). This response does not occur in the VDR knockout mouse. When this attenuation of the Th1 response does not occur, various autoimmune disorders, such as multiple sclerosis, systemic lupus erythematosus, rheumatoid arthritis, and inflammatory bowel disease are more prevalent (Holick 2008). The pro-inflammatory cytokines produced following Th1 activation, such as IL-2 and INF-γ, then activate B cells, macrophages, cell-mediated immunity, and NK-cells. Also, TNF-α, TNF-β, and granulocyte macrophage colony-stimulating factor are produced (Torre et al. 2002). Recently it has been shown that Th17, a subset of T-helper cells producing interleukin 17, plays a key role in the tissue inflammation and injury associated with autoimmune disorders (Harrington et al. 2005). $1,25(OH)_2D$ also attenuates the activity of Th17 cells (Bouillon et al. 2008).

6.3.4   IMMUNOGLOBULINS

The influence of vitamin D on immunoglobulin production is indirect because there are no VDRs in B cells. However, the effect of 1,25(OH)2D on Th1, Th2, Th17, and regulatory T cells (T-reg) is important in modulating B cell actions. Because vitamin D up-regulates the genetic expression of several antimicrobial peptides (AMPs), a suggested immunologic role for these AMPs (other than damaging the lipoprotein membrane of microbes) is to prime the adaptive immune system to produce antigen-specific T lymphocytes and immunoglobulins (Zasloff 2002, 2006).

6.4   THE VITAMIN D RECEPTOR (VDR) KNOCKOUT MOUSE

In some ways the development of a VDR knockout mouse (VDR$^{-/-}$) has been one of the most remarkable tools in assessing the role of the vitamin and its active metabolite, 1,25(OH)$_2$D, on immunologic and anti-infectious function. The impact of the loss of VDR is quite complex and generally influences all forms of immunity. Notable differences are evident dependent on whether the VDR$^{-/-}$ is built upon a C57B1/6J or a BALB/C initial strain (Yu et al. 2011). There are several excellent recent reviews of the immune defects in these mice, and the reader is referred to these (Bouillon et al. 2008; Cantorna 2011). This review will emphasize large immunologic changes found by this knockout technique.

VDR is expressed in activated CD4+ and CD8+ cells and in certain antigen-presenting cells (such as dendritic cells and macrophages; Veldman et al. 2000). A biphasic feature is evident in that VDR is up-regulated early in inflammation and then down-regulated in antigen-presenting cells in later stages in relation to 1,25(OH)$_2$D synthesis. This down-regulation is greater in dendritic cells that contain the vitamin D catabolic pathway, which involves 24 vitamin D hydroxylase, than in macrophages that lack this pathway. Indeed, intracellular 1,25(OH)$_2$D appears to remain elevated in macrophages because the major catabolic pathway for its further metabolism is missing. High levels of 1,25(OH)$_2$D are also evident in the macrophages of human subjects with granulomatous disease and can result in hypercalcemia (Dusso et al. 1997).

Because the major action of 1,25(OH)$_2$D (or at least one action) on the immune system is mediated through VDRs, the knockout mouse reveals a phenotype that is more proinflammatory, with less suppression of TNFα, IL1β, and IL6, and less production of Th-2 pathways, as demonstrated by lower IL-10 production (Cantorna 2011). 1,25(OH)$_2$D suppresses Th1 and Th17 cells; other Th1 cytokines whose production and/or expression are diminished are IL-2 and INFγ in T cells. IL-17, produced by Th17 cells, is diminished in a uveitis model (Tang et al. 2009). 1,25(OH)$_2$D also up-regulates the production of the Th2-associated cytokine IL-4. Addition of the active vitamin metabolite to purified CD4+ T cells both inhibits Th1 cytokines and increases IL-4 in vitro (Boonstra et al. 2001).

In the VDR$^{-/-}$ mouse, two populations of T cells are especially blocked at a more immature form: namely, iNKT cells and TCRαβ CD8+αα T cells (Cantorna 2011; Yu and Cantorna 2008). Otherwise there is no effect of the knockout on the number of CD4+, CD8+, CD4+/CD8+ double positive, and CD4+/CD8+ double negative cells, or on the number of T-reg cells (Cantorna 2011). Yu et al. showed that VDR$^{-/-}$ mice on either C57TBL/GJ or BALB/C background failed to generate an airway hyperreactivity response, and that this was related to a failure of maturation of invariant NKF (iNKF) cells (Yu et al. 2011). Wild-type iNKF cells restored the airway hyperreactivity response when administered to KO mice. Th2 cells from these mice produced less IL-13 in BALB/C mice and IL4 in those mice on the C57BL/GJ background.

An additional mechanism exposed in the VDR KO mouse is the defect in Th1-related production of INFγ, presumably because of defective IL-18 production by macrophages and reduced STAT4 expression in T cells, per se (O'Kelly et al. 2002). Another important aspect of the immune defects in VDR$^{-/-}$ mice is that many of them can be corrected by normalization of extracellular calcium concentrations following a high calcium and lactose diet (Bouillon et al. 2008). Finally, myelopoiesis appears normal in the VDR KO mouse (O'Kelly et al. 2002).

6.5 SPECTRUM OF RESPIRATORY INFECTIONS

An increased incidence or prevalence of several respiratory infections has been noted in populations that are vitamin D deficient or have rickets. This is not a new finding, but it is a persistent one (Table 6.2).

6.5.1 PNEUMONIA

Vitamin D deficiency is associated with higher rates of pneumonia infection of both viral and bacterial origin. In vitamin D–deficient Ethiopian children under the age of five, a thirteen-fold increased risk of pneumonia was evident compared to vitamin D–sufficient case control subjects (Muhe et al. 1997). Half of the children hospitalized for pneumonia in a series from Yemen were rachitic (Banajeh et al. 1997). Forty-three percent of rachitic children in Kuwait had pneumonia (Lubani et al. 1989). An association of subclinical vitamin D deficiency with marked acute lower respiratory infection was also evident in Indian children younger than five years old (Wayse et al. 2004). As mentioned previously, an association between serum vitamin D concentrations <40 nmol/L and acute respiratory tract infection was found as well in young Finnish military recruits (Laaksi et al. 2007). In a prospective cohort study of 198 healthy adults, monthly serum 25(OH)D was measured and participants were evaluated for any form of respiratory tract infection by investigators blinded to 25(OH)D status (Sabetta et al. 2010). Maintenance of a serum concentration of 38 ng/mL or higher afforded a reduction in the incidence of acute viral respiratory tract infections, especially during the fall and winter.

6.5.2 INFLUENZA

Numerous studies suggest a link between vitamin D and influenza (Ginde et al. 2009a). Epidemiologic evidence even suggests that vitamin D status may play a role in the seasonality of influenza. In 1981, Hope-Simpson made the suggestion that a protective seasonal stimulus against influenza in the spring and summer was increased solar radiation (Hope-Simpson 1983, 1992). Long appreciated is the fact that active influenza infections are clustered in winter regardless of the hemisphere; the notion that crowding accounts for this is not currently tenable. Modern workers,

TABLE 6.2
The Anti-Infectious Properties of Vitamin D:
Types of Infections Associated with Deficiency

Bacterial pneumonia

Viral pneumonia

Tuberculosis

Reactive airway disease (viral induced?) and asthma

Respiratory syncytial virus

Lower respiratory infections in cystic fibrosis patients

Urinary tract infections

working indoors, have similar crowding in all seasons. When attenuated influenza virus was given to volunteers, infection occurred mainly in winter (Pfleiderer et al. 2001). Further evidence suggests that influenza peaks during the months following the winter solstice when 25(OH)D values are at a minimum and disappears following the summer solstice (Cannell et al. 2006). Vitamin D status also appears to have a linear association with seasonal infections and lung function in the 1958 British birth cohort groups (Berry et al. 2011). These observations need to be evaluated in clinical trials before dosing with vitamin D can be advised. However, in a randomized, double-blind, placebo-controlled trial among 167 children, those receiving 1200 IU vitamin D were 42% less likely to become infected with seasonal influenza relative to placebo-controlled subjects (Urashima et al. 2010).

Another randomized controlled trial of vitamin D supplementation was undertaken in 162 adult subjects with a biweekly questionnaire to record symptoms of upper respiratory tract infections. While dosing with 2000 IU vitamin D daily in the treated group increased circulating 25(OH)D values, no differences in respiratory tract symptoms or their severity were noted relative to controls (Li-Ng et al. 2009).

In contrast, there has been criticism of the vitamin D–based model of seasonal influenza. Using a large Health Professionals Follow-up Study Data Set, a group of epidemiologists found that the vitamin D–forced model was inferior to absolute humidity- or school calendar–based models of seasonal forcing variables (Shaman et al. 2011). A criticism of this particular study is that mainly males were evaluated. In a review of the thirteen existing human-controlled intervention trials, nine were conducted using a vigorous double-blind design (Yamshchikov et al. 2009). In part because of the clinical heterogeneity of the baseline populations (some children, some adults, some elderly), a need for larger, more carefully designed studies was identified. The authors conclude that no mandate for vitamin D supplementation to prevent respiratory infection currently exists.

6.5.3 RESPIRATORY SYNCYTIAL VIRUS

Respiratory syncytial virus (RSV) infection is universal in children and in a certain proportion results in a viral pneumonia requiring hospital admission. Infection with this virus represents one of the main reasons that young children are hospitalized. Recent epidemiologic evidence indicates that the incidence of RSV relates to 25(OH)D concentration (Belderbos et al. 2011). Using cord blood values, vitamin D deficiency is associated with an increased risk of RSV lower respiratory tract infection in the first year of life. Infected children had a cord blood concentration of 65 nmol/L versus 84 nmol/L in noninfected children ($p < 0.009$).

Neonates born with 25(OH)D values <50 nmol/L had a six-fold increased risk of RSV infection compared to those with 25(OH)D values >75 nmol/L. Of interest, the effect of vitamin D on RSV-infected human airway epithelial cells in vitro is to reduce RSV induction of NFκB-linked cytokines and chemokines (Hansdottir et al. 2010). Vitamin D appears to induce IκBα, an inhibitor of NFκB.

Muscle wasting and cachexia are found in adults with chronic obstructive pulmonary disease. Cachexia is worse in vitamin D–deficient patients who may show persistence of RSV into adult life (Herzog and Cunningham-Rundles 2011). Prospective trials of vitamin D supplementation for populations at risk for RSV infection are probably indicated but require careful design.

6.5.4 ASTHMA

One theory of the recognized increase in the prevalence of asthma and other allergic conditions is the rising prevalence of vitamin D deficiency (Litonjua and Weiss 2007). A recent article by Finklea et al. extensively reviews the evidence for the association between maternal vitamin D status and childhood asthma (Finklea et al. 2011). They review the results of seventeen clinical studies in pregnant mothers and their offspring, in childhood and as adults. As might be expected, results are variable, with the major finding that vitamin D deficiency is more common in asthmatics and that asthma control was more difficult in vitamin D–deficient subjects. Vitamin D supplementation of a cohort of pregnant mothers resulted in a lower risk of asthma in three- and five-year-old children (Camargo et al. 2007; Erkkola et al. 2009). In several studies, however, there was no association between asthma and vitamin D status (Devereux et al. 2010; Hughes et al. 2010).

When vitamin D status has been evaluated in subjects with allergic rhinitis, the prevalence of deficiency was greater (Arshi et al. 2012). One caution is that both allergic rhinitis and vitamin D deficiency are prevalent conditions, and the association may be fortuitous.

6.6 VITAMIN D AND TUBERCULOSIS

Perhaps no infectious agent has been more examined relative to vitamin D status than *Mycobacterium tuberculosis*, the cause of TB. The original recent observations concerning the role of vitamin D in infection and immunity relate to epidemiologic observations, in vitro response (Nnoaham and Clarke 2008), and findings in the VDR-null mouse (Bouillon et al. 2008). The innate immune system involving Toll receptors on the macrophage has come into sharp focus as the anti-TB pathway (Adams 2006; Hewison et al. 2007; Krutzik et al. 2008). This topic has been extensively reviewed, and hence this section will focus upon recent observations (Finklea et al. 2011).

In a study of African immigrants in Melbourne, Australia, higher vitamin D concentrations were associated with a lower probability of any *M. tuberculosis* infection and a lower probability of past infection (Gibney et al. 2008). In the Finklea et al. review, they list eleven case-control studies of vitamin D status and prevalence of TB, eight of which showed an association between infection and lower 25(OH)D values (Finklea et al. 2011). In three additional retrospective studies of patients with TB, over 80% of participants had low 25(OH)D values (Yamshchikov et al. 2010). These observations have been made worldwide, including series from Kenya, Thailand, and Greenland (Davies et al. 1988; Davies et al. 1987; Nielsen et al. 2010).

The presumed mechanism of anti-TB activity is the local macrophage production of cathelicidin (LL-37), a highly specific antimicrobial protein (peptide) whose synthesis is stimulated by intracellular $25(OH)_2D$ (Adams 2006).

It is clear that the cells of the immune system contain the processes required to convert 25(OH)D to active 1,25(OH)$_2$D, and the latter compound activates pathways that lead to antimicrobial responses within macrophages and antigen-presenting cells (Hewison 2012). Another important factor in the killing of *M. tuberculosis* is vitamin D–induced production and release of interferon-γ-induced AMP expression, autophagy, phagosome-lysosome fusion, and antimicrobial activity. The recent report of Martineau et al. (2011a) described a significant association of vitamin D deficiency with susceptibility to TB and reported that this association is greatest in HIV-infected versus noninfected individuals in South Africa. These findings stress the need for prospective trials of vitamin D therapy to prevent or treat TB. The fact that there was a seasonal influence in these South African subjects has led to recommendations to initiate these trials (Realegeno and Modlin 2011). Further evidence in vitro shows that respiratory epithelial cells, per se, can convert 25(OH)D to 1,25(OH)$_2$D and locally increase expression of vitamin D–regulated genes that influence innate immunity (Hansdottir et al. 2008). This feasible intrapulmonary capacity to increase cathelicidin production and synthesis of the TLR co-receptor CD14 offers at least two mechanisms by which the extent of TB can be limited by enhanced host defenses, especially in the lung.

Vitamin D was used to treat TB long before antibiotic therapy. Both cod liver oil and sunshine were prescribed as therapeutics (Bennett 1853; Brincourt 1969; Dowling and Prosser Thomas 1946; Ellman and Anderson 1948). Recent trials of the efficacy of vitamin D in patients with TB have had varied results (Finklea et al. 2011). In a large trial of 365 patients in Guinea-Bissau with active TB who received anti-tuberculosis therapy, no differences were seen between vitamin D–treated subjects and controls (Wejse et al. 2009).

A small trial in children that employed 1,25(OH)$_2$D therapy resulted in clinical improvement in those receiving the drug (Morcos et al. 1998). In a large double-blind, randomized trial in which patients were treated with placebo or 25(OH)D, a single dose of vitamin D appeared to enhance host defenses (Martineau et al. 2007). In another large trial using vitamin D, the response was greatest depending on the genotype of the Taq I VDR polymorphism (Martineau et al. 2011b).

Skepticism concerning the anti-infective role of vitamin D focuses upon the fact that much of the anti-TB effect is an in vitro phenomenon, rather than in vivo (Bruce et al. 2010). As Bruce et al. note, there is little evidence that vitamin D affects the course of human infection. It is conflicting information such as this that indicates the need for prospective trials in groups of patients with TB.

The role of vitamin D as a therapeutic agent in nontuberculous mycobacterial lung diseases is unclear. A recent review of susceptibility factors in these disorders examines the evidence concerning vitamin D (Sexton and Harrison 2008). Among the genetically determined defects in cell-mediated immunity and various other host-defense mechanisms the authors review are the roles of VDR polymorphisms and circulating vitamin D status. They could find no evidence for VDR polymorphisms playing a role in susceptibility to several of the nontuberculous mycobacteria agents.

6.7   VITAMIN D AND CYSTIC FIBROSIS

Mutations in the cystic fibrosis (CF) transmembrane conductance regulator lead to a multiorgan disorder in which thickened secretions impair the clearance of mucus from airways. Combined with reduced bacterial killing, this can lead to an ultimately fatal respiratory failure (O'Sullivan and Freedman 2009). Malabsorption of fat results in vitamin D deficiency in CF patients. With treatment resulting in longer lifespan of these patients, into adulthood, vitamin D status has become even more important (Boyle et al. 2005; Hall et al. 2010; Rovner et al. 2007). In some adult CF centers, more than 90% of patients have 25(OH)D concentrations less than 30 µg/L. Several studies have shown a correlation between reduced bone mineral density in CF subjects and reduced forced expiratory volume in 1 second (FEV1), indicative of poor lung function (Elkin et al. 2001; Haworth et al. 2002; Henderson and Madsen 1996; Stephenson et al. 2007). In another study, serum IgG concentrations were universally correlated with serum 25(OH)D values, suggesting an association between vitamin D status and lung inflammation (Pincikova et al. 2011). There exists a need for randomized controlled trials in CF patients that can correlate vitamin D status, lung function, and inflammation over time. The Cystic Fibrosis Foundation is stressing the need for trials to determine if vitamin D therapy will influence pulmonary exacerbations or clearance of bacterial agents in this group of patients.

6.8   URINARY TRACT INFECTIONS AND BARRIER FUNCTION OF THE BLADDER EPITHELIUM

The notion that vitamin D–associated processes might contribute to the barrier function of the kidney was proposed by Zasloff (Zasloff 2007). His hypothesis is that the bladder is sterile (which is remarkable considering the proximity of the urethral meatus to the rectum) because of proteins and peptides, including alpha and beta defenses and cathelicidin. Microbes commonly found in genital flora contact uroepithelium and activate innate immune responses through Toll-like receptor 4. Since cathelicidin is a vitamin D–dependent protein (as its gene responds to vitamin D), Zasloff hypothesized that susceptibility to urinary tract infections may depend on vitamin D status.

That urinary bladder epithelium was capable of the expression of cathelicidin was evident in a study of bladder tissue from women who underwent vitamin D supplementation for three months. Following supplementation, bladder epithelium did not express the antimicrobial peptide in the basal state, but did so when bladder biopsies were infected with uropathogenic *E. coli* (Hertting et al. 2010). Bladder cells in culture expressed greater amounts of cathelicidin regardless of the presence of microorganisms after 25(OH)D dosing.

Following urinary tract obstruction, there occurs an up-regulation of interleukin 1 family cytokines in local dendritic cells, which then enhances intrarenal Th-17 activation and pathogenic interstitial inflammation (Pindjakova et al. 2012). It has also been shown that converging pathways of the immune system lead to the overproduction of IL-17, the product of Th-17 cells, in the absence of vitamin D signaling (Bruce et al. 2011). All of these tantalizing results suggest that prospective trials of vitamin D therapy should be undertaken in subjects at risk for UTI.

6.9 FUTURE TRIALS

The fundamental issue relative to vitamin D status and infection is that many of the observations in man are the conclusions of epidemiologic associations and retrospective observations. In order to determine if vitamin D status has a general or quite specific anti-infectious property, both double-blind and prospective trials will need to be undertaken. This is consistent with the view expressed in the 2010 Institute of Medicine report (Slomski 2011).

6.10 CONCLUSIONS: THE EPIDEMIOLOGIC ASSOCIATION

In order to assert an epidemiologic association between vitamin D status and immunity and/or infection, the criteria of A. Bradford Hill should be upheld. Among these are strength of association, consistency, specificity, temporality, biological gradient, plausibility, coherence, experiment, and analogy (Hill 1965). Several of these characteristics have been fully stated, but others require comment. Vitamin D deficiency is associated with all-cause mortality rates in the United States (Ginde et al. 2009c). These findings have been found in other studies and are consistent. Many autoimmune disorders, infections, allergic reactions, and possibly cardiovascular and oncologic disorders in man are more common at higher latitudes and among subjects with the lowest quartiles of serum vitamin D values (Holick 2007). As mentioned previously, associations with several lower respiratory tract infections have been described on several continents. Vitamin D deficiency appears to occur prior to onset of these infections, which are more frequent in winter months. A biological gradient is observed in the role of the time of exposure to ultraviolet B light in the determination of vitamin D status. The role of the vitamin in affecting the immune system, especially in the VDR$^{-/-}$ mouse, is highly plausible considering what is known about the biologic action of this secosteroid hormone. The theory that vitamin D deficiency is associated with increased infections, especially TB, is coherent within the role of vitamin D in innate immunity, especially since *M. tuberculosis* attaches to TLR1/2 and activates a cascade that increases cathelicidin synthesis and killing of the microorganism (Walker and Modlin 2009). As stated previously, what is now required are prospective experimental trials, ideally double-blinded in nature. There is an analogy to be made in that rickets and severe infections were common between 1600 and 1930 as a consequence of the extensive use of coal as a fuel (which blocked ultraviolet radiation from the sun) and the confinement of children indoors as child laborers (Chesney 2012b). We now spend more time indoors and limit our sun exposure to such an extent that we are far more likely to be vitamin D deficient. The analogy is that we are reestablishing the environmental conditions that were prevalent during the great epidemics of rickets in past centuries.

REFERENCES

Adams, J. S. 2006. Vitamin D as a defensin. *J Musculoskelet Neuronal Interact* 6: 344–6.
Adams, J. S., Hewison, M. 2008. Unexpected actions of vitamin D: New perspectives on the regulation of innate and adaptive immunity. *Nat Clin Pract Endocrinol Metab* 4: 80–90.

Adams, J. S., Hewison, M. 2011. Update in vitamin D. *J Clin Endocrinol Metab* 95: 471–8.

Arneil, G. C. 1975. Nutritional rickets in children in Glasgow. *Proc Nutr Soc* 34: 101–9.

Arshi, S., Ghalehbaghi, B., Kamrava, S. K., Aminlou, M. 2012. Vitamin D serum levels in allergic rhinitis: Any difference from normal population? *Asia Pac Allergy* 2: 45–8.

Banajeh, S. M., al-Sunbali, N. N., al-Sanahani, S. H. 1997. Clinical characteristics and outcome of children aged under 5 years hospitalized with severe pneumonia in Yemen. *Ann Trop Paediatr* 17: 321–6.

Belderbos, M. E., Houben, M. L., Wilbrink, B. et al. 2011. Cord blood vitamin D deficiency is associated with respiratory syncytial virus bronchiolitis. *Pediatrics* 127: e1513–20.

Bennett, J. H. 1853 *The Pathology and Treatment of Pulmonary Tuberculosis*. 1st ed. Southerland and Knox, Edinburgh.

Berry, D. J., Hesketh, K., Power, C., Hypponen, E. 2011. Vitamin D status has a linear association with seasonal infections and lung function in British adults. *Br J Nutr* 106: 1433–40.

Boonstra, A., Barrat, F. J., Crain, C. et al. 2001. 1Alpha,25-dihydroxyvitamin D_3 has a direct effect on naive CD4(+) T cells to enhance the development of Th2 cells. *J Immunol* 167: 4974–80.

Bouillon, R., Carmeliet, G., Verlinden, L. et al. 2008. Vitamin D and human health: Lessons from vitamin D receptor null mice. *Endocr Rev* 29: 726–76.

Boyle, M. P., Noschese, M. L., Watts, S. L. et al. 2005. Failure of high-dose ergocalciferol to correct vitamin D deficiency in adults with cystic fibrosis. *Am J Respir Crit Care Med* 172: 212–7.

Brincourt, J. 1969. [Liquefying effect on suppurations of an oral dose of calciferol]. *Presse Med* 77: 467–70.

Bruce, D., Ooi, J. H., Yu, S., Cantorna, M. T. 2010. Vitamin D and host resistance to infection? Putting the cart in front of the horse. *Exp Biol Med (Maywood)* 235: 921–7.

Bruce, D., Yu, S., Ooi, J. H., Cantorna, M. T. 2011. Converging pathways lead to overproduction of IL-17 in the absence of vitamin D signaling. *Int Immunol* 23: 519–28.

Camargo, C. A., Jr., Rifas-Shiman, S. L., Litonjua, A. A. et al. 2007. Maternal intake of vitamin D during pregnancy and risk of recurrent wheeze in children at 3 y of age. *Am J Clin Nutr* 85: 788–95.

Cannell, J. J., Vieth, R., Umhau, J. C. et al. 2006. Epidemic influenza and vitamin D. *Epidemiol Infect* 134: 1129–40.

Cantorna, M. T. 2011. Why do T cells express the vitamin D receptor? *Ann N Y Acad Sci* 1217: 77–82.

Chesney, R. W. 2010. Vitamin D and the Magic Mountain: The anti-infectious role of the vitamin. *J Pediatr* 156: 698–703.

Chesney, R. W. 2012a. New thoughts concerning the epidemic of rickets: Was the role of alum overlooked? *Pediatr Nephrol* 27: 3–6.

Chesney, R. W. 2012b. Theobald Palm and his remarkable observation: How the sunshine vitamin came to be recognized. *Nutrients* 4: 42–51.

Chromek, M., Brauner, A. 2008. Antimicrobial mechanisms of the urinary tract. *J Mol Med (Berl)* 86: 37–47.

Chromek, M., Slamova, Z., Bergman, P. et al. 2006. The antimicrobial peptide cathelicidin protects the urinary tract against invasive bacterial infection. *Nat Med* 12: 636–41.

Davies, P. D., Church, H. A., Bovornkitti, S., Charumilind, A., Byrachandra, S. 1988. Altered vitamin D homeostasis in tuberculosis. *Int Med Thailand* 4: 45–7.

Davies, P. D., Church, H. A., Brown, R. C., Woodhead, J. S. 1987. Raised serum calcium in tuberculosis patients in Africa. *Eur J Respir Dis* 71: 341–4.

Devereux, G., Wilson, A., Avenell, A., McNeill, G., Fraser, W. D. 2010. A case-control study of vitamin D status and asthma in adults. *Allergy* 65: 666–7.

Dowling, G. B., Prosser Thomas, E. W. 1946. Treatment of lupus vulgaris with calciferol. *Lancet* 1: 919–22.

Dusso, A. S., Kamimura, S., Gallieni, M. et al. 1997. Gamma-interferon-induced resistance to 1,25-(OH)2 D3 in human monocytes and macrophages: A mechanism for the hypercalcemia of various granulomatoses. *J Clin Endocrinol Metab* 82: 2222–32.

Edlefsen, G. 1901. Zur Aetiologie der Rachitis. *Deutsch Aerzte-Zeitg* 509(39): 64.

Edlefsen, G. 1902. Ueber die Entstehungsursachen de Rachitis und ihre Verwanschaft mit gewissen Infecktionskrankheiten. *Deutsch Aerzte-Zeitg* 169: 200.

Elkin, S. L., Fairney, A., Burnett, S. et al. 2001. Vertebral deformities and low bone mineral density in adults with cystic fibrosis: A cross-sectional study. *Osteoporos Int* 12: 366–72.

Ellman, P., Anderson, K. H. 1948. Calciferol in tuberculous peritonitis with disseminated tuberculosis. *Br Med J* 1: 394.

Erkkola, M., Kaila, M., Nwaru, B. I. et al. 2009. Maternal vitamin D intake during pregnancy is inversely associated with asthma and allergic rhinitis in 5-year-old children. *Clin Exp Allergy* 39: 875–82.

Fagan, D. L., Prehn, J. L., Adams, J. S., Jordan, S. C. 1991. The human myelomonocytic cell line U-937 as a model for studying alterations in steroid-induced monokine gene expression: Marked enhancement of lipopolysaccharide-stimulated interleukin-1 beta messenger RNA levels by 1,25-dihydroxyvitamin D3. *Mol Endocrinol* 5: 179–86.

Findlay, L. 1915. The etiology of rickets: A statistical study of the home conditions of 400 to 500 rachitic children. *Lancet* 185: 956–60.

Finklea, J. D., Grossmann, R. E., Tangpricha, V. 2011. Vitamin D and chronic lung disease: A review of molecular mechanisms and clinical studies. *Adv Nutr* 2: 244–53.

Gibney, K. B., MacGregor, L., Leder, K. et al. 2008. Vitamin D deficiency is associated with tuberculosis and latent tuberculosis infection in immigrants from sub-Saharan Africa. *Clin Infect Dis* 46: 443–6.

Ginde, A. A., Mansbach, J. M., Camargo, C. A., Jr. 2009a. Association between serum 25-hydroxyvitamin D level and upper respiratory tract infection in the Third National Health and Nutrition Examination Survey. *Arch Intern Med* 169: 384–90.

Ginde, A. A., Mansbach, J. M., Camargo, C. A., Jr. 2009b. Vitamin D, respiratory infections, and asthma. *Curr Allergy Asthma Rep* 9: 81–7.

Ginde, A. A., Scragg, R., Schwartz, R. S., Camargo, C. A., Jr. 2009c. Prospective study of serum 25-hydroxyvitamin D level, cardiovascular disease mortality, and all-cause mortality in older U.S. adults. *J Am Geriatr Soc* 57: 1595–603.

Hall, W. B., Sparks, A. A., Aris, R. M. 2010. Vitamin D deficiency in cystic fibrosis. *Int J Endocrinol* 2010: 218691.

Hansdottir, S., Monick, M. M. 2011. Vitamin D effects on lung immunity and respiratory diseases. *Vitam Horm* 86: 217–37.

Hansdottir, S., Monick, M. M., Hinde, S. L. et al. 2008. Respiratory epithelial cells convert inactive vitamin D to its active form: potential effects on host defense. *J Immunol* 181: 7090–9.

Hansdottir, S., Monick, M. M., Lovan, N. et al. 2010. Vitamin D decreases respiratory syncytial virus induction of NF-kappaB-linked chemokines and cytokines in airway epithelium while maintaining the antiviral state. *J Immunol* 184: 965–74.

Harrington, L. E., Hatton, R. D., Mangan, P. R. et al. 2005. Interleukin 17-producing CD4+ effector T cells develop via a lineage distinct from the T helper type 1 and 2 lineages. *Nat Immunol* 6: 1123–32.

Haworth, C. S., Selby, P. L., Horrocks, A. W. et al. 2002. A prospective study of change in bone mineral density over one year in adults with cystic fibrosis. *Thorax* 57: 719–23.

Henderson, R. C., Madsen, C. D. 1996. Bone density in children and adolescents with cystic fibrosis. *J Pediatr* 128: 28–34.

Hertting, O., Holm, A., Luthje, P. et al. 2010. Vitamin D induction of the human antimicrobial peptide cathelicidin in the urinary bladder. *PLoS One* 5: e15580.

Herzog, R., Cunningham-Rundles, S. 2011. Immunologic impact of nutrient depletion in chronic obstructive pulmonary disease. *Curr Drug Targets* 12: 489–500.

Hewison, M. 2012. Vitamin D and immune function: An overview. *Proc Nutr Soc* 71: 50–61.

Hewison, M., Burke, F., Evans, K. N. et al. 2007. Extra-renal 25-hydroxyvitamin D3-1alpha-hydroxylase in human health and disease. *J Steroid Biochem Mol Biol* 103: 316–21.

Hill, A. B. 1965. The environment and disease: Association or causation? *Proc R Soc Med* 58: 295–300.

Holick, M. F. 2007. Vitamin D deficiency. *N Engl J Med* 357: 266–81.

Holick, M. F. 2008. Vitamin D: A D-lightful health perspective. *Nutr Rev* 66: S182–94.

Holt, L. E., Howland, J. 1911. Rickets (Rachitis). In: Holt L. E. (ed.) *The Diseases of Infancy and Childhood*, 6th ed. Appleton, New York: 241–60.

Hope-Simpson, R. E. 1983. Recognition of historic influenza epidemics from parish burial records: A test of prediction from a new hypothesis of influenzal epidemiology. *J Hyg (Lond)* 91: 293–308.

Hope-Simpson, R. E. 1992. *The Transmission of Epidemic Influenza*. Plenum Press, New York.

Hughes, A. M., Lucas, R. M., Ponsonby, A. L. et al. 2010. The role of latitude, ultraviolet radiation exposure and vitamin D in childhood asthma and hayfever: An Australian multicenter study. *Pediatr Allergy Immunol* 22: 327–33.

Imazeki, I., Matsuzaki, J., Tsuji, K., Nishimura, T. 2006. Immunomodulating effect of vitamin D3 derivatives on type-1 cellular immunity. *Biomed Res* 27: 1–9.

Iovane, A., Forte, S. 1907. Contributo sperimentale allo studio della etiologia e patogenesi del rachitismo. *Pediatria* 15: 641.

Jenner, W. 1895. *Clinical Lectures and Essays on Rickets, Tuberculosis, Abdominal Tumours and Other Subjects*. Rivington, Percival, London.

Khajavi, A., Amirhakimi, G. H. 1977. The rachitic lung: Pulmonary findings in 30 infants and children with malnutritional rickets. *Clin Pediatr (Phila)* 16: 36–8.

Koch, J. 1914. Ueber experimentelle Rachitis. *Berl klin Wochenschr* 51: 773.

Krutzik, S. R., Hewison, M., Liu, P. T. et al. 2008. IL-15 links TLR2/1-induced macrophage differentiation to the vitamin D-dependent antimicrobial pathway. *J Immunol* 181: 7115–20.

Laaksi, I., Ruohola, J. P., Tuohimaa, P. et al. 2007. An association of serum vitamin D concentrations <40 nmol/L with acute respiratory tract infection in young Finnish men. *Am J Clin Nutr* 86: 714–7.

Lemire, J. M. 1992. Immunomodulatory role of 1,25-dihydroxyvitamin D3. *J Cell Biochem* 49: 26–31.

Li-Ng, M., Aloia, J. F., Pollack, S. et al. 2009. A randomized controlled trial of vitamin D3 supplementation for the prevention of symptomatic upper respiratory tract infections. *Epidemiol Infect* 137: 1396–404.

Litonjua, A. A., Weiss, S. T. 2007. Is vitamin D deficiency to blame for the asthma epidemic? *J Allergy Clin Immunol* 120: 1031–5.

Lubani, M. M., al-Shab, T. S., al-Saleh, Q. A. et al. 1989. Vitamin-D-deficiency rickets in Kuwait: The prevalence of a preventable disease. *Ann Trop Paediatr* 9: 134–9.

Marshall, T. G. 2008. Vitamin D discovery outpaces FDA decision making. *Bioessays* 30: 173–82.

Martineau, A. R., Nhamoyebonde, S., Oni, T. et al. 2011a. Reciprocal seasonal variation in vitamin D status and tuberculosis notifications in Cape Town, South Africa. *Proc Natl Acad Sci U S A* 108: 19013–7.

Martineau, A. R., Timms, P. M., Bothamley, G. H. et al. 2011b. High-dose vitamin D(3) during intensive-phase antimicrobial treatment of pulmonary tuberculosis: A double-blind randomised controlled trial. *Lancet* 377: 242–50.

Martineau, A. R., Wilkinson, R. J., Wilkinson, K. A. et al. 2007. A single dose of vitamin D enhances immunity to mycobacteria. *Am J Respir Crit Care Med* 176: 208–13.

Mathieu, C., Adorini, L. 2002. The coming of age of 1,25-dihydroxyvitamin D(3) analogs as immunomodulatory agents. *Trends Mol Med* 8: 174–9.

Mathieu, C., van Etten, E., Decallonne, B. et al. 2004. Vitamin D and 1,25-dihydroxyvitamin D3 as modulators in the immune system. *J Steroid Biochem Mol Biol* 89-90: 449–52.

Mellanby, E. 1919. An experimental investigation on rickets. *Lancet* I: 407–12.

Modlin, R. L., Cheng, G. 2004. From plankton to pathogen recognition. *Nat Med* 10: 1173–4.

Morcos, M. M., Gabr, A. A., Samuel, S. et al. 1998. Vitamin D administration to tuberculous children and its value. *Boll Chim Farm* 137: 157–64.

Morpurgo, B. 1902. Durch Infection hervorgerufene malacische und rachitische Skelettveranderungen an jungen weissen Ratten. *Centralb f allg Path u path Anat* 13: 113.

Moussu, G. 1903. Anatomie et physiologie pathologiques de la cachexie osseuse du porc. *Bull Soc centr de méd vét* n. s. 21: 303.

Muhe, L., Lulseged, S., Mason, K. E., Simoes, E. A. 1997. Case-control study of the role of nutritional rickets in the risk of developing pneumonia in Ethiopian children. *Lancet* 349: 1801–4.

Nielsen, N. O., Skifte, T., Andersson, M. et al. 2010. Both high and low serum vitamin D concentrations are associated with tuberculosis: a case-control study in Greenland. *Br J Nutr* 104: 1487–91.

Nnoaham, K. E., Clarke, A. 2008. Low serum vitamin D levels and tuberculosis: A systematic review and meta-analysis. *Int J Epidemiol* 37: 113–9.

O'Kelly, J., Hisatake, J., Hisatake, Y. et al. 2002. Normal myelopoiesis but abnormal T lymphocyte responses in vitamin D receptor knockout mice. *J Clin Invest* 109: 1091–9.

O'Sullivan, B. P., Freedman, S. D. 2009. Cystic fibrosis. *Lancet* 373: 1891–904.

Park, E. A. 1923. The etiology of rickets. *Physiol Rev* 3: 106–63.

Pfleiderer, M., Lower, J., Kurth, R. 2001. Cold-attenuated live influenza vaccines, a risk-benefit assessment. *Vaccine* 20: 886–94.

Pichler, J., Gerstmayr, M., Szepfalusi, Z. et al. 2002. 1 alpha,25(OH)2D3 inhibits not only Th1 but also Th2 differentiation in human cord blood T cells. *Pediatr Res* 52: 12–18.

Pincikova, T., Nilsson, K., Moen, I. E. et al. 2011. Inverse relation between vitamin D and serum total immunoglobulin G in the Scandinavian Cystic Fibrosis Nutritional Study. *Eur J Clin Nutr* 65: 102–9.

Pindjakova, J., Hanley, S. A., Duffy, M. M. et al. 2012. Interleukin-1 accounts for intrarenal Th17 cell activation during ureteral obstruction. *Kidney Int* 81: 379–90.

Prehn, J. L., Fagan, D. L., Jordan, S. C., Adams, J. S. 1992. Potentiation of lipopolysaccharide-induced tumor necrosis factor-alpha expression by 1,25-dihydroxyvitamin D3. *Blood* 80: 2811–16

Realegeno, S., Modlin, R. L. 2011. Shedding light on the vitamin D-tuberculosis-HIV connection. *Proc Natl Acad Sci U S A* 108: 18861–2.

Richards, J. B., Valdes, A. M., Gardner, J. P. et al. 2007. Higher serum vitamin D concentrations are associated with longer leukocyte telomere length in women. *Am J Clin Nutr* 86: 1420–5.

Rovner, A. J., Stallings, V. A., Schall, J. I., Leonard, M. B., Zemel, B. S. 2007. Vitamin D insufficiency in children, adolescents, and young adults with cystic fibrosis despite routine oral supplementation. *Am J Clin Nutr* 86: 1694–9.

Sabetta, J. R., DePetrillo, P., Cipriani, R. J. et al. 2010. Serum 25-hydroxyvitamin D and the incidence of acute viral respiratory tract infections in healthy adults. *PLoS One* 5: e11088.

Sexton, P., Harrison, A. C. 2008. Susceptibility to nontuberculous mycobacterial lung disease. *Eur Respir J* 31: 1322–33.

Shaman, J., Jeon, C. Y., Giovannucci, E., Lipsitch, M. 2011. Shortcomings of vitamin D-based model simulations of seasonal influenza. *PLoS One* 6: e20743.

Slomski, A. 2011. IOM endorses vitamin D, calcium only for bone health, dispels deficiency claims. *JAMA* 305: 453–4, 6.

Stephenson, A., Brotherwood, M., Robert, R. et al. 2007. Cholecalciferol significantly increases 25-hydroxyvitamin D concentrations in adults with cystic fibrosis. *Am J Clin Nutr* 85: 1307–11.

Tang, J., Zhou, R., Luger, D. et al. 2009. Calcitriol suppresses antiretinal autoimmunity through inhibitory effects on the Th17 effector response. *J Immunol* 182: 4624–32.

Torre, D., Speranza, F., Giola, M. et al. 2002. Role of Th1 and Th2 cytokines in immune response to uncomplicated *Plasmodium falciparum* malaria. *Clin Diagn Lab Immunol* 9: 348–51.

Urashima, M., Segawa, T., Okazaki, M. et al. 2010. Randomized trial of vitamin D supplementation to prevent seasonal influenza A in schoolchildren. *Am J Clin Nutr* 91: 1255–60.

Veldman, C. M., Cantorna, M. T., DeLuca, H. F. 2000. Expression of 1,25-dihydroxyvitamin D(3) receptor in the immune system. *Arch Biochem Biophys* 374: 334–8.

Walker, V. P., Modlin, R. L. 2009. The vitamin D connection to pediatric infections and immune function. *Pediatr Res* 65: 106R–13R.

Wayse, V., Yousafzai, A., Mogale, K., Filteau, S. 2004. Association of subclinical vitamin D deficiency with severe acute lower respiratory infection in Indian children under 5 y. *Eur J Clin Nutr* 58: 563–7.

Wejse, C., Gomes, V. F., Rabna, P. et al. 2009. Vitamin D as supplementary treatment for tuberculosis: A double-blind, randomized, placebo-controlled trial. *Am J Respir Crit Care Med* 179: 843–50.

Yamshchikov, A. V., Desai, N. S., Blumberg, H. M., Ziegler, T. R., Tangpricha, V. 2009. Vitamin D for treatment and prevention of infectious diseases: a systematic review of randomized controlled trials. *Endocr Pract* 15: 438–49.

Yamshchikov, A. V., Kurbatova, E. V., Kumari, M. et al. 2010. Vitamin D status and antimicrobial peptide cathelicidin (LL-37) concentrations in patients with active pulmonary tuberculosis. *Am J Clin Nutr* 92: 603–11.

Yu, S., Cantorna, M. T. 2008. The vitamin D receptor is required for iNKT cell development. *Proc Natl Acad Sci U S A* 105: 5207–12.

Yu, S., Zhao, J., Cantorna, M. T. 2011. Invariant NKT cell defects in vitamin D receptor knockout mice prevents experimental lung inflammation. *J Immunol* 187: 4907–12.

Zasloff, M. 2002. Antimicrobial peptides of multicellular organisms. *Nature* 415: 389–95.

Zasloff, M. 2006. Fighting infections with vitamin D. *Nat Med* 12: 388–90.

Zasloff, M. 2007. Antimicrobial peptides, innate immunity, and the normally sterile urinary tract. *J Am Soc Nephrol* 18: 2810–6.

7 Vitamin D, Vitamin D Binding Protein, and Cardiovascular Disease

Diane Berry and Elina Hyppönen

CONTENTS

7.1 INTRODUCTION

Nearly all of the circulating vitamin D metabolites are bound to serum proteins, mostly to vitamin D binding protein (VDBP). Vitamin D is primarily obtained through sunlight-induced synthesis, with typically smaller amounts coming from diet. Also, genetic variations affect steps in the vitamin D metabolic pathways, with heritability estimates for 25-hydroxyvitamin D (25(OH)D, the status indicator) from family studies ranging from 29% to 80% (Hunter et al. 2001, Shea et al. 2009, Wjst et al. 2006). In our recent meta-analysis done in the context of the SUNLIGHT (Study of Underlying Genetic Determinants of Vitamin D and Highly Related Traits) Consortium, the gene coding the VDBP (group-specific component or Gc globulin, *GC*) was seen to exert the strongest effect on 25(OH)D concentrations (Wang et al. 2010). Differences between the carriers of *GC* risk allele and others were ~8 nmol/L, an effect size similar to that seen for the use of vitamin D supplementation in these types of population studies (Hyppönen and Power 2007). In this chapter, we provide a short overview on vitamin D metabolism and the possible importance of vitamin D on cardiovascular disease, with a particular focus on evidence suggesting a role for vitamin D binding protein in mediating/altering these associations.

7.2   VITAMIN D METABOLISM

Vitamin D is a pro-hormone that comes in two forms, cholecalciferol (vitamin D_3) and ergocalciferol (vitamin D_2) (Haddad and Hahn 1973). Most of vitamin D is obtained as vitamin D_3, through synthesis in the skin after exposure to UVB radiation (typically from sunlight) of wavelengths 290–315 nm (Holick 1995). The synthesis converts 7-dehydrocholesterol (DHC-7) to pre-vitamin D_3, which is then quickly converted by heat induction to vitamin D_3. Overexposure to UVB radiation does not cause vitamin D toxicity as excess pre-vitamin D_3 and vitamin D_3 is inactivated by radiation (Holick et al. 1981, Webb et al. 1989). Vitamin D_3 can also be obtained from animal-based food products, with oily fish (e.g., salmon, sardines, and mackerel) being the best natural source, and egg yolk and meat containing smaller quantities. Vitamin D_2 is the plant form of vitamin D and can occur naturally in some types of mushrooms (Lamberg-Allardt 2006).

After entering the circulation, most of the vitamin D (whether obtained from skin synthesis or diet) is transported to the liver for the hydroxylation to 25-hydroxyvitamin D (25(OH)D) (Figure 7.1). The 25-hydroxylation can also occur in some extra-hepatic target tissues (DeLuca 2008). Circulating concentrations of 25(OH)D increase in proportion to vitamin D intake as 25-hydroxylation in the liver has little or no regulation (Horst et al. 2005). The half-life of 25(OH)D is relatively long (15 days) (Jones 2008b), and hence, total 25(OH)D is commonly used to indicate nutritional vitamin D status (Horst et al. 2005, Jones 2008a). For hormonal activation, 25(OH)D undergoes a further hydroxylation, which occurs mainly in the kidneys and to some extent in extra-renal tissues, producing the active form of 1,25-dihydroxyvitamin D (1,25(OH)$_2$D, calcitriol) (Horst et al. 2005, Shultz et al. 1983). Circulating concentrations of 1,25(OH)$_2$D are tightly regulated, its half-life is only 10–20 hours. Degradation of 25(OH)D and 1,25(OH)$_2$D is initiated by 24-hydroxylation, leading to inactive metabolites and eventually extraction from the system (DeLuca 2008).

7.2.1   GENETIC EFFECTS ON VITAMIN D METABOLISM

Genes *CYP2R1*, *CYP27B1*, and *CYP24A1* from the cytochrome P450 superfamily of enzymes are primarily responsible for the successive hydroxylation and degradation of vitamin D sterols (Schuster 2011). Genetic variants from *CYP2R1*, *CYP27B1*, and *CYP24A1* have been found to be associated with 25(OH)D concentrations in genome-wide association studies (GWAS) and candidate gene studies (Wang et al. 2010, Ahn et al. 2010, Hyppönen et al. 2009). Furthermore, genetic variants from *DHCR7*, which encodes a reductase catalyzing the conversion of 7-DHC to cholesterol and hence removes a substrate from vitamin D metabolic pathway, were also found to be associated with 25(OH)D concentrations in GWAS meta-analyses (Ahn et al. 2010, Wang et al. 2010). *CYP2R1* expresses the microsomal enzyme of

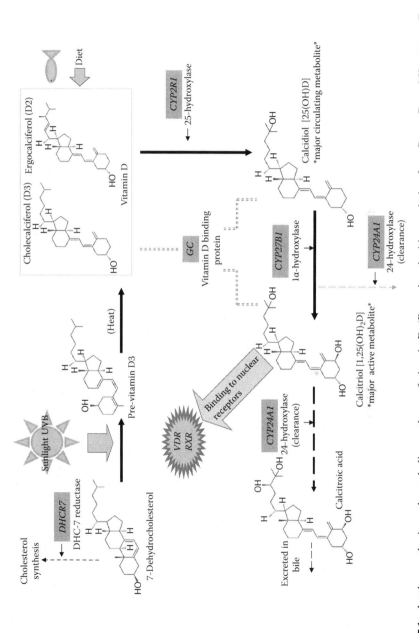

FIGURE 7.1 Intake, synthesis, and metabolism pathway of vitamin D. (Reproduced with permission from Berry, D. and Hyppönen, E., *Curr Opin Nephrol Hypertens*, 20, 331–6, 2011.)

25-hydroxylase primarily responsible for the hepatic hydroxylation of vitamin D. The renal hydroxylation to $1,25(OH)_2D$ by 1α-hydroxylase is encoded by *CYP27B1*. The clearance of most vitamin D metabolites initiated by the 24-hydroxylation is encoded by *CYP24A1*. The hormonal effects of $1,25(OH)_2D$ are mediated through binding with nuclear vitamin D receptor (VDR), which in turn forms a heterodimer structure with the retinoid-X receptor (RXR) (Pike and Shevde 2005). The complex then binds with the regulatory element in the target gene. High-resolution mapping has identified some 2,800 VDR binding sites in the genome, highlighting the potential for a wide range of influences across different metabolic pathways (Ramagopalan et al. 2010).

7.2.2   THE VDBP IN THE METABOLISM PATHWAY OF VITAMIN D

VDBP is a protein carrier of vitamin D–related sterols, including vitamin D_2 and D_3, 25(OH)D, and $1,25(OH)_2D$ (Daiger et al. 1975). The binding protein has a slightly higher binding affinity with the major circulating form of 25(OH)D than with the active form $1,25(OH)_2D$ or with vitamin D_2 and D_3 (Bouillon et al. 1980) (Kawakami et al. 1979). The majority of the 25(OH)D metabolite is transported in circulation bound by VDBP (83%–85%; high affinity), with a much smaller proportion bound by albumin (12%–15%; low affinity) (Bikle et al. 1986b). As with other circulating hormones, only a very small proportion of the vitamin D metabolites are free or unbound (for 25(OH)D it is 0.04%). For the 1α-hydroxylation to $1,25(OH)_2D$ to occur in the proximal tubules of the kidney, it is the VDBP-bound 25(OH)D that undergoes megalin/cubulin mediated endocytosis, thereby preventing its excretion (Nykjaer et al. 2001, Nykjaer et al. 1999).

There is evidence from animal studies to suggest that VDBP has a slightly higher affinity with vitamin D_3 compared with D_2 (Hollis 1984). A recent systematic review and meta-analysis found that 25(OH)D concentrations were higher after supplementing with a vitamin D_3 than with a D_2 (Tripkovic 2012), which in part may be due to vitamin D_3's high binding affinity with VDBP.

7.2.2.1   Synthesis of Vitamin D Binding Protein

The VDBP is primarily synthesized in the liver, although there is some evidence from animal studies suggesting secondary synthesis sites for VDBP, such as the kidney (McLeod and Cooke 1989). The initiation of the synthesis appears to be self-contained within the hepatocytes; however, in secondary sites it is suspected that the gene expression is probably driven by different determinants (Hiroki et al. 2006). The turnover of VDBP in the liver is high, as its half-life (2.5 days) is relatively short compared to its major ligand 25(OH)D (Horst et al. 2005, Kawakami et al. 1981).

Women typically have slightly higher VDBP concentrations than men regardless of ethnicity (Bolland et al. 2007, Gressner et al. 2009, Jorgensen et al. 2004), as concentrations are influenced by female sex hormones (Cheema et al. 1989). VDBP concentrations have been found to be positively correlated with 25(OH)D levels in some (Bolland et al. 2007, Lauridsen et al. 2005, Saadi et al. 2006), but not in all studies (Reid et al. 2011, Wood et al. 2011).

There is ongoing debate whether free 25(OH)D concentrations have greater bio-availability in certain cell types compared to the metabolite that is ligand bound (Chun et al. 2012, Zella et al. 2008). Free 25(OH)D can be calculated as:

$$\text{Free 25(OH)D} = \frac{\text{Total 25(OH)D}}{1 + K_{a\,\text{alb}} \times \text{albumin} + K_{a\,\text{VDBP}} \times \text{VDBP}},$$

where $K_{a\,\text{alb}}$ and $K_{a\,\text{VDBP}}$ are the affinity constants for albumin ($K_{a\,\text{alb}} = 6 \times 10^5$ mol/L) and VDBP ($K_{a\,\text{VDBP}} = 7 \times 10^8$ mol/L) (Bikle et al. 1986a). The free fraction of 25(OH)D is calculated as the ratio of free 25(OH)D over total 25(OH)D. A high negative correlation between the free fraction of 25(OH)D and VDBP (correlation score r ranges from -0.77 to -0.99) has been reported, whereas it is unclear whether a correlation between free 25(OH)D and VDBP exists (Bikle et al. 1986a, Bolland et al. 2007).

7.2.2.2 Group-Specific Component (*GC*) Gene

The *GC* gene is part of an albumin multi-gene family that includes albumin, alpha-fetoprotein, and alpha-albumin/afamin (Cooke and David 1985, Nishio and Dugaiczyk 1996). The gene family is located on chromosome 4 within the 4q12-q13 region; albumin, α-fetoprotein, and α-albumin/afamin are located next to one another, and *GC* is apart some 1.6M away (Song et al. 1999). Of the four genes, *GC* is the most divergent in terms of structure, which suggests that it is the oldest (Witke et al. 1993). *GC* has thirteen exons and is more than twice as long as the other genes in the family (Witke et al. 1993).

Initially identified by immune-electrophoresis of human serum, there are two autosomal and codominant alleles of *GC* (Gc1 and Gc2) (Hirschfeld 1959, Hirschfeld et al. 1960). Phenotype differences in the mobility of the VDBP in sera are related to the genetic inheritance of the alleles (Cleve and Bearn 1961). There are two common genetic subtypes of Gc1, notably that of Gc1s (slow) and Gc1f (fast), named after mobility differences in protein bands observed by isofo-cusing electrophoresis (Constans and Viau 1977). *GC* can also be subdivided to three common isoforms (Gc1s, Gc1f, Gc2) based on the amino acids 432 and 436 (Genome build 37.3, in earlier versions it was 416 and 420) in the coding region for exon 11 of the gene (Braun et al. 1992). Since the isoforms have different iso-electric points, they can be measured by isoelectric focusing (Constans and Viau 1977). Isoform Gc1s has a glutamic acid at amino acid position 432, while both Gc1f and Gc2 have aspartic acid in that position. Gc2 has a lysine reside at amino acid 436, whereas Gc1s and Gc1f have a threonine reside at this position (Braun et al. 1992). The differences in amino acids have been shown to alter the binding affinities of the isoforms with 25(OH)D; Gc1f has the highest affinity, then Gc1s, and Gc2 has the lowest affinity (Arnaud and Constans 1993). The nonsynonymous single nucleotide polymorphisms (SNPs) (rs7041 = Asp432Glu with minor allele frequency (MAF) of 0.43 in Hapmap CEU population, rs4588 = Thr436Lys with MAF of 0.27 in Hapmap CEU, Genome build 37.3) can be used to infer the haplo-types relating to the isoforms (Fang et al. 2009).

Different studies have presented the *GC* genetic variants in several ways. The isoforms can be presented as diplotypes (Gc1f-1f, Gc1s-1s, Gc1f-1s, Gc1s-2, Gc1f-2, and Gc2-2), or grouped as Gc1-1 (Gc1f-1f, Gc1s-1s, Gc1f-1s), Gc1-2 (Gc1s-2, Gc1f-2), and Gc2-2 (which can be referred as DBP phenotypes; DBP1-1, DBP2-1, and DBP2-2 when measured from serum), all of which can be determined by electrophoretic techniques or more recently genotyping (Fang et al. 2009, Lauridsen et al. 2001). The SNP rs4588 is equivalent to comparing Gc1s and Gc1f with Gc2 as the difference is the transversion of threonine to lysine.

7.2.3 ASSOCIATIONS OF *GC* VARIANTS WITH 25(OH)D AND **VDBP** CONCENTRATIONS

Of all the SNPs found to be associated with 25(OH)D in the GWAS, individual *GC* SNPs (in particular those in high LD with nonsynonymous SNP rs4588) explain the most variation in vitamin D status (Wang et al. 2010). The minor alleles of rs4588 (or rs2282679 in high-linkage disequilibrium with rs4588 [$r^2 = 0.98$] and typically genotyped in GWA platforms) and rs7041 are associated with lower 25(OH)D concentrations, compared with the major alleles (Wang et al. 2010). Out of the three major isoforms in studies where participants are of European ancestry, carriers of Gc1s have higher 25(OH)D concentrations than noncarriers, whereas carriers of Gc2 have lower 25(OH)D concentrations than noncarriers. This pattern of 25(OH) D concentrations across *GC* haplotypes can clearly be observed in the British 1958 birth cohort that consists of 5,330 participants of European ancestry (Table 7.1). In the Twins UK study (Wang et al. 2010), the minor allele of *GC* SNP rs2282679 was associated with lower VDBP levels compared with the major allele. Likewise, in individuals of European ancestry, VDBP concentrations have been reported to be the highest for those carrying Gc1-1, the lowest in individuals carrying Gc2-2, and within the middle of distribution in individuals carrying Gc2-1 (Lauridsen et al. 2001, Speeckaert et al. 2010).

7.3 VITAMIN D AND CARDIOVASCULAR DISEASE

CVD is the leading cause of death in the United Kingdom, and during 2008 it accounted for a third of deaths across all ages (Scarborough et al. 2010). CVD is also one of the main causes of premature death, killing 28% of men and 20% of women before reaching the age of 75 years (Scarborough et al. 2010). Since the 1970s, there has been a steady decline in mortality rates for CVD, and this is attributed to medical treatment and changes in lifestyle, especially in smoking habits (Unal et al. 2004). Vitamin D deficiency has been proposed as a risk factor for CVD and related metabolic abnormalities, and as reviewed in the following, much of this evidence comes from observational studies.

7.3.1 MECHANISMS

There are several plausible mechanisms through which vitamin D intake or status could affect cardiovascular health and related risk factors (Figure 7.2). VDRs are

TABLE 7.1

Copies of the *GC* Genotypes (Haplotype Pairs) with Mean 25(OH)D Concentrations in the 1958BC

Gc Genotype / n / 25(OH)D, nmol geometric mean (95% CI)	First Haplotype			
Second Haplotype	*Gc1s*	*Gc1f*	*Gc2*	*Gcx*
Gc1s	1600	0	0	0
	57.1	–	–	–
	(55.9, 58.4)			
Gc1f	902	132	0	0
	56.4	52.7	–	–
	(54.7, 58.1)	(48.4, 57.5)		
Gc2	1785	444	438	11
	52.0	51.8	49.2	35.9
	(50.9, 53.1)	(49.7, 54.1)	(47.2, 51.3)	(25.9, 49.8)
Gcx	14	0	0	4
	51.8	–	–	41.2
	(40.9, 65.7)			(16.5, 102.7)

Source: Berry, D. J., *Analysing the association of vitamin D status on selected cardiovascular risk markers using seasonal and genetic variations.* PhD thesis, University College London, 2012.

Note: Haplotypes estimated by SimHap (Carter et al. 2008) from SNPs rs7041 and rs4588; *Gcx* has glutamic and lysine acids at amino acid positions 432 and 436, respectively.

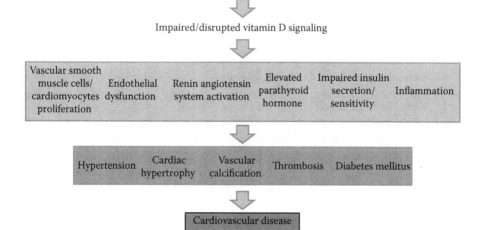

FIGURE 7.2 Proposed mechanisms of vitamin D deficiency with CVD.

expressed throughout the human circulatory system, including endothelial cells (Suzuki et al. 2009), vascular smooth muscle cells within blood vessels (Wu-Wong et al. 2007), and cardiac myocytes (O'Connell and Simpson 1996). These tissues have also exhibited expression of 1α-hydroxylase, suggesting that they are capable of extra-renal synthesis of active vitamin D (Chen et al. 2008, Somjen et al. 2005). Somewhat controversially, it has been recently questioned whether VDR exists in the muscle tissue (including cardiac, endothelial, and vascular smooth muscle) (Bischoff-Ferrari 2012, Wang and DeLuca 2011), therefore prompting a call for further research.

Experiments in animals have found that proliferation of vascular smooth muscle cells can be inhibited by calcitriol administration (Mitsuhashi et al. 1991). An overactive renin-angiotensin system (RAS) can impair renal function and deteriorate cardiovascular health (Li 2012), and down-regulation of RAS activity is one of the key mechanisms proposed for calcitriol (Li et al. 2002). Evidence to support this mechanism has been primarily obtained from animal experiments; for example, treatment with calcitriol has been shown to down-regulate RAS and to improve cardiac function in 1α-hydroxylase knockout mice (Zhou et al. 2008), and in salt-sensitive rats with cardiac hypertrophy (Bae et al. 2011, Choi et al. 2011). However, a recent randomized controlled trial in patients with chronic kidney disease did not find improvements in left ventricular mass index or diastolic function by treatment with paricalcitol (active vitamin D analogue) (Thadhani et al. 2012).

The regulation of local RAS activity in the pancreas has also been postulated as a mechanism for which vitamin D deficiency may impair insulin secretion and sensitivity (Cheng et al. 2011). However, calcitriol may also have a direct influence on pancreatic β cells (Wolden-Kirk et al. 2011) as the pancreas expresses both VDR (Pike et al. 1980) and 1α-hydroxylase (Bland et al. 2004). Vitamin D response elements exist in the human insulin gene promoter (Maestro et al. 2003), and transcriptional activation of the human insulin gene was observed after treatment with calcitriol, further supporting a possible role for calcitriol in regulating the insulin response (Maestro et al. 2002).

Calcitriol could also affect the risk of cardiovascular disease by influences on coagulation and by helping to maintain a normal antithrombotic homeostasis (Ohsawa et al. 2000). Compared to the wild type, VDR knockout mice have displayed greater tendency for thrombus formation in the liver and kidney (Aihara et al. 2004). Calcitriol has important influences on immune function (Penna and Adorini 2000) and may help to down-regulate vascular inflammation (Equils et al. 2006, Suzuki et al. 2009). In animal studies, calcitriol was observed to down-regulate inflammation induced by plasminogen activator inhibitor (a major fibrinolytic inhibitor) (Chen et al. 2011) and to reduce the development of atherosclerotic lesions by induction of regulatory T cells and immature dendritic cells (Takeda et al. 2010). In patients with type 2 diabetes, treatment with calcitriol has also been reported to reduce cholesterol uptake by macrophages and to inhibit foam cell formation (Oh et al. 2009).

Vitamin D may have a dual role in the development of vascular calcification, with both deficient and excessive dosages displaying adverse effects (Rodriguez et al. 2011, Shroff et al. 2008). Vitamin D–related vascular calcification has typically been observed in conjunction with treatment with the active calcitriol (or related analogues) in patients with chronic kidney disease (Briese et al. 2006, Goldsmith et al. 1997). However, excessive intake of vitamin D (in toxic quantities) can also lead to soft-tissue calcification (including vascular) and bone resorption, causing renal and heart damage (Selby et al. 1995, Shephard and Deluca 1980).

There is relatively little research on the role of VDBP in CVD. However, VDBP could potentially have a direct effect on cardiovascular health through influences on inflammation and immune function (Kew and Webster 1988, Yamamoto and Homma 1991). It may also affect the risk of thrombotic events by binding to actin (Laing and Cooke 2005), thereby increasing the clearance rate of actin from the circulation (Bogaerts et al. 2005, Dueland et al. 1990). Actin filaments are released from the vascular smooth muscle cells after their death caused by tissue damage (Janmey et al. 1992). Elevated concentrations of actin in the blood can increase blood viscosity to dangerous levels of resistance and can cause clot formation by aggregating platelets (Bogaerts et al. 2005). Furthermore, an in vitro experiment found that in endothelial cells under stress, such as in a situation of vascular injury, the release of VDBP can facilitate the response of vascular smooth muscle cells (Raymond et al. 2005).

7.3.2 Observational Studies

There have been several recent meta-analyses investigating the association of vitamin D status with cardio-metabolic disorders (hypertension, diabetes mellitus, dyslipidemia) and morbidity/mortality related to CVD. Most of the meta-analyses have pooled together cross-sectional and prospective studies and evaluated disorders or CVD events between high and low 25(OH)D concentrations. The inclusion criteria vary for the published meta-analyses, thus indirectly commenting on the quality and heterogeneity of the studies published in the area (Pittas et al. 2010, Sokol et al. 2011).

A review of prospective studies investigating the association between vitamin D status and CVD events found ten studies that were of sufficient quality to be meta-analyzed (Sokol et al. 2011). Risk of CVD events was higher for individuals with 25(OH)D <50 nmol/L, compared with individuals with 25(OH)D ≥50 nmol/L, although there was evidence of heterogeneity among the studies. After dropping two studies that contributed to the heterogeneity and only including studies that had adjusted for season, the risk did somewhat attenuate but remained statistically significant. A subsequent large prospective study on two American cohorts (Nurse's Health Study and Health Professional Follow-Up Study) followed participants for 22 years and 20 years, respectively, and consisted of 74,272 women and 44,592 men (Sun et al. 2011). Findings from that study were also to some extent supportive of a beneficial association, suggesting that men with a dietary daily intake of vitamin D ≥600 IU have a 16% lower risk of CVD, compared with men who had a daily intake <100 IU.

A recent meta-analysis on hypertension included four prospective and fourteen cross-sectional studies (Burgaz et al. 2011). Most studies had some participants, if not all, with European ancestry and had adjusted for an indicator of body weight and season. The risk of hypertension was 27% lower in the highest 25(OH)D category compared with the lowest. A lower incidence of hypertension for the highest 25(OH)D category versus the lowest was also reported in an earlier meta-analysis that focused solely on prospective studies (Pittas et al. 2010).

Evidence of a protective association between 25(OH)D and incidence risk of type 2 diabetes has recently come from a systematic review and meta-analysis of nine prospective studies and two unpublished prospective studies (Forouhi et al. 2012). The meta-analyzed prospective studies all adjusted for a measure of body weight or obesity, except for one study, which had the largest contribution to the meta-analysis. Incidence of type 2 diabetes was 41% lower for individuals with 25(OH)D concentrations in the top quartile, compared with those with 25(OH)D concentrations in the bottom quartile. After excluding the large study where body weight had not been controlled for, the estimate for reduction in type 2 diabetes risk by higher compared with lower 25(OH)D concentrations was slightly attenuated to 37%.

A study reviewing the association of vitamin D on serum lipids identified some twenty-two cross-sectional studies published between 2000 and 2010, although only thirteen studies adjusted for an obesity indicator (Jorde and Grimnes 2011). Four out of the possible thirteen studies reported a significant positive association between 25(OH)D and high-density lipoprotein after adjusting for an obesity measure. For triglycerides, five studies from a possible eleven observed a significant negative association of 25(OH)D with triglycerides after adjusting for an obesity measure. Although no meta-analysis was performed, the authors concluded that high 25(OH)D levels were associated with a favorable lipid profile (Jorde and Grimnes 2011).

A challenge with observational studies evaluating the association between vitamin D status and cardiovascular risk arises from the strong association between obesity and low 25(OH)D concentrations. Adipose tissue is an active endocrine organ, and obesity is known to lead to a chronic state of low-grade inflammation likely to confer adverse influences on the risk of CVD disease (Rocha and Libby 2009). Furthermore, obesity is also a known risk factor for vitamin D deficiency (Cheng et al. 2010, Forrest and Stuhldreher 2011, Jorde et al. 2010, Martins et al. 2007). It appears likely that the association is due to volumetric expansion (Drincic et al. 2012), or by increased sequestration of vitamin D to adipose tissues leading to lowering of 25(OH)D concentrations (Earthman et al. 2012). Hence, when evaluating the association between 25(OH)D and CVD or related risk factors using information from observational studies, it is paramount to consider obesity and to exclude associations arising through unrelated independent effects of excess adiposity. However, at the same time, low vitamin D status per se may mediate some of the adversity related to obesity, demonstrating the need for randomized controlled trials or alternative study designs (such as Mendelian randomization (Davey Smith and Ebrahim 2003)) and the inherent uncertainty with simple observational associations.

7.3.3 Intervention Studies

The recent Cochrane review of "Vitamin D supplementation for prevention of mortality in adults" included fifty randomized trials with 94,148 participants (Bjelakovic et al. 2011). The trials were initially designed to investigate predominantly the effects of vitamin D supplementation on markers of bone health and strength (such as number of falls and fractures), but all studies had also reported on mortality. The majority of participants included in the trials were elderly women and the mean age across the trials was 74 years. In the meta-analysis, vitamin D_3 supplementation reduced the risk of all-cause mortality, while supplementation with vitamin D_2, alfacalcidol or calcitriol, was not associated with mortality risk. Seven trials were available for analyses on CVD mortality, with 41,906 participants in total. However, vitamin D supplementation was not associated with CVD mortality. An alternative meta-analysis on randomized trials supplementing with vitamin D (that excluded trials intervening with calcitriol or an analogue), analyzed the influences on the risk of incident myocardial infarction and stroke (Elamin et al. 2011). Similarly, this analysis found no difference in either event between the treatment groups.

The largest study by far included in the Cochrane Review was Women's Health Initiative (WHI) that had some 36,000 women between the ages of 50 and 79 years at baseline, who were randomly assigned with daily calcium and vitamin D_3 supplements or a placebo and followed for a mean of seven years (Jackson et al. 2006). In this study alone, no reduction in mortality (cardiovascular or all-cause) from supplementation was observed. A substudy of some 750 women in the WHI were scored for coronary artery calcification after the trial, and there was no change in the score between those supplemented and not (Manson et al. 2010).

Despite the evidence from observational studies, meta-analyses of randomized trials on risk factors of cardiovascular disease have mostly found no improvement in the cardiometabolic profile after intervention. For hypertension, a recent review deemed fourteen randomized trials supplementing with vitamin D to be of sufficient quality to be meta-analyzed (Elamin et al. 2011). The conclusion was there was no evidence of a change in systolic or diastolic blood pressure as a result of vitamin D supplementation. The review also included meta-analyses of randomized trials on lipids (total cholesterol, triglycerides, low- and high-density lipoproteins), and there was no significant difference between the groups supplemented with vitamin D and not. For measures of glycemia in healthy individuals with normal glucose tolerance, a meta-analysis of five trials found no difference in plasma glucose levels after supplementation (Pittas et al. 2010). However, there is some evidence that for individuals with glucose intolerance, vitamin D supplementation may lead to improvements in insulin resistance (Mitri et al. 2011).

7.3.4 VBDP STUDIES: GENETIC AND MEASURED SERUM CONCENTRATIONS

There are few epidemiological studies directly investigating the association of serum VDBP or genetic variants of *GC* with risk factors/markers of cardiovascular disease. Most epidemiological studies have tended to report associations that have not been adjusted for potential confounders and have been fairly limited in study size, and none to our knowledge has assessed the risk of cardiovascular mortality. However, two recent studies have reported on quantity of calcification in coronary arteries (CAC), which has been shown to be a strong predictor of cardiovascular disease beyond standard risk factors (Detrano et al. 2008). Both studies found no associations between *GC* SNPs and CAC (Shen et al. 2010, Young et al. 2011). The larger of the two studies included one discovery study and a further two replication datasets, but no association was observed between eight *GC* tag SNPs and quantity of CAC at the discovery stage that consisted of a population-based sample of 697 Amish individuals (Shen et al. 2010).

The most common risk factor of CVD investigated with genotypes/phenotypes of VDBP is adiposity (Dorjgochoo et al. 2011, Jiang et al. 2007, Taes et al. 2006). More recently, the GIANT consortium meta-analyses results from GWAS on measures of adiposity (BMI and waist-hip ratio adjusted for BMI) have become publically available. The discovery meta-analyses consisted of up to 120,000 individuals of European ancestry, and no association was found between the nonsynonymous *GC* SNPs of rs7041 and rs2282679 and the adiposity measures (p-values $\geq$ 0.43) (Heid et al. 2010, Speliotes et al. 2010). The largest candidate gene study to date has been in some 7,000 Chinese women, where again no association was observed between the *GC* SNPs and BMI (Dorjgochoo et al. 2011). Phenotypes of VDBP were also found not to be associated with body composition (fat mass and lean mass) in unadjusted models (Taes et al. 2006). In contrast, a study using data from 1,873 individuals of U.S. European ancestry from 405 families did observe associations between genetic variations in *GC* (SNP and haplotype) and age- and sex-adjusted measures of fat mass, BMI, and percentage of fat mass (Jiang et al. 2007). The size and direction of the associations was not reported in this study, but the *GC* variants were reported to explain from 1.19% to 1.42% variation of percentage fat mass, which appears somewhat large given the amount of variation the top *GC* SNP explains in 25(OH)D concentrations (as reported in the GWAS (Wang et al. 2010)).

Studies investigating the association between serum concentrations of VDBP and adiposity have typically been relatively small and provided mixed results (Bolland et al. 2007, Powe et al. 2011, Taes et al. 2006, Winters et al. 2009). A strong negative correlation between BMI and VDBP concentrations, and positive correlation between bioavailable 25(OH)D and BMI, was observed in forty-nine young adults of mixed ethnic origin (Powe et al. 2011). Another study that consisted of women with European and African ancestry found no differences in VDBP levels between obese (BMI > 30 m²) and nonobese participants (Winters et al. 2009). In elderly men of European ancestry, the correlation between BMI and VDBP was positive (Taes et al. 2006). A later study stratifying by sex suggested a positive correlation between VDBP and BMI in men and a negative correlation in women, but neither of the correlations was statistically significant (Bolland et al. 2007). However, in the same study, free 25(OH)D concentrations were negatively correlated with BMI in women but not correlated in men.

Genetic-related variants of *GC* (phenotypes of DBP) have only been found to be associated with total cholesterol and high-density lipoproteins in a study of cystic fibrosis patients (Speeckaert et al. 2008). In the discovery meta-analysis consisting of some 100,000 individuals of European ancestry from the Global Lipids Genetic Consortium, *GC* SNPs rs7041 and rs2282679 were not associated with low- and high-density lipoproteins, total cholesterol, and triglycerides (p-value $\geq$ 0.11) (Teslovich et al. 2010). Serum VDBP concentrations have been correlated with triglycerides ($r = 0.21$), but not with high- and low-density lipoproteins in a twin study of some 1,500 individuals, which were representative of the U.K. population (Arora et al.

2011). Furthermore, in the same study, serum VDBP was positively correlated with C-reactive protein and insulin levels. Analyses in 211 men aged 71 years and older found positive associations of triglycerides and low-density lipoprotein with serum VDBP and a negative cholesterol association after adjusting for albumin, BMI and 1,25(OH)$_2$D3 (Speeckaert et al. 2010).

There are limited studies on T2D and VDBP, and all have been focused on genetic variations in *GC*. The frequency of diplotypes Gc1f-1f and Gc1s-2 was found to vary among Japanese T2D cases and controls, although the frequency of all other *GC* diplotypes was similar (Hirai et al. 1998). Furthermore, alleles of the *GC* diplotypes have been associated with serum markers of glucose tolerance (fasting insulin and HOMA-R) in healthy Japanese individuals (Hirai et al. 2000). In studies with participants of European ancestry, no evidence has been observed that either non-synonymous SNP (rs4588 and rs7041) was associated with the disease risk of T2D (Klupa et al. 1999, Malecki et al. 2002, Ye et al. 2001).

7.4   CONCLUSIONS

The 2011 Institute of Medicine report on dietary reference intakes on vitamin D concluded that beyond skeletal health, the role of vitamin D is so far inconclusive (Institute of Medicine 2011). The strongest promise that some of the proposed extraskeletal health benefits may turn out to be causal, comes from the recent Cochrane review reporting reductions in all-cause mortality risk by the intake of vitamin D supplements even at fairly modest dosages (Bjelakovic et al. 2011). However, despite accumulating evidence for an association of vitamin D status and intake with CVD and risk factors of CVD from observational studies, the Cochrane review found no significant reductions in mortality caused by cardiovascular disease. Whether that reflects a true lack of effect, that benefits of vitamin D intake are restricted to particular subgroups, or issues relating to dosage, timing, duration, or some other methodological aspect remain to be established. As reviewed, there are various mechanisms by which higher vitamin D intakes might be beneficial in relation to cardiovascular risk while a possible direct role of VDBP remains to be established.

REFERENCES

Ahn, J., Yu, K., Stolzenberg-Solomon, R., et al. 2010. Genome-wide association study of circulating vitamin D levels. *Hum Mol Genet,* 19: 2739–45.

Aihara, K., Azuma, H., Akaike, M., et al. 2004. Disruption of nuclear vitamin D receptor gene causes enhanced thrombogenicity in mice. *J Biol Chem,* 279: 35798–802.

Arnaud, J., & Constans, J. 1993. Affinity differences for vitamin D metabolites associated with the genetic isoforms of the human serum carrier protein (DBP). *Hum Genet,* 92: 183–8.

Arora, P., Garcia-Bailo, B., Dastani, Z., et al. 2011. Genetic polymorphisms of innate immunity-related inflammatory pathways and their association with factors related to type 2 diabetes. *BMC Med Genet,* 12: 95.

Bae, S., Yalamarti, B., Ke, Q., et al. 2011. Preventing progression of cardiac hypertrophy and development of heart failure by paricalcitol therapy in rats. *Cardiovasc Res,* 91: 632–9.

Berry, D., & Hyppönen, E. 2011. Determinants of vitamin D status: Focus on genetic variations. *Curr Opin Nephrol Hypertens,* 20: 331–6.

Berry, D. J. 2012. *Analysing the association of vitamin D status on selected cardiovascular risk markers using seasonal and genetic variations.* PhD thesis, University College London.

Bikle, D. D., Gee, E., Halloran, B., et al. 1986a. Assessment of the free fraction of 25-hydroxyvitamin D in serum and its regulation by albumin and the vitamin D-binding protein. *J Clin Endocrinol Metab,* 63: 954–9.

Bikle, D. D., Halloran, B. P., Gee, E., Ryzen, E., & Haddad, J. G. 1986b. Free 25-hydroxyvitamin D levels are normal in subjects with liver disease and reduced total 25-hydroxyvitamin D levels. *J Clin Invest,* 78: 748–52.

Bischoff-Ferrari, H. A. 2012. Relevance of vitamin D in muscle health. *Rev Endocr Metab Disord,* 13: 71–7.

Bjelakovic, G., Gluud, L. L., Nikolova, D., et al. 2011. Vitamin D supplementation for prevention of mortality in adults. *Cochrane Database Syst Rev:* CD007470.

Bland, R., Markovic, D., Hills, C. E., et al. 2004. Expression of 25-hydroxyvitamin D3-1alpha-hydroxylase in pancreatic islets. *J Steroid Biochem Mol Biol,* 89-90: 121–5.

Bogaerts, I., Vergoven, C., Van Baelen, H., & Bouillon, R. 2005. New aspects of DBP. *In:* Feldman, D., Glorieux, F. H., & Pike, J. W. (eds.). *Vitamin D.* Amsterdam: Elsevier Academic Press.

Bolland, M. J., Grey, A. B., Ames, R. W., et al. 2007. Age-, gender-, and weight-related effects on levels of 25-hydroxyvitamin D are not mediated by vitamin D binding protein. *Clin Endocrinol (Oxf),* 67: 259–64.

Bouillon, R., Van Baelen, H., & De Moor, P. 1980. Comparative study of the affinity of the serum vitamin D-binding protein. *J Steroid Biochem,* 13: 1029–34.

Braun, A., Bichlmaier, R., & Cleve, H. 1992. Molecular analysis of the gene for the human vitamin-D-binding protein (group-specific component): Allelic differences of the common genetic GC types. *Hum Genet,* 89: 401–6.

Briese, S., Wiesner, S., Will, J. C., et al. 2006. Arterial and cardiac disease in young adults with childhood-onset end-stage renal disease-impact of calcium and vitamin D therapy. *Nephrol Dial Transplant,* 21: 1906–14.

Burgaz, A., Orsini, N., Larsson, S. C., & Wolk, A. 2011. Blood 25-hydroxyvitamin D concentration and hypertension: a meta-analysis. *J Hypertens,* 29: 636–45.

Carter, K. W., Mccaskie, P. A., & Palmer, L. J. 2008. SimHap GUI: An intuitive graphical user interface for genetic association analysis. *BMC Bioinformatics,* 9: 557.

Cheema, C., Grant, B. F., & Marcus, R. 1989. Effects of estrogen on circulating "free" and total 1,25-dihydroxyvitamin D and on the parathyroid-vitamin D axis in postmenopausal women. *J Clin Invest,* 83: 537–42.

Chen, S., Glenn, D. J., Ni, W., et al. 2008. Expression of the vitamin D receptor is increased in the hypertrophic heart. *Hypertension,* 52: 1106–12.

Chen, Y., Kong, J., Sun, T., et al. 2011. 1,25-Dihydroxyvitamin D suppresses inflammation-induced expression of plasminogen activator inhibitor-1 by blocking nuclear factor-kappaB activation. *Arch Biochem Biophys,* 507: 241–7.

Cheng, Q., Li, Y. C., Boucher, B. J., & Leung, P. S. 2011. A novel role for vitamin D: Modulation of expression and function of the local renin-angiotensin system in mouse pancreatic islets. *Diabetologia,* 54: 2077–81.

Cheng, S., Massaro, J. M., Fox, C. S., et al. 2010. Adiposity, cardiometabolic risk, and vitamin D status: The Framingham Heart Study. *Diabetes,* 59: 242–248.

Choi, J. H., Ke, Q., Bae, S., et al. 2011. Doxercalciferol, a pro-hormone of vitamin D, prevents the development of cardiac hypertrophy in rats. *J Card Fail,* 17: 1051–8.

Chun, R. F., Peercy, B. E., Adams, J. S., & Hewison, M. 2012. Vitamin D binding protein and monocyte response to 25-hydroxyvitamin D and 1,25-dihydroxyvitamin D: Analysis by mathematical modeling. *PLoS One,* 7: e30773.

Cleve, H., & Bearn, A. G. 1961. Inherited variations in human serum proteins: Studies on the group-specific component. *Ann N Y Acad Sci,* 94: 218–24.

Constans, J., & Viau, M. 1977. Group-specific component: Evidence for two subtypes of the Gc1 gene. *Science,* 198: 1070–1.

Cooke, N. E., & David, E. V. 1985. Serum vitamin D-binding protein is a third member of the albumin and alpha fetoprotein gene family. *J Clin Invest,* 76: 2420–4.

Daiger, S. P., Schanfield, M. S., & Cavalli-Sforza, L. L. 1975. Group-specific component (Gc) proteins bind vitamin D and 25-hydroxyvitamin D. *Proc Natl Acad Sci U S A,* 72: 2076–80.

Davey Smith, G., & Ebrahim, S. 2003. "Mendelian randomization": Can genetic epidemiology contribute to understanding environmental determinants of disease? *Int J Epidemiol,* 32: 1–22.

Deluca, H. F. 2008. Evolution of our understanding of vitamin D. *Nutr Rev,* 66: S73–S87.

Detrano, R., Guerci, A. D., Carr, J. J., et al. 2008. Coronary calcium as a predictor of coronary events in four racial or ethnic groups. *N Engl J Med,* 358: 1336–45.

Dorjgochoo, T., Shi, J., Gao, Y. T., et al. 2011. Genetic variants in vitamin D metabolism-related genes and body mass index: Analysis of genome-wide scan data of approximately 7000 Chinese women. *Int J Obes (Lond),* 36(9): 1252–5.

Drincic, A. T., Armas, L. A., Van Diest, E. E., & Heaney, R. P. 2012. Volumetric dilution, rather than sequestration best explains the low vitamin D status of obesity. *Obesity (Silver Spring),* 20(7): 1444–8.

Dueland, S., Blomhoff, R., & Pedersen, J. I. 1990. Uptake and degradation of vitamin D binding protein and vitamin D binding protein-actin complex in vivo in the rat. *Biochem J,* 267: 721–5.

Earthman, C. P., Beckman, L. M., Masodkar, K., & Sibley, S. D. 2012. The link between obesity and low circulating 25-hydroxyvitamin D concentrations: Considerations and implications. *Int J Obes (Lond),* 36(3): 387–96.

Elamin, M. B., Abu Elnour, N. O., Elamin, K. B., et al. 2011. Vitamin D and cardiovascular outcomes: A systematic review and meta-analysis. *J Clin Endocrinol Metab,* 96: 1931–42.

Equils, O., Naiki, Y., Shapiro, A. M., et al. 2006. 1,25-Dihydroxyvitamin D inhibits lipopoly-saccharide-induced immune activation in human endothelial cells. *Clin Exp Immunol,* 143: 58–64.

Fang, Y., Van Meurs, J. B., Arp, P., et al. 2009. Vitamin D binding protein genotype and osteoporosis. *Calcif Tissue Int,* 85: 85–93.

Forouhi, N. G., Ye, Z., Rickard, A. P., et al. 2012. Circulating 25-hydroxyvitamin D concentration and the risk of type 2 diabetes: Results from the European Prospective Investigation into Cancer (EPIC)-Norfolk cohort and updated meta-analysis of prospective studies. *Diabetologia,* 55(8): 2173–82.

Forrest, K. Y., & Stuhldreher, W. L. 2011. Prevalence and correlates of vitamin D deficiency in US adults. *Nutr Res,* 31: 48–54.

Goldsmith, D. J., Covic, A., Sambrook, P. A., & Ackrill, P. 1997. Vascular calcification in long-term haemodialysis patients in a single unit: A retrospective analysis. *Nephron,* 77: 37–43.

Gressner, O. A., Gao, C., Siluschek, M., Kim, P., & Gressner, A. M. 2009. Inverse association between serum concentrations of actin-free vitamin D-binding protein and the histopathological extent of fibrogenic liver disease or hepatocellular carcinoma. *Eur J Gastroenterol Hepatol,* 21: 990–5.

Haddad, J. G., Jr., & Hahn, T. J. 1973. Natural and synthetic sources of circulating 25-hydroxyvitamin D in man. *Nature,* 244: 515–7.

Heid, I. M., Jackson, A. U., Randall, J. C., et al. 2010. Meta-analysis identifies 13 new loci associated with waist-hip ratio and reveals sexual dimorphism in the genetic basis of fat distribution. *Nat Genet,* 42: 949–60.

Hirai, M., Suzuki, S., Hinokio, Y., et al. 1998. Group specific component protein genotype is associated with NIDDM in Japan. *Diabetologia,* 41: 742–3.

Hirai, M., Suzuki, S., Hinokio, Y., et al. 2000. Variations in vitamin D-binding protein (group-specific component protein) are associated with fasting plasma insulin levels in Japanese with normal glucose tolerance. *J Clin Endocrinol Metab,* 85: 1951–3.

Hiroki, T., Song, Y. H., Liebhaber, S. A., & Cooke, N. E. 2006. The human vitamin D-binding protein gene contains locus control determinants sufficient for autonomous activation in hepatic chromatin. *Nucleic Acids Res,* 34: 2154–65.

Hirschfeld, J. 1959. Immune-electrophoretic demonstration of qualitative differences in human sera and their relation to the haptoglobins. *Acta Pathol Microbiol Scand,* 47: 160–8.

Hirschfeld, J., Jonsson, B., & Rasmuson, M. 1960. Inheritance of a new group-specific system demonstrated in normal human sera by means of an immuno-electrophoretic technique. *Nature,* 185: 931–2.

Holick, M. F. 1995. Environmental factors that influence the cutaneous production of vitamin D. *Am J Clin Nutr,* 61: 638S–645S.

Holick, M. F., Maclaughlin, J. A., & Doppelt, S. H. 1981. Regulation of cutaneous previtamin D3 photosynthesis in man: Skin pigment is not an essential regulator. *Science,* 211: 590–3.

Hollis, B. W. 1984. Comparison of equilibrium and disequilibrium assay conditions for ergocalciferol, cholecalciferol and their major metabolites. *J Steroid Biochem,* 21: 81–6.

Horst, R., Reinhardt, T., & Satyanarayana Reddy, G. 2005. Vitamin D metabolism. *In:* Feldman, D., Glorieux, F. H., & Pike, J. W. (eds.) *Vitamin D.* 2nd ed. Amsterdam: Elsevier Academic Press.

Hunter, D., De Lange, M., Snieder, H., et al. 2001. Genetic contribution to bone metabolism, calcium excretion, and vitamin D and parathyroid hormone regulation. *J Bone Miner Res,* 16: 371–8.

Hyppönen, E., Berry, D. J., Wjst, M., & Power, C. 2009. Serum 25-hydroxyvitamin D and IgE – a significant but nonlinear relationship. *Allergy,* 64: 613–20.

Hyppönen, E., & Power, C. 2007. Hypovitaminosis D in British adults at age 45 y: Nationwide cohort study of dietary and lifestyle predictors. *Am J Clin Nutr,* 85: 860-8.

Institute of Medicine. 2011. *Dietary reference intakes for calcium and vitamin D.* Ross, A. C., Taylor, C. L., Yaktine, A. L., & Del Valle, H. B. (eds.). Washington, DC.

Jackson, R. D., Lacroix, A. Z., Gass, M., et al. 2006. Calcium plus vitamin D supplementation and the risk of fractures. *N Engl J Med,* 354: 669–83.

Janmey, P. A., Lamb, J. A., Ezzell, R. M., Hvidt, S., & Lind, S. E. 1992. Effects of actin filaments on fibrin clot structure and lysis. *Blood,* 80: 928–36.

Jiang, H., Xiong, D. H., Guo, Y. F., et al. 2007. Association analysis of vitamin D–binding protein gene polymorphisms with variations of obesity-related traits in Caucasian nuclear families. *Int J Obes (Lond),* 31: 1319–24.

Jones, G. 2008a. Vitamin D. *In:* Kiple, K. F., & Conee Ornelas, K. (eds.) *The Cambridge world history of food* (pp. 763–768). Cambridge, UK: Cambridge University Press.

Jones, G. 2008b. Pharmacokinetics of vitamin D toxicity. *Am J Clin Nutr,* 88: 582S–586S.

Jorde, R., & Grimnes, G. 2011. Vitamin D and metabolic health with special reference to the effect of vitamin D on serum lipids. *Prog Lipid Res,* 50: 303–12.

Jorde, R., Sneve, M., Emaus, N., Figenschau, Y., & Grimnes, G. 2010. Cross-sectional and longitudinal relation between serum 25-hydroxyvitamin D and body mass index: The Tromso study. *Eur J Nutr,* 49: 401–7.

Jorgensen, C. S., Christiansen, M., Norgaard-Pedersen, B., et al. 2004. Gc globulin (vitamin D-binding protein) levels: An inhibition ELISA assay for determination of the total concentration of Gc globulin in plasma and serum. *Scand J Clin Lab Invest,* 64: 157–66.

Kawakami, M., Blum, C. B., Ramakrishnan, R., Dell, R. B., & Goodman, D. S. 1981. Turnover of the plasma binding protein for vitamin D and its metabolites in normal human subjects. *J Clin Endocrinol Metab,* 53: 1110–6.

Kawakami, M., Imawari, M., & Goodman, D. S. 1979. Quantitative studies of the interaction of cholecalciferol (vitamin D3) and its metabolites with different genetic variants of the serum binding protein for these sterols. *Biochem J,* 179: 413–23.

Kew, R. R., & Webster, R. O. 1988. Gc-globulin (vitamin D-binding protein) enhances the neutrophil chemotactic activity of C5a and C5a des Arg. *J Clin Invest,* 82: 364–9.

Klupa, T., Malecki, M., Hanna, L., et al. 1999. Amino acid variants of the vitamin D–binding protein and risk of diabetes in white Americans of European origin. *Eur J Endocrinol,* 141: 490–3.

Laing, C. J., & Cooke, N. K. 2005. Vitamin D binding protein. *In:* Feldman, D., Glorieux, F. H., & Pike, J. W. (eds.) *Vitamin D* (pp.117–134). Amsterdam: Elsevier Academic Press.

Lamberg-Allardt, C. 2006. Vitamin D in foods and as supplements. *Prog Biophys Mol Biol,* 92: 33–8.

Lauridsen, A. L., Vestergaard, P., Hermann, A. P., et al. 2005. Plasma concentrations of 25-hydroxy-vitamin D and 1,25-dihydroxy-vitamin D are related to the phenotype of Gc (vitamin D-binding protein): A cross-sectional study on 595 early postmenopausal women. *Calcif Tissue Int,* 77: 15–22.

Lauridsen, A. L., Vestergaard, P., & Nexo, E. 2001. Mean serum concentration of vitamin D-binding protein (Gc globulin) is related to the Gc phenotype in women. *Clin Chem,* 47: 753–6.

Li, Y. C. 2012. Vitamin D: roles in renal and cardiovascular protection. *Curr Opin Nephrol Hypertens,* 21: 72–9.

Li, Y. C., Kong, J., Wei, M., et al. 2002. 1,25-Dihydroxyvitamin D(3) is a negative endocrine regulator of the renin-angiotensin system. *J Clin Invest,* 110: 229–38.

Maestro, B., Davila, N., Carranza, M. C., & Calle, C. 2003. Identification of a vitamin D response element in the human insulin receptor gene promoter. *J Steroid Biochem Mol Biol,* 84: 223–30.

Maestro, B., Molero, S., Bajo, S., Davila, N., & Calle, C. 2002. Transcriptional activation of the human insulin receptor gene by 1,25-dihydroxyvitamin D(3). *Cell Biochem Funct,* 20: 227–32.

Malecki, M. T., Klupa, T., Wanic, K., et al. 2002. Vitamin D binding protein gene and genetic susceptibility to type 2 diabetes mellitus in a Polish population. *Diabetes Res Clin Pract,* 57: 99–104.

Manson, J. E., Allison, M. A., Carr, J. J., et al. 2010. Calcium/vitamin D supplementation and coronary artery calcification in the Women's Health Initiative. *Menopause,* 17: 683–91.

Martins, D., Wolf, M., Pan, D., et al. 2007. Prevalence of cardiovascular risk factors and the serum levels of 25-hydroxyvitamin D in the United States: Data from the Third National Health and Nutrition Examination Survey. *Arch Intern Med,* 167: 1159–65.

McLeod, J. F., & Cooke, N. E. 1989. The vitamin D-binding protein, alpha-fetoprotein, albumin multigene family: Detection of transcripts in multiple tissues. *J Biol Chem,* 264: 21760–9.

Mitri, J., Muraru, M. D., & Pittas, A. G. 2011. Vitamin D and type 2 diabetes: A systematic review. *Eur J Clin Nutr,* 65: 1005–15.

Mitsuhashi, T., Morris, R. C., Jr., & Ives, H. E. 1991. 1,25-dihydroxyvitamin D3 modulates growth of vascular smooth muscle cells. *J Clin Invest,* 87: 1889–95.

Nishio, H., & Dugaiczyk, A. 1996. Complete structure of the human alpha-albumin gene, a new member of the serum albumin multigene family. *Proc Natl Acad Sci U S A,* 93: 7557–61.

Nykjaer, A., Dragun, D., Walther, D., et al. 1999. An endocytic pathway essential for renal uptake and activation of the steroid 25-(OH) vitamin D3. *Cell,* 96: 507–15.

Nykjaer, A., Fyfe, J. C., Kozyraki, R., et al. 2001. Cubilin dysfunction causes abnormal metabolism of the steroid hormone 25(OH) vitamin D(3). *Proc Natl Acad Sci U S A,* 98: 13895–900.

O'Connell, T. D., & Simpson, R. U. 1996. Immunochemical identification of the 1,25-dihydroxyvitamin D3 receptor protein in human heart. *Cell Biol Int,* 20: 621–4.

Oh, J., Weng, S., Felton, S. K., et al. 2009. 1,25(OH)2 vitamin D inhibits foam cell formation and suppresses macrophage cholesterol uptake in patients with type 2 diabetes mellitus. *Circulation,* 120: 687–98.

Ohsawa, M., Koyama, T., Yamamoto, K., et al. 2000. 1Alpha,25-dihydroxyvitamin D(3) and its potent synthetic analogs downregulate tissue factor and upregulate thrombomodulin expression in monocytic cells, counteracting the effects of tumor necrosis factor and oxidized LDL. *Circulation,* 102: 2867–72.

Penna, G., & Adorini, L. 2000. 1 Alpha,25-dihydroxyvitamin D3 inhibits differentiation, maturation, activation, and survival of dendritic cells leading to impaired alloreactive T cell activation. *J Immunol,* 164: 2405–11.

Pike, J. W., Gooze, L. L., & Haussler, M. R. 1980. Biochemical evidence for 1,25-dihydroxyvitamin D receptor macromolecules in parathyroid, pancreatic, pituitary, and placental tissues. *Life Sci,* 26: 407–14.

Pike, J. W., & Shevde, N. K. 2005. The vitamin D receptor. *In:* Feldman, D., Glorieux, F. H., & Pike, J. W. (eds.) *Vitamin D* (pp.167–192). Amsterdam: Elsevier Academic Press.

Pittas, A. G., Chung, M., Trikalinos, T., et al. 2010. Systematic review: Vitamin D and cardiometabolic outcomes. *Ann Intern Med,* 152: 307–14.

Powe, C. E., Ricciardi, C., Berg, A. H., et al. 2011. Vitamin D–binding protein modifies the vitamin D-bone mineral density relationship. *J Bone Miner Res,* 26: 1609–16.

Ramagopalan, S. V., Heger, A., Berlanga, A. J., et al. 2010. A ChIP-seq defined genome-wide map of vitamin D receptor binding: Associations with disease and evolution. *Genome Res,* 20: 1352–60.

Raymond, M. A., Desormeaux, A., Labelle, A., et al. 2005. Endothelial stress induces the release of vitamin D-binding protein, a novel growth factor. *Biochem Biophys Res Commun,* 338: 1374–82.

Reid, D., Toole, B. J., Knox, S., et al. 2011. The relation between acute changes in the systemic inflammatory response and plasma 25-hydroxyvitamin D concentrations after elective knee arthroplasty. *Am J Clin Nutr,* 93: 1006–11.

Rocha, V. Z., & Libby, P. 2009. Obesity, inflammation, and atherosclerosis. *Nat Rev Cardiol,* 6: 399–409.

Rodriguez, M., Martinez-Moreno, J. M., Rodriguez-Ortiz, M. E., Munoz-Castaneda, J. R., & Almaden, Y. 2011. Vitamin D and vascular calcification in chronic kidney disease. *Kidney Blood Press Res,* 34: 261–8.

Saadi, H. F., Nagelkerke, N., Benedict, S., et al. 2006. Predictors and relationships of serum 25 hydroxyvitamin D concentration with bone turnover markers, bone mineral density, and vitamin D receptor genotype in Emirati women. *Bone,* 39: 113–43.

Scarborough, P., Bhatnagar, P., Wickramasinghe, K., et al. 2010. *Mortality* (Online). London: British Heart Foundation.

Schuster, I. 2011. Cytochromes P450 are essential players in the vitamin D signaling system. *Biochim Biophys Acta,* 1814: 186–99.

Selby, P. L., Davies, M., Marks, J. S., & Mawer, E. B. 1995. Vitamin D intoxication causes hypercalcaemia by increased bone resorption which responds to pamidronate. *Clin Endocrinol (Oxf),* 43: 531–6.

Shea, M. K., Benjamin, E. J., Dupuis, J., et al. 2009. Genetic and non-genetic correlates of vitamins K and D. *Eur J Clin Nutr,* 63: 458–464.

Shen, H., Bielak, L. F., Ferguson, J. F., et al. 2010. Association of the vitamin D metabolism gene CYP24A1 with coronary artery calcification. *Arterioscler Thromb Vasc Biol,* 30: 2648–54.

Shephard, R. M., & Deluca, H. F. 1980. Plasma concentrations of vitamin D3 and its metabolites in the rat as influenced by vitamin D3 or 25-hydroxyvitamin D3 intakes. *Arch Biochem Biophys,* 202: 43-53.

Shroff, R., Egerton, M., Bridel, M., et al. 2008. A bimodal association of vitamin D levels and vascular disease in children on dialysis. *J Am Soc Nephrol,* 19: 1239–46.

Shultz, T. D., Fox, J., Heath, H., III, & Kumar, R. 1983. Do tissues other than the kidney produce 1,25-dihydroxyvitamin D3 in vivo? A reexamination. *Proc Natl Acad Sci U S A* 80: 1746–50.

Sokol, S. I., Tsang, P., Aggarwal, V., Melamed, M. L., & Srinivas, V. S. 2011. Vitamin D status and risk of cardiovascular events: Lessons learned via systematic review and meta-analysis. *Cardiol Rev,* 19: 192–201.

Somjen, D., Weisman, Y., Kohen, F., et al. 2005. 25-hydroxyvitamin D3-1alpha-hydroxylase is expressed in human vascular smooth muscle cells and is upregulated by parathyroid hormone and estrogenic compounds. *Circulation,* 111: 1666–71.

Song, Y. H., Naumova, A. K., Liebhaber, S. A., & Cooke, N. E. 1999. Physical and meiotic mapping of the region of human chromosome 4q11-q13 encompassing the vitamin D binding protein DBP/Gc-globulin and albumin multigene cluster. *Genome Res,* 9: 581–7.

Speeckaert, M. M., Taes, Y. E., De Buyzere, M. L., et al. 2010. Investigation of the potential association of vitamin D binding protein with lipoproteins. *Ann Clin Biochem,* 47: 143–50.

Speeckaert, M. M., Wehlou, C., Vandewalle, S., et al. 2008. Vitamin D binding protein, a new nutritional marker in cystic fibrosis patients. *Clin Chem Lab Med,* 46: 365–70.

Speliotes, E. K., Willer, C. J., Berndt, S. I., et al. 2010. Association analyses of 249,796 individuals reveal 18 new loci associated with body mass index. *Nat Genet,* 42: 937–48.

Sun, Q., Shi, L., Rimm, E. B., et al. 2011. Vitamin D intake and risk of cardiovascular disease in U.S. men and women. *Am J Clin Nutr,* 94: 534–42.

Suzuki, Y., Ichiyama, T., Ohsaki, A., et al. 2009. Anti-inflammatory effect of 1alpha,25-dihydroxyvitamin D(3) in human coronary arterial endothelial cells: Implication for the treatment of Kawasaki disease. *J Steroid Biochem Mol Biol,* 113: 134–8.

Taes, Y. E., Goemaere, S., Huang, G., et al. 2006. Vitamin D binding protein, bone status and body composition in community-dwelling elderly men. *Bone,* 38: 701–7.

Takeda, M., Yamashita, T., Sasaki, N., et al. 2010. Oral administration of an active form of vitamin D3 (calcitriol) decreases atherosclerosis in mice by inducing regulatory T cells and immature dendritic cells with tolerogenic functions. *Arterioscler Thromb Vasc Biol,* 30: 2495–503.

Teslovich, T. M., Musunuru, K., Smith, A. V., et al. 2010. Biological, clinical and population relevance of 95 loci for blood lipids. *Nature,* 466: 707–13.

Thadhani, R., Appelbaum, E., Pritchett, Y., et al. 2012. Vitamin D therapy and cardiac structure and function in patients with chronic kidney disease: The PRIMO randomized controlled trial. *JAMA,* 307: 674–84.

Tripkovic, L. L., Hart, K., Smith, C., et al. 2012. Comparison of vitamin D2 vs. vitamin D3 supplementation in raising serum 25(OH)D status: A systematic review and meta-analysis. *Am J Clin Nutr,* 95(6): 1357–64.

Unal, B., Critchley, J. A., & Capewell, S. 2004. Explaining the decline in coronary heart disease mortality in England and Wales between 1981 and 2000. *Circulation,* 109: 1101–7.

Wang, T. J., Zhang, F., Richards, J. B., et al. 2010. Common genetic determinants of vitamin D insufficiency: A genome-wide association study. *Lancet,* 376: 180–8.

Wang, Y., & Deluca, H. F. 2011. Is the vitamin D receptor found in muscle? *Endocrinology,* 152: 354–63.

Webb, A. R., Decosta, B. R., & Holick, M. F. 1989. Sunlight regulates the cutaneous production of vitamin D3 by causing its photodegradation. *J Clin Endocrinol Metab,* 68: 882–7.

Winters, S. J., Chennubhatla, R., Wang, C., & Miller, J. J. 2009. Influence of obesity on vitamin D-binding protein and 25-hydroxy vitamin D levels in African American and white women. *Metabolism,* 58: 438–42.

Witke, W. F., Gibbs, P. E., Zielinski, R., et al. 1993. Complete structure of the human Gc gene: Differences and similarities between members of the albumin gene family. *Genomics,* 16: 751–4.

Wjst, M., Altmuller, J., Faus-Kessler, T., et al. 2006. Asthma families show transmission disequilibrium of gene variants in the vitamin D metabolism and signalling pathway. *Respir Res,* 7: 60.

Wolden-Kirk, H., Overbergh, L., Christesen, H. T., Brusgaard, K., & Mathieu, C. 2011. Vitamin D and diabetes: Its importance for beta cell and immune function. *Mol Cell Endocrinol,* 347: 106–20.

Wood, A. M., Bassford, C., Webster, D., et al. 2011. Vitamin D-binding protein contributes to COPD by activation of alveolar macrophages. *Thorax,* 66: 205–10.

Wu-Wong, J. R., Nakane, M., Ma, J., Ruan, X., & Kroeger, P. E. 2007. VDR-mediated gene expression patterns in resting human coronary artery smooth muscle cells. *J Cell Biochem,* 100: 1395–405.

Yamamoto, N., & Homma, S. 1991. Vitamin D3 binding protein (group-specific component) is a precursor for the macrophage-activating signal factor from lysophosphatidylcholine-treated lymphocytes. *Proc Natl Acad Sci U S A,* 88: 8539–43.

Ye, W. Z., Dubois-Laforgue, D., Bellanne-Chantelot, C., Timsit, J., & Velho, G. 2001. Variations in the vitamin D-binding protein (Gc locus) and risk of type 2 diabetes mellitus in French Caucasians. *Metabolism,* 50: 366–9.

Young, K. A., Snell-Bergeon, J. K., Naik, R. G., et al. 2011. Vitamin D deficiency and coronary artery calcification in subjects with type 1 diabetes. *Diabetes Care,* 34: 454-8.

Zella, L. A., Shevde, N. K., Hollis, B. W., Cooke, N. E., & Pike, J. W. 2008. Vitamin D-binding protein influences total circulating levels of 1,25-dihydroxyvitamin D3 but does not directly modulate the bioactive levels of the hormone in vivo. *Endocrinology,* 149: 3656–67.

Zhou, C., Lu, F., Cao, K., et al. 2008. Calcium-independent and 1,25(OH)2D3-dependent regulation of the renin-angiotensin system in 1alpha-hydroxylase knockout mice. *Kidney Int,* 74: 170–9.

8 Vitamins E and C
Effects on Matrix Components in the Vascular System

Jean-Marc Zingg, Mohsen Meydani, and Angelo Azzi

CONTENTS

8.1 INTRODUCTION

Connective tissue in the vascular system consists of vascular smooth muscle cells (VSMC) and the interstitial extracellular matrix (ECM). VSMC are located in the media of the vascular wall surrounded by their own basement membranes and embedded within an interstitial matrix consisting mainly of type I fibrillar collagen, the glycoprotein fibronectin, tenascins, and dermatan and chondroitin sulfate proteoglycans (Raines 2000). Two major types of VSMC have been described, and their aberrant regulation is at the basis of atherosclerosis development: the contractile phenotype of VSMC is essential for hemodynamic stability and maintenance of the vascular tone, whereas the synthetic phenotype is capable of repairing the injured vessel wall by migrating, proliferating, and elaborating an appropriate ECM (reviewed in Alexander and Dzau 2000, Zingg and Azzi 2007). In addition to maintaining the vascular structure, connective tissue supports against mechanical stress generated by pulsatile blood flow, which results in hydrostatic pressure, cyclic tension, and wall shear stresses. Connective tissue cells respond to these stressors, adapt to them, and convert them into changes of the extracellular compartment (Chiquet et al. 2003).

Injuries to the endothelium caused by excess mechanical and shear forces as a consequence of hypertension, or in areas of turbulent flow, can initiate the recruitment of inflammatory cells (monocytes/macrophages, T cells, mast cells) and trigger the proliferation and migration of VSMC into the area of vascular damage, thus contributing to the thickening of the vascular wall. This process is regulated by an intricate interplay of pro- and anti-inflammatory cytokines, chemokines, and growth factors, whose balance decides whether the damaged tissue will stabilize or will progress into more advanced lesions. Moreover, these events are influenced by changes in the extracellular environment mainly synthesized by the VSMC in response to endothelial damage. The synthesized ECM proteins enable VSMC to migrate, proliferate, and deposit further extracellular matrix in a typical wound-healing reaction. Since VSMC respond to ECM components such as collagen I, fibronectin, or heparin by reducing their proliferation (Christen et al. 1999, Kazi et al. 2002, Koyama et al. 1996), the amount and composition of the matrix produced by mechanical damage will result in an altered response of the vascular tissue (Raines 2000). Whereas these events are thought to be essential for normal vascular tissue homeostasis, excessive proliferative and secretory activity as a result of chronic inflammation will contribute to the formation of atherosclerotic and restenotic lesions.

Less is known whether the composition and stability of the ECM in the vascular wall can be influenced by dietary components such as cholesterol, vitamins (e.g., vitamins A, C, D, and E), phytochemicals, micronutrients, or different fatty acids (e.g., omega-3 or omega-6 unsaturated fatty acids). At the molecular level, these compounds may influence the ECM by means of their antioxidant properties, their ability to modulate signal transduction, and their capacity to regulate gene expression (e.g., of ECM genes in VSMC). Moreover, these dietary components can regulate the expression of integrin receptors at the cell surface and thus modulate ECM-triggered signal transition and gene expression, leading to altered cellular responses important for cardiovascular physiology, including cell adhesion, proliferation, survival/apoptosis, migration, invasion, and contraction (Breyer and Azzi 2001, Munteanu et al. 2004, Samandari et al. 2006).

Peroxidation of lipids in the subendothelial space is believed to be a central event in the development of atherosclerosis. Although vitamins E (α-tocopherol) and C (L-ascorbic acid) are well known by their ability to scavenge free radicals in the lipid and the water phase, respectively, their ability to prevent this process and the progression of *in vivo* lesions remains uncertain (Munteanu et al. 2004). The clearest *in vivo* evidence on protective effects of vitamin E on the cardiovascular system has been provided in animal models of atherosclerosis (reviewed by Munteanu and Zingg 2007). In humans, although some epidemiological studies have given indications that vitamin E protects against atherosclerosis progression, clinical intervention trials have reported contradictory results (reviewed in Munteanu and Zingg 2007). The reasons of this discrepancy may possibly lie in the result of the enrolment in clinical studies of patients with already manifested symptoms of atherosclerosis or with increased risk of cardiovascular disease (Brigelius-Flohe et al. 2002, Ricciarelli et al. 2001). Nevertheless, low plasma concentrations of vitamins E and C and β-carotene have been reported in epidemiological studies of cardiovascular disease (Ames et al. 1993).

The combination of vitamins E and C has been suggested to be more potent in preventing atherosclerosis, which may be due in part to regeneration of vitamin E by vitamin C (Abudu et al. 2004, Antoniads et al. 2003, Chan 1993, Kagan et al. 1992, May 1999). In these studies, the prevention of atherosclerosis by vitamins E and C is thought to be mainly the result of chemically preventing lipid peroxidation in low-density lipoprotein (LDL) thus reducing the formation of foam cells and consequent lesion development (Upston et al. 2003). There is less known on the ability of these vitamins to modulate other molecular pathways that are involved in the development of CVD, for example, by affecting VSMC and the ECM. Therefore, this review focuses on the molecular activities of vitamins E and C on the cells of the connective tissue in the vasculature, which are important for the maintenance of an intact vascular wall as well as in the repair of atherosclerotic lesions during disease development.

8.2 MOLECULAR ACTIVITIES OF VITAMIN E ON ECM COMPONENTS IN THE VASCULAR SYSTEM

Only a few studies have addressed the possibility that vitamin E might directly regulate the properties of the connective tissue and thus influence the maintenance and repair of the vascular wall (Table 8.1).

The ECM in the vasculature is mainly produced by VSMC; thus, the ability of vitamin E to influence ECM may be monitored by its regulatory effects on these cells (reviewed in Villacorta et al. 2007, Zingg and Azzi 2007). Since in the normal vasculature VSMC growth has ceased and VSMC gene expression is reduced to levels required predominantly to maintain contractile function without producing significant amounts of ECM, this question may be more relevant for de-differentiating migrating VSMC. Upon injury, intimal VSMC and a proportion of the medial cells "de-differentiate" as they migrate into the intima and continue proliferating without showing evidence for re-differentiation up to 60 days (Kocher et al. 1991, Orlandi et al. 1994). These changes in VSMC occurring after tissue damage are similar to the ones observed during culturing of intimal cells that de-differentiate to an epithelioid phenotype (Neuville et al. 1997). VSMC undergo relatively large changes upon plaque formation, as monitored by comparing VSMC of the media and the plaque on gene expression arrays (Mulvihill et al. 2004).

Thus, the question arises whether the phenotypic change seen in VSMC during atherosclerosis can be influenced, for example, by inducing a phenotypic change in VSMC toward more differentiated nonproliferating and contractile properties that can be followed by monitoring the expression and formation of an alpha-smooth muscle (α-SM) actin cytoskeleton. In fact, more potent differentiating compounds, such as retinoic acids (RA), increase α-SM actin and heavy chain myosin formation, and differentiation is also promoted by all-trans-retinoic acid (atRA) in rat aortic VSMC by increasing of protein kinase C alpha (PKCα) and α-SM actin expression (Axel et al. 2001, Haller et al. 1995), events that may also be important for the remodeling of blood vessels needed to limit restenosis after balloon angioplasty (Wiegman et al. 2000). Similarly, vitamin D analogs have been proposed to influence differentiation of mesangial cells, which share features with VSMC in which they mechanically support the capillary wall and acquire a VSMC phenotype in a

TABLE 8.1
Vitamin E Influences the Extracellular Matrix in the Vascular System

Activity of Vitamin E	Possible Molecular Mechanisms Involved	References
Prevention of lipid peroxidation, of oxLDL formation and consequent reduction of foam cell formation	Chain-breaking antioxidant activity, chemical scavenging of reactive oxygen and nitrogen species Inhibition of protein kinase C alpha (PKCα) with consequent inhibition of superoxide production by NADPH oxidase	(Cachia et al. 1998, Greenwald and Moy 1980, Witztum and Steinberg 1991)
Inhibition of vascular smooth muscle cell (VSMC) proliferation with consequent reduction of thickening of the vascular wall (intima-media thickness (IMT)	Inhibition of protein kinase C alpha (PKCα) via activation of protein phosphatase 2A (PP2A) or via direct binding to PKCα, activation of cellular release of transforming growth factor beta (TGF-β)	(McCary et al. 2012, Ozer et al. 1995, Ricciarelli et al. 1998, Salonen et al. 2000, Tasinato et al. 1995)
Prevention of altered vascular blood flow, extracellular matrix deposition, basement membrane thickening, and neovascularization during diabetes and insulin resistance	Inhibition of high glucose–induced protein kinase C (PKC) activation by activation of diacylglycerol kinase with consequent diminution of diacylglycerol and PKC activation	(Koya et al. 1997, Rask-Madsen and King 2005, Way et al. 2001)
Reduction of collagen accumulation in type II diabetic rat model	Normalization of TGF-β1 levels and of the medial vessel area Inhibition of medial VSMC proliferation and reduction of plasma and tissue levels of 8-iso-PGF$_2\alpha$ and overexpression of TGF-β1 and TGF-β1 receptor	(Shinomiya et al. 2002)
Stabilization of plaque with a fibrous stable cap by increasing ECM deposition	Induction of connective tissue growth factor (CTGF) expression leading to enhanced deposition of ECM (collagen, fibronectin) in the intima and the stabilization of the plaques	(Grotendorst et al. 1996, Orbe et al. 2003, Villacorta et al. 2003)
Improvement of plaque stability by inhibiting ECM degradation	Reduction of the expression of matrix metalloproteinase 3 (MMP-3) and of ruptured plaques Inhibition of collagenase I (matrix metalloproteinase 1, MMP-1) in VSMC without affecting the expression of its natural inhibitor, tissue inhibitor of metalloproteinase 1 (TIMP1) and MMP-19	(Li et al. 2004, Mauch et al. 2002, Ricciarelli et al. 1999)
Stimulation of wound repair	Preservation of endothelial cell migration and the normal wound repair process by restoring Rac translocation to the plasma membrane Induction of CTGF and stimulation of wound closure of mesangial cell monolayers Promotion of the formation of filopodia and lamellipodia, β-actin reorganization, and activation of focal adhesion kinase (FAK), praxillin, and Rac	(Blom et al. 2001, Chen et al. 2001, Ghosh et al. 2002)

transforming growth factor beta (TGF-β) type II receptor-mediated manner (Abe et al. 1999). Moreover, 1,25-dihydroxyvitamin D3 increases VSMC proliferation by stimulating vascular endothelial growth factor (VEGF) expression via a vitamin D response element in its promoter (Cardus et al. 2006, Cardus et al. 2009).

Although vitamin E inhibits VSMC proliferation and influences the expression of a number of genes (reviewed in Villacorta et al. 2007, Zingg and Azzi 2004, 2007), it is unknown whether this is the consequence of changing their differentiation state or rather reflects the specific modulation of individual genes (Zingg and Azzi 2004). At least the expression of the VSMC differentiation marker α-SM actin in the media of the aortas of cholesterol-fed rabbits was not altered by vitamin E (Sirikci et al. 1996). However, a transient up-regulation of α-tropomyosin, a fibrillar protein implicated in stabilization of the actin cytoskeleton, by α-tocopherol both at mRNA and protein levels was observed in rat VSMC (Aratri et al. 1999). Up-regulation by α-tocopherol may restore decreased amounts of tropomyosin in VSMC of the media and those that have migrated into the intima after balloon catheter denudation of rat carotid artery (Kocher et al. 1991), with consequent phenotypic modulation of VSMC (Kashiwada et al. 1997).

VSMC proliferation and matrix metalloproteinase (MMP)-2 expression as a result of high levels of leptin is reduced by α-tocopherol, which in turn may reduce the adverse effects of obesity and diabetes on the cardiovascular system (Li et al. 2005). In vitamin E–deficient rats, the total amounts of uronic acid, as an index of ECM glycosaminoglycan representing hyaluronic acid (HA), heparan sulfate (HS), dermatan sulfate (DS), and chondroitin sulfate (CS), decreased significantly as a result of decreased biosynthesis especially of the sulfated DS and CS (Iwama et al. 1985). A regulatory role of vitamin E on ECM gene expression has also been observed in diabetic patients, in which α-tocopherol reduced high glucose–induced collagen and fibronectin deposition by inhibiting specifically protein kinase C beta and gamma (PKCβ and PKCγ) (Koya and King 1998, Park et al. 1999). Likewise, increased collagen deposition and ventricular stiffness in the heart of young Wistar rats induced by high dietary fructose is prevented by α-tocopherol or a tocotrienol-rich fraction (Patel et al. 2011). A reduction of ECM production may also be important during liver fibrosis and contribute to the recently observed preventive effects of vitamin E against nonalcoholic steatohepatitis (NASH) in adults without diabetes (DiSario et al. 2007, Sanyal et al. 2010).

The expression of the connective tissue growth factor (CTGF), which stimulates extracellular matrix deposition, is up-regulated by α-tocopherol in primary VSCM (Villacorta et al. 2003). However, CTGF is mainly detected in α-SM actin-positive VSMC, which therefore has an already differentiated, contractile, and nonproliferating phenotype (Luscher 1997, Mulvihill et al. 2004, Oemar and Oemar et al. 1997). Nevertheless, it has been postulated that CTGF mediates myofibroblast differentiation, although CTGF-treated cells do not increase α-SM actin content (Folger et al. 2001). However, this may be more related to a role of CTGF in wound repair, a phenomenon that apparently does not require α-SM actin. Thus, it remains to be demonstrated whether induction of CTGF in atherosclerotic plaques or in thoracic aortic dissections is *per se* cause or consequence of the pathologic process (Wang et al. 2006). The induction of connective tissue growth factor (CTGF) and subsequently collagen expression by α-tocopherol (Villacorta et al. 2003), as well as inhibition of collagenase (MMP-1) expression (Ricciarelli et al. 1999), as a result of inhibition of protein kinase C (PKC) activity (Ricciarelli et al. 1998) may lead to acceleration of wound repair processes as well as to stabilization of the fibrous cap. In fact, in a porcine model of atherosclerosis, vitamins E and C were proposed to stabilize atherosclerotic plaques after angioplasty in hypercholesterolemic pigs by increasing collagen expression and by reducing hypercholesterolemia-induced MMP-1 expression (Orbe et al. 2003). Vitamin E inhibits the MMP-1 promoter via inhibition of binding of a Fos/Jun transcription factor complex to the AP-1 response element that is strongly activated by PKCα after stimulation by phorbol 12-myristate 13-acetate (PMA). On the other hand, in VSMC not stimulated with PMA, α-tocopherol leads to an early activation of AP-1 binding to DNA (Azzi et al. 1995, Ricciarelli et al. 1999, Stauble et al. 1994).

In other cell types, other members of the vitamin E family and analogs do in fact exhibit differentiating properties (You et al. 2001). Evidence of increased expression of cytoskeletal proteins such as smooth muscle myosin heavy chain, α-actin, and α-tropomyosin by vitamin E has been recently provided in the bronchioles, microvasculature, and alveolar septae of murine lungs (Oommen et al. 2007). In contrast, ozone (O_3) treatment induced a stronger expression of the ECM gene tenascin C in vitamin E–deficient α-TTP$^{-/-}$ mice when compared to wild-type mice (Vasu et al. 2007). Vitamin E and C supplementation of mice increased the number of progenitor endothelial cells as well as affecting the expression of many genes (Fiorito et al. 2008). On the other hand, gene expression profiling in vitamin E–deficient skeletal muscle shows increased levels of muscle structure and ECM genes, possibly resulting from oxidative stress, as reflected by increased expression of oxidative response proteins such as heme oxygenase, ferritin, transferrin, and the glutathione peroxidase precursor, as well as of genes related to anti-inflammatory and fibrotic conditions (Nier et al. 2006).

8.3 MOLECULAR ACTIVITIES OF VITAMIN C ON ECM COMPONENTS IN THE VASCULAR SYSTEM

Vitamin C (L-ascorbic acid) is mostly known for its preventive action against scurvy, but other health benefits have been attributed to it as well (reviewed in Rosenfeld 1997). In clinical and epidemiological studies, dietary intake of 100 mg/day of vitamin C reduces the incidence of mortality from heart disease, stroke, and cancer (Carr and Frei 1999), and the combined treatment with vitamins C and E showed some benefits in transplant-associated atherosclerosis (Fang et al. 2002, Liu and Meydani 2002). However, despite numerous studies investigating possible associations between vitamin C intake and the risk of developing cardiovascular disease (CVD), no clear evidence of significant benefits in prevention of CVD were observed in meta-analyses (Bjelakovic et al. 2007, Marchioli 1999, Ness et al. 1999). In a long-term study, vitamin C and E supplementation tended to reduce F2-isoprostanes but failed to alter oxidized LDL (oxLDL) formation or autoantibodies to oxLDL (Kinlay et al. 2004). Nevertheless, in small clinical studies, combined vitamins C and E improved endothelial function in patients with risk factors for atherosclerosis such as diabetes mellitus, smoking, hypertension, or hypercholesterolemia (Antoniades et al. 2003). Several animal studies indicate that vitamin C can slow the progression of experimental atherosclerosis (reviewed in Padayatty et al. 2003), and again, the combination of vitamin E and C supplementation synergistically reduces atherosclerotic lesion formation (Napoli et al.

2004). In pigs, the myocardial microvascular architecture is changed by hypercholesterolemia, possibly also as a result of an altered ECM, and supplementation with 100 IU/kg vitamin E and 1 g vitamin C daily attenuated these events (Zhu et al. 2004).

At the molecular level, both metabolic and antioxidant functions may contribute to the possible reduction of CVD risk by L-ascorbic acid (reviewed in Lynch et al. 1996, Naidu 2003). Similar to α-tocopherol, L-ascorbic acid can act as a chemical antioxidant and prevent the oxidative modification of LDL to atherogenic oxLDL (Carr and Frei 1999, Frei et al. 1989, Jialal et al. 1990), although in human intervention studies no clear changes in biomarkers of oxidation were observed (reviewed in Padayatty et al. 2003). In studies conducted *in vitro*, physiological concentrations of L-ascorbic acid strongly inhibit LDL oxidation by vascular endothelial cells (Martin and Frei 1997). L-ascorbic acid acts as a synergistic antioxidant together with α-tocopherol to prevent LDL oxidation (Sato et al. 1990), although it is best prevented by tocopherols (Jialal and Fuller 1995; reviewed in Frei et al. 1996). In addition, L-ascorbic acid reacts with the α-tocopheroxyl radical to regenerate α-tocopherol (Buettner 1993), and the ascorbate free radical (one-electron oxidized form) and dehydroascorbic acid (two-electron oxidized form) are enzymatically recycled to L-ascorbic acid (May et al. 2004). The ascorbate free radical can be reduced by NADH-dependent cytochrome $b5$ reductases or thioredoxin reductase (May et al. 1998), and dehydroascorbic acid can be reduced directly by glutathione (GSH) and GSH-dependent enzymes (Washburn and Wells 1999) or by NADPH-dependent dehydroascorbate reductase (Del Bello et al. 1994). In addition to prevention of oxLDL formation, L-ascorbic acid reduces free radicals by adaptively increasing endogenous antioxidant gene expression and protects against apoptosis of VSMC induced by moderately oxLDL (Siow et al. 1999b). Further activities of L-ascorbic acid that may influence atherosclerosis include its effect as a hypotensive agent, on cholesterol metabolism by modulating the conversion of cholesterol to bile acids (Ginter et al. 1982), and on plasma triglyceride levels via modulation of lipoprotein lipase activity (reviewed in Carr and Frei 1999, Villacorta et al. 2007). L-ascorbic acid also prevents leukocyte-endothelium interaction, an important step in initiating atherosclerosis; leukocyte-endothelial cell interactions induced by cigarette smoke or oxidized LDL are inhibited by L-ascorbic acid *in vivo* (Lehr et al. 1994, Lehr et al. 1995, Lehr et al. 1997).

The ECM is influenced by L-ascorbic acid in many different manners (Table 8.2), most importantly by acting as an essential cofactor in the oxidoreduction function by regenerating prosthetic metal ions in several hydroxylases, monooxygenases, and dioxygenases, enzymes that are involved in the synthesis of collagen, carnitine, and neurotransmitters (Arrigoni and DeTullio 2002, Levine 1986, Padh 1991).

In these enzymes, L-ascorbic acid accelerates hydroxylation reactions by maintaining the active center of metal ions in a reduced state for optimal activity (Arrigoni and DeTullio 2002), which may help in maintaining arterial wall integrity by stimulating biosynthesis of collagen and glycosaminoglycans. In particular, L-ascorbic acid is required for hydroxylation of proline residues in procollagen to hydroxyproline by acting as a cofactor for the enzyme prolyl hydroxylase, leading to stabilization of the collagen triple helical structure, and it also increases procollagen secretion (reviewed in Tinker and Rucker 1985).

By stimulating collagen synthesis, L-ascorbic acid also plays a critical role in wound repair and in the healing/regeneration process (Hellman and Burns 1958). This

TABLE 8.2
Vitamin C Influences the Extracellular Matrix in the Vascular System

Activity of Vitamin C	Molecular Mechanisms Involved	References
Prevention of lipid peroxidation; prevention of oxLDL formation; prevention of oxidation of α-tocopherol (sparing effect)	Chain-breaking antioxidant activity, chemical scavenging of reactive oxygen and nitrogen species.	(Lynch et al. 1996, Naidu 2003)
Influence of vascular tone by increasing NO	Nitric oxide, a central regulator of vascular tone and homeostasis by inducing VSMC relaxation, is increased in human umbilical vein endothelial by α-tocopherol, which may be related to an increase of eNOS phosphorylation. Coincubation with L-ascorbic acid amplified the effects of α-tocopherol on eNOS phosphorylation and NO formation, which is possibly related to the regeneration of oxidized α-tocopherol by L-ascorbic acid.	(Halpner et al. 1998, Heller et al. 2004b, Heller et al. 2004a)
Maintenance of collagen	L-ascorbic acid deficiency results in reduced hydroxylation of proline and lysine, thus destabilizing the collagen triple helix, and the impaired collagen synthesis leads mainly to defective connective tissue.	(Tinker and Rucker 1985)
Modulation of apoptosis	Prevention of apoptosis and protection against plaque instability in advanced atherosclerosis. Apoptosis of VSMC may contribute to the plaque "necrotic" core, cap rupture, and thrombosis. In human VSMC, moderately oxidized LDL, with its high lipid hydroperoxide content, rather than mildly or highly oxidized LDL, causes apoptosis.	(Siow et al. 1999b, Siow et al. 1999a)
L-ascorbic acid is a cofactor for hydroxylation involved in carnitine synthesis	L-ascorbic acid is essential for the synthesis of liver carnitine (β-hydroxy-epsilon-N-trimethyl-L-lysine), required for transport and transfer of fatty acids into mitochondria for ATP production. In rabbits, daily administration of L-carnitine completely prevents the progression of atherosclerotic lesions induced by hypercholesterolemia in both aorta and coronaries, whereas D-carnitine increased the progression of atherosclerotic lesions with the appearance of foam cells and apparent intimal plaques.	(Sayed-Ahmed et al. 2001, Hulse et al. 1978)

is demonstrated in mice unable to synthesize L-ascorbic acid due to a knockout of the gene L-gulono-gamma-lactone oxidase (Gulo$^{-/-}$), which has aortic wall damage and altered vascular integrity as evidenced by the disruption of elastic laminae, VSMC proliferation, and focal endothelial desquamation of the luminal surface (Maeda et al. 2000). Chronic vitamin C deficiency in Gulo$^{-/-}$ApoE$^{-/-}$ double-knockout mice does not influence the initiation or progression of atherosclerotic plaques, but collagen deposition is compromised, leading to a type of plaque morphology that is potentially vulnerable to rupture, due to lower amounts of collagen, larger necrotic cores within the plaques, and reduced fibroproliferation and neovascularization in the aortic adventitia (Nakata and Maeda 2002). Similar to that, the combined deficiency of vitamins E and C in guinea pigs causes severe central nervous system damage, with primary damage of the blood vessels (Burk et al. 2006). In vitamin C–deficient guinea pigs, poor wound healing was related to defective interstitial procollagen expression and blood vessel formation (Kipp et al. 1995). In the neonatal rat aorta, L-ascorbic acid stimulates collagen but inhibits elastin deposition (Quaglino et al. 1991).

In human arterial endothelial cells, L-ascorbic acid, alone or in combination with α-tocopherol–induced proliferation and DNA synthesis and antagonized the antiproliferative effects of oxLDL, but in VSMC the proliferation is inhibited (Ulrich-Merzenich et al. 2002). The ECM may contribute to the inhibition of DNA synthesis and proliferation, since VSMC plated on extracellular matrices deposited by VSMC in the presence of 0.1–1 mM L-ascorbic acid had an up to 50% lower proliferation rate than on matrices from L-ascorbic acid–deficient cells (Ivanov et al. 1997). The inhibitory effect is not specific for the biological active isomer of L-ascorbic acid, and isoascorbate and D-ascorbic acid are even more effective in reducing cell growth than L-ascorbic acid (Alcain and Buron 1994).

L-ascorbic acid influences *in vitro* differentiation of several mesenchyme-derived cell types, stabilizes the differentiated state, and cooperates with other agents to induce differentiation in a leukemia cell line (Alcain and Buron 1994, Arrigoni and DeTullio 2002). VSMC differentiation and the expression of smooth muscle–specific genes including h1-calponin, h-caldesmon, SM22α, and α-actin are induced by L-ascorbic acid after treatment of mouse TBR-B bone marrow stromal cells (Arakawa et al. 2000). The treatment of embryonic stem (ES) cells with L-ascorbic acid can induce differentiation into cardiac cells with spontaneous and rhythmic contractile activity expressing sarcomeric myosin, α-actinin, GATA4, Nkx2.5, α-MHC, β-MHC, and ANF (Takahashi et al. 2003). L-ascorbic acid stimulated chondrocyte terminal differentiation and the expression of collagen type X, possibly by increasing vitamin D levels and vitamin D receptor expression (Farquharson et al. 1998). Similar to that, the formation of a mineralized ECM during osteoblast differentiation is dependent on the presence of L-ascorbic acid (Franceschi et al. 1994, Franceschi and Iyer 1992).

Since many transcription factors and enzymes involved in signal transduction are redox regulated, L-ascorbic acid may modulate their activity, influencing the expression of ECM genes (Kunsch and Medford 1999, Sen and Packer 1996, Sun and Oberley 1996); in addition, several genes relevant for the synthesis of collagen matrix are modulated by L-ascorbic acid at the level of gene transcription, hydroxylation, secretion, and mRNA stabilization via direct L-ascorbic acid/RNA complex formation (Arrigoni and DeTullio 2002).

In the absence of L-ascorbic acid, cultured VSMC from fetal rat hearts express relatively more elastin and less collagen (DeClerck and Jones 1980), as a result from the combined, marked stabilization of collagen mRNA, the lesser stability of elastin mRNA, and the significant repression of elastin gene transcription (Davidson et al. 1997). In cultured human skin fibroblasts, the levels of mRNA for pro-collagens pro α1(I), pro α2(I), and pro α1(III) are elevated in the presence of L-ascorbic acid, whereas the level of mRNA for fibronectin is either unchanged (Geesin et al. 1988, Pinnel et al. 1987) or decreased (Peterszegi et al. 2002), suggesting that *cis*-regulatory elements are responsible for transcriptional activation by L-ascorbic acid or by ascorbate 2-phosphate (Duarte et al. 2009, Kurata et al. 1993).

8.4 ROLE OF VITAMINS E AND C BINDING PROTEINS ON ECM

Since the cells of the vascular system are in constant immediate contact with the bloodstream, the plasma level of α-tocopherol and L-ascorbic acid may directly influence the ECM after being transported across the endothelium. The plasma level of α-tocopherol is mainly determined by the proteins that recognize specifically α-tocopherol during uptake and distribution (Figure 8.1), whereas the other

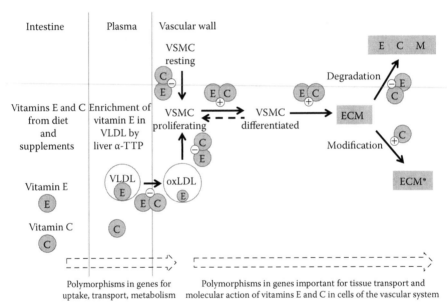

FIGURE 8.1 Regulatory role of vitamins E and C on the extracellular matrix (ECM) of the vascular system. Vitamins E and C are taken up from the circulation into the vascular wall and influence VSMC proliferation, differentiation, de-differentiation, and ECM production, modification, and degradation. The bioavailability of vitamins E and C depends on several proteins that are involved in the uptake, transport, and distribution of vitamins E and C from the diet into plasma and tissues. Moreover, the bioactivity of vitamins E and C depends on specific proteins that are able to bind and interact with these vitamins in cells and tissues. This is the case not only for α-tocopherol, which is enriched in plasma by the liver α-TTP protein via selective incorporation of this analogue into VLDL, but also for all the other tocopherol and tocotrienol analogues that may have a role at much lower concentration in plasma and tissues, possibly after binding to specific proteins. Polymorphisms in these genes are important determinants for plasma and tissue levels of vitamins E and C and their molecular action (reviewed in Zingg et al. 2008a), and thus may influence the ECM produced by the VSMC in the vascular wall. *modified ECM; E C M = degraded ECM.

tocopherol analogues are recognized for preferential metabolism (Table 8.3) (reviewed in Mustacich et al. 2007, Rigotti 2007, Zingg et al. 2008a).

The main protein influencing plasma vitamin E levels, the α-tocopherol transfer protein (α-TTP), mediates the enrichment of α-tocopherol by the liver and its delivery to the blood flow incorporated in VLDL (Brigelius-Flohe and Traber 1999, Hentati et al. 1996, Sato et al. 1991, Sato et al. 1993). α-TTP deficiency leads to vitamin E deficiency in the body with increased atherosclerotic lesion formation in

TABLE 8.3
Vitamin E Binding Proteins

Vitamin E Binding Proteins	Molecular Activity	References
Alpha-tocopherol transfer protein (α-TTP)	Binding of tocopherols and selective enrichment of α-tocopherol in plasma, with consequent higher tissue concentrations	(Hentati et al. 1996, Sato et al. 1991, Sato et al. 1993)
Niemann-Pick-C1-like 1	Binding of tocopherols and intracellular transport	(Narushima et al. 2008)
Tocopherol associated protein 1, TAP1 (SEC14L2)	Binding of tocopherols and intracellular transport; possible role in cellular signaling; possible role in plasma lipid levels	(Kempna et al. 2003, Ni et al. 2005, Neuzil et al. 2006, Porter 2003, Wright et al. 2009, Zimmer et al. 2000, Zingg et al. 2008b)
Tocopherol associated protein 2, TAP2 (SEC14L3)	Binding of tocopherols and intracellular transport; possible role in cellular signaling	(Kempna et al. 2003, Zingg et al. 2008b)
Tocopherol associated protein 3, TAP3 (SEC14L4)	Binding of tocopherols and intracellular transport; possible role in cellular signaling	(Kempna et al. 2003, Kempna et al. 2010, Zingg et al. 2008b)
Afamin	Binding and transport of α-tocopherol in cerebrospinal fluid	(Voegele et al. 2002)
Albumin	Binding of α-tocopherol in plasma and possible role in tissue distribution	(Fanali et al. 2013, Oram et al. 2001)
Phospholipid transfer protein (PLTP)	Binding of tocopherols and transfer between lipoproteins and tissues	(Jiang et al. 2002, Kostner et al. 1995)
Protein kinase C alpha (PKCα)	Binding of α-tocopherol and inhibition of PKCα	(McCary et al. 2012, Ricciarelli et al. 1998)
Cyclooxygenase-2 (COX-2)	Binding of γ-tocopherol and inhibition of COX-2	(Jiang et al. 2000, Jiang et al. 2001, Jiang and Ames 2003)
Phospholipase A2 (PLA2)	Binding of α-tocopherol and inhibition of PLA2	(Chandra et al. 2002)
Lipoxygenase (LOX)	Binding of α-tocopherol and inhibition of 5-, and 15-LOX	(Grossman and Waksman 1984, Reddanna et al. 1985)
	Binding of α-tocotrienol and inhibition of 12-LOX	(Khanna et al. 2003)

apoE$^{-/-}$ knockout mice (Terasawa et al. 2000). In these vitamin E-deficient animals, vitamin E supplementation prevents atherosclerosis independent of reducing lipid oxidation in the vessel wall (Suarna et al. 2006).

Vitamin E absorption from micelles in the intestine is a second important determinant of the plasma vitamin E level and it is mediated by scavenger receptor BI (SR-BI), which absorbs vitamin E into intestinal enterocytes (Reboul et al. 2006; reviewed in Iqbal and Hussain 2009). SR-BI also plays a role in the uptake of HDL-associated vitamin E across the brain capillary endothelial cells forming the blood-brain barrier, in hepatocytes, and in type II pneumocytes (Goti et al. 1998, Goti et al. 2000, Goti et al. 2001, Kolleck et al. 1999, Mardones et al. 2002a, Mardones and Rigotti 2004). SR-BI-deficient mice have defective tissue uptake of α-tocopherol that may contribute to the reproductive, cardiovascular, and neurodegenerative pathologies observed with these animals (Mardones et al. 2002b).

When transported by lipoproteins in plasma, vitamin E exchange between lipoproteins and between lipoproteins and cells is enhanced by the plasma phospholipid transfer protein (PLTP; Kostner et al. 1995, Lemaire-Ewing et al. 2010). PLTP-deficient mice show reduced vitamin E content in tissues such as brain and elevated lipofuscin, cholesterol oxides, and cellular peroxides (Desrumaux et al. 2005) and increased macrophage cholesterol accumulation (Ogier et al. 2007) but delayed formation of conjugated dienes resulting from increased levels of vitamin E in circulating apoB-lipoprotein–containing lipoproteins at the expense of the vascular wall (Jiang et al. 2002). To what degree an altered plasma lipoprotein/tissue distribution of vitamin E as a result of low PLTP levels contributes to the increased risk for peripheral atherosclerosis observed in nondiabetic, nonsmoking individuals remains to be determined (Schgoer et al. 2008, Schneider et al. 2004).

Other tocopherol binding proteins, the tocopherol-associated proteins TAP1, TAP2, and TAP3 (also named SEC14L2, SEC14L3, and SEC14L4, respectively) (Kempna et al. 2003, Neuzil et al. 2006, Porter 2003, Zimmer et al. 2000, Zingg et al. 2008b), Niemann-Pick-C1-like 1 (Narushima et al. 2008), afamin (Voegele et al. 2002), and albumin (Oram et al. 2001, Fanali et al. 2013) have been described, but their role in vitamin E tissue transport, activity, and metabolism, and influence on atherosclerotic risk remains to be determined (reviewed in Zingg et al. 2008a,b).

As recently exemplified by Milman et al. (2008) and Blum et al. (2010), diabetic patients with the haptoglobin 2-2 allele have reduced plasma vitamin E and C levels as a result of increased oxidation and consequent increased cardiovascular risk. Thus, certain polymorphisms in genes relevant for the bioavailability of vitamin E and C (uptake, distribution, oxidation, metabolism, molecular action) may be important determinants for the preventive effects of these vitamins against atherosclerosis (reviewed in Zingg et al. 2008a, Zingg 2012). However, in the case of the haptoglobin 2-2 allele, molecular mechanisms in addition to free radical scavenging may play a role, since the combination α-tocopherol with L-ascorbic acid did not have any further protective effect and even mitigated the effect of α-tocopherol on HDL oxidation (Asleh and Levy 2010). Likewise, polymorphisms in the apolipoprotein E gene not only influence the plasma lipid profile, inflammatory cytokine expression, and the level of free radicals, but also the plasma and tissue distribution of vitamin E, with apoE4 allele being associated with a lower retention of vitamin E in tissues but higher levels in the circulation (Huebbe et al. 2010).

Polymorphisms in several other genes such as α-TTP, hTAPs, CD36, or SR-BI scavenger receptors, and ABC transporters have been suggested to play a role for the different responsiveness of different individuals to α-tocopherol (Doring et al. 2004, Jialal and Devaraj 2003, Zingg et al. 2008a). However, whereas a specific role of α-TTP in determining the plasma vitamin E concentration has been well established, for the TAP proteins this is so far only supported *in vitro* (Neuzil et al. 2006, Ni et al. 2005, Zingg et al. 2008b). Therefore, polymorphisms in vitamin E–binding proteins may not only influence the plasma vitamin E levels but also in tissues and furthermore influence the molecular action of vitamin E on cells in tissues (Kempna et al. 2010, Lecompte et al. 2011, Zingg et al. 2008a, Zingg et al. 2008b, Zingg and Azzi 2009). However, as described in the following, a possible influence of genetic polymorphisms on ECM in the cardiovascular tissue can currently only be deduced from research done with other tissues. In a recent study, Wright et al. (2009) have investigated whether polymorphisms in α-TTP or hTAP1 are associated with elevated prostate cancer risk resulting from altered plasma vitamin E concentrations. However, since these proteins are also expressed in the prostate (Morley et al. 2010, Ni et al. 2007, Zingg et al. 2008b), α-TTP and hTAP1 polymorphisms may not only determine the plasma vitamin E levels but also influence other cellular events in prostate or breast cancer cells, such as cell proliferation, signal transduction, and gene expression (e.g., via phosphatidylinositol-kinases Kempna et al. 2003), or by influencing the biosynthesis of cholesterol/steroids (e.g., via influencing cholesterol/steroid biosynthesis by regulating squalene epoxidase or HMG-CoA reductase) (Johnykutty et al. 2009, Kempna et al. 2010, Ni et al. 2005, Zingg 2007a, Zingg and Azzi 2009).

Polymorphisms in the *CD36/FAT* scavenger receptor gene have recently been found to influence plasma vitamin E levels and possibly also in tissues (Lecompte et al. 2011), which may similarly in patients with CD36 deficiency influence the risk for cardiovascular disease (Ma et al. 2004, Yuasa-Kawase et al. 2012). CD36 is involved not only in the uptake of oxLDL into monocytes/macrophages and VSMC but also in fatty acid transport, signal transduction, and gene expression (Eyre et al. 2007, Stuart et al. 2005, Triantafilou et al. 2006). Since CD36 expression is negatively regulated by vitamin E (Munteanu et al. 2006, Ricciarelli et al. 2000), these polymorphisms as well as the gene variants resulting from alternative splicing may influence the responsiveness to vitamin E (Andersen et al. 2006, Cheung et al. 2007, Ma et al. 2004, Zingg et al. 2002) and the risk for atherosclerosis (Love-Gregory et al. 2008, Love-Gregory et al. 2011) and metabolic syndrome (Noel et al. 2010).

In contrast to vitamin E, which is highly hydrophobic and requires specific proteins to be transported throughout the body, L-ascorbic acid does not, since it is hydrophilic. However, to pass across hydrophobic layers such as the plasma membrane, specific transporters have been implicated in the transport of vitamin C (sodium coupled vitamin C transporters 1 and 2 [*SVCT1/SLC23A1*, *SVCT2/ SLC23A2*] or the dehydroascorbate transporters [*GLUT1*, *GLUT3*]), tissue distribution, and metabolism (dehydroascorbate reductase [*DHAR*], e.g., the omega class of glutathione transferases *GSTO1-1* or *GSTO2-2*), and their cellular expression levels could contribute to the individual vitamin C response (Erichsen et al. 2006, Hediger 2002, Na et al. 2006, Seno et al. 2004, Takanaga et al. 2004, Whitbread et al. 2005). Polymorphisms in these genes may influence not only the vitamin C but also the vitamin E bioavailability and bioactivity in the ECM, although it was recently shown that for glutathione-*S*-transferases, certain polymorphisms influenced the level of vitamin C but not of vitamin E (Horska et al. 2011). In the peripheral nervous system, reduced expression of various collagen types (IV, V, and XXVIII) and laminin 2 was observed in SVCT2$^{+/-}$ heterozygote mice independent on the level of hydroxyproline, showing that the tissue L-ascorbic acid level is important for the expression of ECM genes (Gess et al. 2011). Possibly, analogues of L-ascorbic acid, such as disodium isostearyl 2-O-L-ascorbyl phosphate or ascorbic acid 2-phosphate that are slowly hydrolyzed in tissues and that are more potent in stimulating collagen expression and other EMC genes could show preventive effects on atherosclerosis by influencing the ECM in the vascular system (Duarte et al. 2009, Shibayama et al. 2008).

8.5   VITAMIN E ANALOGUES AND THEIR INFLUENCE ON ECM

The focus of vitamin E research has long been to relate a specific disease risk with the plasma levels of α-tocopherol, since only this analogue of vitamin E is enriched in plasma (to an average concentration of about 23 μM). However, in recent years specific cellular activities of the other natural vitamin E analogues (β-, γ-, δ-tocopherols, α-, β-, γ-, δ-tocotrienols, α-tocopheryl quinone, α-tocopheryl phosphate) have been discovered, which may act at much lower concentration as a result of their specific interaction with enzymes and proteins in cells and thus possibly also influence the ECM (Zingg 2007a,b). In human keratinocytes, γ-tocopherol and to a lower level also α-, β-, and δ-tocopherols up-regulate PPARγ, leading to the induction of transglutaminase expression, an enzyme that plays an important physiological role in hemostasis, wound healing, and the assembly and remodeling of the ECM (DePascale et al. 2006, Wodzinska 2005). In human tenon's fibroblasts, the α-, γ-, and δ-tocotrienols inhibit cell proliferation, cell migration, and the synthesis of collagen and thus may have an antifibrotic and antiscarring effect in filtrating glaucoma surgery (Tappeiner et al. 2010). Likewise, in human intestinal fibroblasts, tocotrienols reduce fibrosis and ECM production, such as procollagen 1 and 3, and laminin γ-1, whereas MMP-3 expression was increased (Luna et al. 2011a,b). In addition to affecting the same molecular targets as α-tocopherol, these vitamin E analogues may change gene expression by additional mechanisms, such as by activating the pregnenolone X receptor (PXR), which induces the *CYP3A4* and *CYP4F2* genes required for their metabolism, and polymorphisms in these genes may therefore be important determinants for their action by influencing the bioavailability and bioactivity of these vitamin E analogues in the body (Brigelius-Flohe 2003, Landes et al. 2003).

The vascular endothelial growth factor (VEGF) plays not only an important role during angiogenesis and vasculogenesis but also in vascular permeability

(Wong et al. 2009) and in the production and stability of the ECM, for example, by increasing fibronectin secretion (Kazi et al. 2004) or the production of MMP-1, -3, and -9 from VSMC (Wang and Keiser 1998). VEGF expression is modulated by vitamin E in several *in vitro* and *in vivo* experimental systems, in which either activation (Daghini et al. 2007, Kasimanickam et al. 2010, Kasimanickam et al. 2012, Zhang et al. 2004a) or inhibition of VEGF (Nespereira et al. 2003, Schindler and Mentlein 2006, Tang and Meydani 2001) has been observed. VEGF is expressed in the normal human aorta and veins in VSMC, and its expression is increased in vessels narrowed by atherosclerotic plaques, suggesting functions in addition to promoting angiogenesis such as the maintenance and repair of the endothelium, which may indirectly inhibit VSMC proliferation and intimal thickening (Couffinhal et al. 1997). Moreover, the amount of ECM may influence the action of the different VEGF isoforms, since they have different affinity to the heparin proteoglycan (Ruhrberg 2003). VEGF expression in monocytes is induced by α-tocopheryl phosphate (αTP) leading to increased *in vitro* angiogenesis in HUVEC, whereas αT is not effective (Zingg et al. 2010). Induction of VEGF by αTP was the consequence of stimulation of the PI3K/Akt signaling pathway, whereas α-tocopherol inhibits activation of this pathway (Kempna et al. 2004, Munteanu et al. 2006, Numakawa et al. 2006), which may involve binding to hTAPs and/or α-TTP (Kempna et al. 2003, Zingg et al. 2010, Zingg et al. 2012). Whether activation of PI3K/Akt by αTP regulates further cellular events in VSMC such as the formation of gap junctional intercellular communication (Chaumontet et al. 2008, Ito et al. 2010) or cortactin movement from the cell membrane to the nucleus (Ceccarelli et al. 2007, Saitoh et al. 2009) remains to be further elucidated. Interestingly, cortactin regulates lysosomal uptake and secretion of ECM components such as fibronectin and ECM-degrading metalloproteinases from invadopodia (Furmaniak-Kazmierczak et al. 2007, Lener et al. 2006, Sung et al. 2011, Sung and Weaver 2011) that are of functional importance for the movement of VSCM and macrophages through the ECM (Van Goethem et al. 2011). Thus, αTP may modulate ECM formation and removal by podosomes and invadopodia by influencing cortactin activity important for cell adhesion, migration, and tissue invasion (Ren et al. 2009).

Moreover, synthetic vitamin E derivatives may also influence the ECM, for example, by affecting similar cellular molecular targets as the natural vitamin E analogues, albeit possibly with a higher activity. The analogue α-tocopheryl succinate inhibits human prostate cancer cell invasiveness by reducing the expression of the ECM-degrading enzyme MMP-9 (Zhang et al. 2004b). Tretinoin tocopheryl, a synthetic composite compound ester linked between α-tocopherol and all-trans-retinoic acid, activates ECM gene expression (elastin, collagen 1, 3, and 6, fibronectin) in dermal fibroblasts that may accelerate wound healing (Mori et al. 1994).

8.6 CONCLUSIONS

Vitamins E and C influence the ECM by chemically reducing the formation of oxidized small molecules, proteins, and lipids, which are possible causes for vascular injury and cellular deregulation, and by influencing redox-dependent signal transduction and gene expression. In addition to their direct antioxidant effects, both vitamins can bind and modulate the activity of specific enzymes and act as their cofactors and regulators with consequent modulation of VSMC signal transduction and gene expression. Although a preventive effect of vitamins E and C against cardiovascular disease (CVD) is not universally accepted, a regulatory role of these vitamins on VSMC proliferation, differentiation, and ECM production may significantly contribute not only to the maintenance of an intact vascular wall but also in the repair of atherosclerotic lesions during disease development. In addition to the availability of vitamins E and C in the daily diet, polymorphisms in genes relevant for the uptake, distribution, metabolism, and activity of these vitamins may be important modulators to maintain an intact ECM in the vascular system.

REFERENCES

ABE, H., IEHARA, N., UTSUNOMIYA, K., KITA, T. & DOI, T. 1999. A vitamin D analog regulates mesangial cell smooth muscle phenotypes in a transforming growth factor-beta type II receptor-mediated manner. *J Biol Chem,* 274, 20874–8.

ABUDU, N., MILLER, J. J., ATTAELMANNAN, M. & LEVINSON, S. S. 2004. Vitamins in human arteriosclerosis with emphasis on vitamin C and vitamin E. *Clin Chim Acta,* 339, 11–25.

ALCAIN, F. J. & BURON, M. I. 1994. Ascorbate on cell growth and differentiation. *J Bioenerg Biomembr,* 26, 393–8.

ALEXANDER, R. W. & DZAU, V. J. 2000. Vascular biology: The past 50 years. *Circulation,* 102, IV112–6.

AMES, B. N., SHIGENAGA, M. K. & HAGEN, T. M. 1993. Oxidants, antioxidants, and the degenerative diseases of aging. *Proc Natl Acad Sci U S A,* 90, 7915–22.

ANDERSEN, M., LENHARD, B., WHATLING, C., ERIKSSON, P. & ODEBERG, J. 2006. Alternative promoter usage of the membrane glycoprotein CD36. *BMC Mol Biol,* 7, 8.

ANTONIADES, C., TOUSOULIS, D., TENTOLOURIS, C., TOUTOUZAS, P. & STEFANADIS, C. 2003. Oxidative stress, antioxidant vitamins, and atherosclerosis. From basic research to clinical practice. *Herz,* 28, 628–38.

ARAKAWA, E., HASEGAWA, K., YANAI, N., OBINATA, M. & MATSUDA, Y. 2000. A mouse bone marrow stromal cell line, TBR-B, shows inducible expression of smooth muscle-specific genes. *FEBS Lett,* 481, 193–6.

ARATRI, E., SPYCHER, S. E., BREYER, I. & AZZI, A. 1999. Modulation of alpha-tropomyosin expression by alpha-tocopherol in rat vascular smooth muscle cells. *FEBS Lett,* 447, 91–4.

ARRIGONI, O. & DE TULLIO, M. C. 2002. Ascorbic acid: Much more than just an antioxidant. *Biochim Biophys Acta,* 1569, 1–9.

ASLEH, R. & LEVY, A. P. 2010. Divergent effects of alpha-tocopherol and vitamin C on the generation of dysfunctional HDL associated with diabetes and the Hp 2-2 genotype. *Antioxid Redox Signal,* 12, 209–17.

AXEL, D. I., FRIGGE, A., DITTMANN, J., RUNGE, H., SPYRIDOPOULOS, I., RIESSEN, R., VIEBAHN, R. & KARSCH, K. R. 2001. All-trans retinoic acid regulates proliferation, migration, differentiation, and extracellular matrix turnover of human arterial smooth muscle cells. *Cardiovasc Res,* 49, 851–62.

AZZI, A., BOSCOBOINIK, D., MARILLEY, D., OZER, N. K., STAUBLE, B. & TASINATO, A. 1995. Vitamin E: A sensor and an information transducer of the cell oxidation state. *Am J Clin Nutr,* 62, 1337S–1346S.

BJELAKOVIC, G., NIKOLOVA, D., GLUUD, L. L., SIMONETTI, R. G. & GLUUD, C. 2007. Mortality in randomized trials of antioxidant supplements for primary and secondary prevention: Systematic review and meta-analysis. *Jama,* 297, 842–57.

BLOM, I. E., VAN DIJK, A. J., WIETEN, L., DURAN, K., ITO, Y., KLEIJ, L., DENICHILO, M., RABELINK, T. J., WEENING, J. J., ATEN, J. & GOLDSCHMEDING, R. 2001. In vitro evidence for differential involvement of CTGF, TGFbeta, and PDGF-BB in mesangial response to injury. *Nephrol Dial Transplant,* 16, 1139–48.

BLUM, S., VARDI, M., BROWN, J. B., RUSSELL, A., MILMAN, U., SHAPIRA, C., LEVY, N. S., MILLER-LOTAN, R., ASLEH, R. & LEVY, A. P. 2010. Vitamin E reduces cardiovascular disease in individuals with diabetes mellitus and the haptoglobin 2-2 genotype. *Pharmacogenomics,* 11, 675–84.

BREYER, I. & AZZI, A. 2001. Differential inhibition by alpha- and beta-tocopherol of human erythroleukemia cell adhesion: Role of integrins. *Free Radic Biol Med,* 30, 1381–9.

BRIGELIUS-FLOHE, R. 2003. Vitamin E and drug metabolism. *Biochem Biophys Res Commun,* 305, 737–40.

BRIGELIUS-FLOHE, R., KELLY, F. J., SALONEN, J. T., NEUZIL, J., ZINGG, J. M. & AZZI, A. 2002. The European perspective on vitamin E: Current knowledge and future research. *Am J Clin Nutr,* 76, 703–16.

BRIGELIUS-FLOHE, R. & TRABER, M. G. 1999. Vitamin E: Function and metabolism. *FASEB J,* 13, 1145–55.

BUETTNER, G. R. 1993. The pecking order of free radicals and antioxidants: Lipid peroxidation, alpha-tocopherol, and ascorbate. *Arch Biochem Biophys,* 300, 535–43.

BURK, R. F., CHRISTENSEN, J. M., MAGUIRE, M. J., AUSTIN, L. M., WHETSELL, W. O., JR., MAY, J. M., HILL, K. E. & EBNER, F. F. 2006. A combined deficiency of vitamins E and C causes severe central nervous system damage in guinea pigs. *J Nutr,* 136, 1576–81.

CACHIA, O., BENNA, J. E., PEDRUZZI, E., DESCOMPS, B., GOUGEROT-POCIDALO, M. A. & LEGER, C. L. 1998. Alpha-tocopherol inhibits the respiratory burst in human monocytes. Attenuation of p47(phox) membrane translocation and phosphorylation. *J Biol Chem,* 273, 32801–5.

CARDUS, A., PANIZO, S., ENCINAS, M., DOLCET, X., GALLEGO, C., ALDEA, M., FERNANDEZ, E. & VALDIVIELSO, J. M. 2009. 1,25-dihydroxyvitamin D3 regulates VEGF production through a vitamin D response element in the VEGF promoter. *Atherosclerosis,* 204, 85–9.

CARDUS, A., PARISI, E., GALLEGO, C., ALDEA, M., FERNANDEZ, E. & VALDIVIELSO, J. M. 2006. 1,25-Dihydroxyvitamin D3 stimulates vascular smooth muscle cell proliferation through a VEGF-mediated pathway. *Kidney Int,* 69, 1377–84.

CARR, A. C. & FREI, B. 1999. Does vitamin C act as pro-oxidant under physiological conditions? *FASEB J,* 13, 1007–24.

CECCARELLI, S., CARDINALI, G., ASPITE, N., PICARDO, M., MARCHESE, C., TORRISI, M. R. & MANCINI, P. 2007. Cortactin involvement in the keratinocyte growth factor and fibroblast growth factor 10 promotion of migration and cortical actin assembly in human keratinocytes. *Exp Cell Res,* 313, 1758–77.

CHAN, A. C. 1993. Partners in defense, vitamin E and vitamin C. *Can J Physiol Pharmacol,* 71, 725–31.

CHANDRA, V., JASTI, J., KAUR, P., BETZEL, C., SRINIVASAN, A. & SINGH, T. P. 2002. First structural evidence of a specific inhibition of phospholipase A2 by alpha-tocopherol (vitamin E) and its implications in inflammation: Crystal structure of the complex formed between phospholipase A2 and alpha-tocopherol at 1.8 A resolution. *J Mol Biol,* 320, 215–22.

CHAUMONTET, C., BEX, V., VERAN, F. & MARTEL, P. 2008. The vitamin E analog tocopherol succinate strongly inhibits gap junctional intercellular communication in rat liver epithelial cells (IAR203). *J Nutr Biochem,* 19, 263–8.

CHEN, C. C., CHEN, N. & LAU, L. F. 2001. The angiogenic factors Cyr61 and connective tissue growth factor induce adhesive signaling in primary human skin fibroblasts. *J Biol Chem,* 276, 10443–52.

CHEUNG, L., ANDERSEN, M., GUSTAVSSON, C., ODEBERG, J., FERNANDEZ-PEREZ, L., NORSTEDT, G. & TOLLET-EGNELL, P. 2007. Hormonal and nutritional regulation of alternative CD36 transcripts in rat liver—a role for growth hormone in alternative exon usage. *BMC Mol Biol,* 8, 60.

CHIQUET, M., RENEDO, A. S., HUBER, F. & FLUCK, M. 2003. How do fibroblasts translate mechanical signals into changes in extracellular matrix production? *Matrix Biol,* 22, 73–80.

CHRISTEN, T., BOCHATON-PIALLAT, M. L., NEUVILLE, P., RENSEN, S., REDARD, M., VAN EYS, G. & GABBIANI, G. 1999. Cultured porcine coronary artery smooth muscle cells. A new model with advanced differentiation. *Circ Res,* 85, 99–107.

COUFFINHAL, T., KEARNEY, M., WITZENBICHLER, B., CHEN, D., MUROHARA, T., LOSORDO, D. W., SYMES, J. & ISNER, J. M. 1997. Vascular endothelial growth factor/vascular permeability factor (VEGF/VPF) in normal and atherosclerotic human arteries. *Am J Pathol,* 150, 1673–85.

DAGHINI, E., ZHU, X. Y., VERSARI, D., BENTLEY, M. D., NAPOLI, C., LERMAN, A. & LERMAN, L. O. 2007. Antioxidant vitamins induce angiogenesis in the normal pig kidney. *Am J Physiol Renal Physiol,* 293, F371–81.

DAVIDSON, J. M., LUVALLE, P. A., ZOIA, O., QUAGLINO, D., JR. & GIRO, M. 1997. Ascorbate differentially regulates elastin and collagen biosynthesis in vascular smooth muscle cells and skin fibroblasts by pretranslational mechanisms. *J Biol Chem,* 272, 345–52.

DE CLERCK, Y. A. & JONES, P. A. 1980. The effect of ascorbic acid on the nature and production of collagen and elastin by rat smooth-muscle cells. *Biochem J,* 186, 217–25.

DE PASCALE, M. C., BASSI, A. M., PATRONE, V., VILLACORTA, L., AZZI, A. & ZINGG, J. M. 2006. Increased expression of transglutaminase-1 and PPARgamma after vitamin E treatment in human keratinocytes. *Arch Biochem Biophys,* 447, 97–106.

DECLERCK, Y. A. & JONES, P. A. 1980. Effect of ascorbic acid on the resistance of the extracellular matrix to hydrolysis by tumor cells. *Cancer Res,* 40, 3228–31.

DEL BELLO, B., MAELLARO, E., SUGHERINI, L., SANTUCCI, A., COMPORTI, M. & CASINI, A. F. 1994. Purification of NADPH-dependent dehydroascorbate reductase from rat liver and its identification with 3 alpha-hydroxysteroid dehydrogenase. *Biochem J,* 304 (Pt 2), 385–90.

DESRUMAUX, C., RISOLD, P. Y., SCHROEDER, H., DECKERT, V., MASSON, D., ATHIAS, A., LAPLANCHE, H., LE GUERN, N., BLACHE, D., JIANG, X. C., TALL, A. R., DESOR, D. & LAGROST, L. 2005. Phospholipid transfer protein (PLTP) deficiency reduces brain vitamin E content and increases anxiety in mice. *FASEB J,* 19, 296–7.

DI SARIO, A., CANDELARESI, C., OMENETTI, A. & BENEDETTI, A. 2007. Vitamin E in chronic liver diseases and liver fibrosis. *Vitam Horm,* 76, 551–73.

DORING, F., RIMBACH, G. & LODGE, J. K. 2004. In silico search for single nucleotide polymorphisms in genes important in vitamin E homeostasis. *IUBMB Life,* 56, 615–20.

DUARTE, T. L., COOKE, M. S. & JONES, G. D. 2009. Gene expression profiling reveals new protective roles for vitamin C in human skin cells. *Free Radic Biol Med,* 46, 78–87.

ERICHSEN, H. C., ENGEL, S. A., ECK, P. K., WELCH, R., YEAGER, M., LEVINE, M., SIEGA-RIZ, A. M., OLSHAN, A. F. & CHANOCK, S. J. 2006. Genetic variation in the sodium-dependent vitamin C transporters, SLC23A1, and SLC23A2 and risk for preterm delivery. *Am J Epidemiol,* 163, 245–54.

EYRE, N. S., CLELAND, L. G., TANDON, N. N. & MAYRHOFER, G. 2007. Importance of the carboxyl terminus of FAT/CD36 for plasma membrane localization and function in long-chain fatty acid uptake. *J Lipid Res,* 48, 528–42.

FANALI, G., FASANO, M., ASCENZI, P., ZINGG, J.-M. & AZZI, A. 2013. α-Tocopherol binding to human serum albumin. *Biofactors.* Epub ahead of print. DOI: 10.1002/biof.1070.

FANG, J. C., KINLAY, S., BELTRAME, J., HIKITI, H., WAINSTEIN, M., BEHRENDT, D., SUH, J., FREI, B., MUDGE, G. H., SELWYN, A. P. & GANZ, P. 2002. Effect of vitamins C and E on progression of transplant-associated arteriosclerosis: A randomised trial. *Lancet,* 359, 1108–13.

FARQUHARSON, C., BERRY, J. L., MAWER, E. B., SEAWRIGHT, E. & WHITEHEAD, C. C. 1998. Ascorbic acid-induced chondrocyte terminal differentiation: The role of the extracellular matrix and 1,25-dihydroxyvitamin D. *Eur J Cell Biol,* 76, 110–8.

FIORITO, C., RIENZO, M., CRIMI, E., ROSSIELLO, R., BALESTRIERI, M. L., CASAMASSIMI, A., MUTO, F., GRIMALDI, V., GIOVANE, A., FARZATI, B., MANCINI, F. P. & NAPOLI, C. 2008. Antioxidants increase number of progenitor endothelial cells through multiple gene expression pathways. *Free Radic Res,* 42, 754–62.

FOLGER, P. A., ZEKARIA, D., GROTENDORST, G. & MASUR, S. K. 2001. Transforming growth factor-beta-stimulated connective tissue growth factor expression during corneal myofibroblast differentiation. *Invest Ophthalmol Vis Sci,* 42, 2534–41.

FRANCESCHI, R. T. & IYER, B. S. 1992. Relationship between collagen synthesis and expression of the osteoblast phenotype in MC3T3-E1 cells. *J Bone Miner Res,* 7, 235–46.

FRANCESCHI, R. T., IYER, B. S. & CUI, Y. 1994. Effects of ascorbic acid on collagen matrix formation and osteoblast differentiation in murine MC3T3-E1 cells. *J Bone Miner Res,* 9, 843–54.

FREI, B., ENGLAND, L. & AMES, B. N. 1989. Ascorbate is an outstanding antioxidant in human blood plasma. *Proc Natl Acad Sci U S A,* 86, 6377–81.

FREI, B., KEANEY, J. F., JR., RETSKY, K. L. & CHEN, K. 1996. Vitamins C and E and LDL oxidation. *Vitam Horm,* 52, 1–34.

FURMANIAK-KAZMIERCZAK, E., CRAWLEY, S. W., CARTER, R. L., MAURICE, D. H. & COTE, G. P. 2007. Formation of extracellular matrix-digesting invadopodia by primary aortic smooth muscle cells. *Circ Res,* 100, 1328–36.

GEESIN, J. C., DARR, D., KAUFMAN, R., MURAD, S. & PINNELL, S. R. 1988. Ascorbic acid specifically increases type I and type III procollagen messenger RNA levels in human skin fibroblast. *J Invest Dermatol,* 90, 420–4.

GESS, B., ROHR, D., FLEDRICH, R., SEREDA, M. W., KLEFFNER, I., HUMBERG, A., NOWITZKI, J., STRECKER, J. K., HALFTER, H. & YOUNG, P. 2011. Sodium-dependent vitamin C transporter 2 deficiency causes hypomyelination and extracellular matrix defects in the peripheral nervous system. *J Neurosci,* 31, 17180–92.

GHOSH, P. K., VASANJI, A., MURUGESAN, G., EPPELL, S. J., GRAHAM, L. M. & FOX, P. L. 2002. Membrane microviscosity regulates endothelial cell motility. *Nat Cell Biol,* 4, 894–900.

GINTER, E., BOBEK, P., KUBEC, F., VOZAR, J. & URBANOVA, D. 1982. Vitamin C in the control of hypercholesterolemia in man. *Int J Vitam Nutr Res Suppl,* 23, 137–52.

GOTI, D., HAMMER, A., GALLA, H. J., MALLE, E. & SATTLER, W. 2000. Uptake of lipoprotein-associated alpha-tocopherol by primary porcine brain capillary endothelial cells. *J Neurochem,* 74, 1374–83.

GOTI, D., HRZENJAK, A., LEVAK-FRANK, S., FRANK, S., VAN DER WESTHUYZEN, D. R., MALLE, E. & SATTLER, W. 2001. Scavenger receptor class B, type I is expressed in porcine brain capillary endothelial cells and contributes to selective uptake of HDL-associated vitamin E. *J Neurochem,* 76, 498–508.

GOTI, D., REICHER, H., MALLE, E., KOSTNER, G. M., PANZENBOECK, U. & SATTLER, W. 1998. High-density lipoprotein (HDL3)-associated alpha-tocopherol is taken up by HepG2 cells via the selective uptake pathway and resecreted with endogenously synthesized apo-lipoprotein B-rich lipoprotein particles. *Biochem J,* 332, 57–65.

GREENWALD, R. A. & MOY, W. W. 1980. Effect of oxygen-derived free radicals on hyaluronic acid. *Arthritis Rheum,* 23, 455–63.

GROSSMAN, S. & WAKSMAN, E. G. 1984. New aspects of the inhibition of soybean lipoxygenase by alpha-tocopherol. Evidence for the existence of a specific complex. *Int J Biochem,* 16, 281–9.

GROTENDORST, G. R., OKOCHI, H. & HAYASHI, N. 1996. A novel transforming growth factor beta response element controls the expression of the connective tissue growth factor gene. *Cell Growth Differ,* 7, 469–80.

HALLER, H., LINDSCHAU, C., QUASS, P., DISTLER, A. & LUFT, F. C. 1995. Differentiation of vascular smooth muscle cells and the regulation of protein kinase C-alpha. *Circ Res,* 76, 21–9.

HALPNER, A. D., HANDELMAN, G. J., HARRIS, J. M., BELMONT, C. A. & BLUMBERG, J. B. 1998. Protection by vitamin C of loss of vitamin E in cultured rat hepatocytes. *Arch Biochem Biophys,* 359, 305–9.

HEDIGER, M. A. 2002. New view at C. *Nat Med,* 8, 445–6.

HELLER, R., HECKER, M., STAHMANN, N., THIELE, J. J., WERNER-FELMAYER, G. & WERNER, E. R. 2004a. Alpha-tocopherol amplifies phosphorylation of endothelial nitric oxide synthase at serine 1177 and its short-chain derivative trolox stabilizes tetrahydrobiopterin. *Free Radic Biol Med,* 37, 620–31.

HELLER, R., WERNER-FELMAYER, G. & WERNER, E. R. 2004b. Alpha-tocopherol and endothelial nitric oxide synthesis. *Ann N Y Acad Sci,* 1031, 74–85.

HELLMAN, L. & BURNS, J. J. 1958. Metabolism of L-ascorbic acid-1-C14 in man. *J Biol Chem,* 230, 923–30.

HENTATI, A., DENG, H. X., HUNG, W. Y., NAYER, M., AHMED, M. S., HE, X., TIM, R., STUMPF, D. A., SIDDIQUE, T. & AHMED 1996. Human alpha-tocopherol transfer protein: Gene structure and mutations in familial vitamin E deficiency. *Ann Neurol,* 39, 295–300.

HORSKA, A., MISLANOVA, C., BONASSI, S., CEPPI, M., VOLKOVOVA, K. & DUSINSKA, M. 2011. Vitamin C levels in blood are influenced by polymorphisms in glutathione S-transferases. *Eur J Nutr,* 50, 437–46.

HUEBBE, P., LODGE, J. K. & RIMBACH, G. 2010. Implications of apolipoprotein E genotype on inflammation and vitamin E status. *Mol Nutr Food Res,* 54, 623–30.

HULSE, J. D., ELLIS, S. R. & HENDERSON, L. M. 1978. Carnitine biosynthesis. Betahydroxylation of trimethyllysine by an alpha-ketoglutarate-dependent mitochondrial dioxygenase. *J Biol Chem,* 253, 1654–9.

IQBAL, J. & HUSSAIN, M. M. 2009. Intestinal lipid absorption. *Am J Physiol Endocrinol Metab,* 296, E1183–94.

ITO, S., HYODO, T., HASEGAWA, H., YUAN, H., HAMAGUCHI, M. & SENGA, T. 2010. PI3K/Akt signaling is involved in the disruption of gap junctional communication caused by v-Src and TNF-alpha. *Biochem Biophys Res Commun,* 400, 230–5.

IVANOV, V. O., IVANOVA, S. V. & NIEDZWIECKI, A. 1997. Ascorbate affects proliferation of guinea-pig vascular smooth muscle cells by direct and extracellular matrix-mediated effects. *J Mol Cell Cardiol,* 29, 3293–303.

IWAMA, M., HONDA, A., OHOHASHI, Y., SAKAI, T. & MORI, Y. 1985. Alterations in glycosaminoglycans of the aorta of vitamin E-deficient rats. *Atherosclerosis,* 55, 115–23.

JIALAL, I. & DEVARAJ, S. 2003. Antioxidants and atherosclerosis: Don't throw out the baby with the bath water. *Circulation,* 107, 926–8.

JIALAL, I. & FULLER, C. J. 1995. Effect of vitamin E, vitamin C and beta-carotene on LDL oxidation and atherosclerosis. *Can J Cardiol,* 11 Suppl G, 97G–103G.

JIALAL, I., VEGA, G. L. & GRUNDY, S. M. 1990. Physiologic levels of ascorbate inhibit the oxidative modification of low density lipoprotein. *Atherosclerosis,* 82, 185–91.

JIANG, Q. & AMES, B. N. 2003. Gamma-tocopherol, but not alpha-tocopherol, decreases proinflammatory eicosanoids and inflammation damage in rats. *FASEB J,* 17, 816–22.

JIANG, Q., CHRISTEN, S., SHIGENAGA, M. K. & AMES, B. N. 2001. Gamma-tocopherol, the major form of vitamin E in the US diet, deserves more attention. *Am J Clin Nutr,* 74, 714–22.

JIANG, Q., ELSON-SCHWAB, I., COURTEMANCHE, C. & AMES, B. N. 2000. Gamma-tocopherol and its major metabolite, in contrast to alpha-tocopherol, inhibit cyclooxygenase activity in macrophages and epithelial cells. *Proc Natl Acad Sci U S A,* 97, 11494–99.

JIANG, X. C., TALL, A. R., QIN, S., LIN, M., SCHNEIDER, M., LALANNE, F., DECKERT, V., DESRUMAUX, C., ATHIAS, A., WITZTUM, J. L. & LAGROST, L. 2002. Phospholipid transfer protein deficiency protects circulating lipoproteins from oxidation due to the enhanced accumulation of vitamin E. *J Biol Chem,* 277, 31850–6.

JOHNYKUTTY, S., TANG, P., ZHAO, H., HICKS, D. G., YEH, S. & WANG, X. 2009. Dual expression of alpha-tocopherol-associated protein and estrogen receptor in normal/benign human breast luminal cells and the downregulation of alpha-tocopherol-associated protein in estrogen-receptor-positive breast carcinomas. *Mod Pathol,* 22, 770–5.

KAGAN, V. E., SERBINOVA, E. A., FORTE, T., SCITA, G. & PACKER, L. 1992. Recycling of vitamin E in human low density lipoproteins. *J Lipid Res,* 33, 385–97.

KASHIWADA, K., NISHIDA, W., HAYASHI, K., OZAWA, K., YAMANAKA, Y., SAGA, H., YAMASHITA, T., TOHYAMA, M., SHIMADA, S., SATO, K. & SOBUE, K. 1997. Coordinate expression of alpha-tropomyosin and caldesmon isoforms in association with phenotypic modulation of smooth muscle cells. *J Biol Chem,* 272, 15396–404.

KASIMANICKAM, R. K., KASIMANICKAM, V. R., HALDORSON, G. J. & TIBARY, A. 2012. Effect of tocopherol supplementation during last trimester of pregnancy on mRNA abundances of interleukins and angiogenesis in ovine placenta and uterus. *Reprod Biol Endocrinol,* 10, 4.

KASIMANICKAM, R. K., KASIMANICKAM, V. R., RODRIGUEZ, J. S., PELZER, K. D., SPONENEBERG, P. D. & THATCHER, C. D. 2010. Tocopherol induced angiogenesis in placental vascular network in late pregnant ewes. *Reprod Biol Endocrinol,* 8, 86.

KAZI, A. S., LOTFI, S., GONCHAROVA, E. A., TLIBA, O., AMRANI, Y., KRYMSKAYA, V. P. & LAZAAR, A. L. 2004. Vascular endothelial growth factor-induced secretion of fibronectin is ERK dependent. *Am J Physiol Lung Cell Mol Physiol,* 286, L539–45.

KAZI, M., LUNDMARK, K., RELIGA, P., GOUDA, I., LARM, O., RAY, A., SWEDENBORG, J. & HEDIN, U. 2002. Inhibition of rat smooth muscle cell adhesion and proliferation by non-anticoagulant heparins. *J Cell Physiol,* 193, 365–72.

KEMPNA, P., REITER, E., AROCK, M., AZZI, A. & ZINGG, J. M. 2004. Inhibition of HMC-1 mast cell proliferation by vitamin E: Involvement of the protein kinase B pathway. *J Biol Chem,* 279, 50700–9.

KEMPNA, P., RICCIARELLI, R., AZZI, A. & ZINGG, J. M. 2010. Alternative splicing and gene polymorphism of the human TAP3/SEC14L4 gene. *Mol Biol Rep,* 37, 3503–8.

KEMPNA, P., ZINGG, J. M., RICCIARELLI, R., HIERL, M., SAXENA, S. & AZZI, A. 2003. Cloning of novel human SEC14p-like proteins: Cellular localization, ligand binding and functional properties. *Free Radic Biol Med,* 34, 1458–1472.

KHANNA, S., ROY, S., RYU, H., BAHADDURI, P., SWAAN, P. W., RATAN, R. R. & SEN, C. K. 2003. Molecular basis of vitamin E action. Tocotrienol modulates 12-lipoxygenase, a key mediator of glutamate-induced neurodegeneration. *J Biol Chem,* 278, 43508–15.

KINLAY, S., BEHRENDT, D., FANG, J. C., DELAGRANGE, D., MORROW, J., WITZTUM, J. L., RIFAI, N., SELWYN, A. P., CREAGER, M. A. & GANZ, P. 2004. Long-term effect of combined vitamins E and C on coronary and peripheral endothelial function. *J Am Coll Cardiol,* 43, 629–34.

KIPP, D., WILSON, S., GOSIEWSKA, A. & PETERKOFSKY, B. 1995. Differential regulation of collagen gene expression in granulation tissue and non-repair connective tissues in vitamin C-deficient guinea pigs. *Wound Repair Regen,* 3, 192–203.

KOCHER, O., GABBIANI, F., GABBIANI, G., REIDY, M. A., COKAY, M. S., PETERS, H. & HUTTNER, I. 1991. Phenotypic features of smooth muscle cells during the evolution of experimental carotid artery intimal thickening. Biochemical and morphologic studies. *Lab Invest,* 65, 459–70.

KOLLECK, I., SCHLAME, M., FECHNER, H., LOOMAN, A. C., WISSEL, H. & RUSTOW, B. 1999. HDL is the major source of vitamin E for type II pneumocytes. *Free Radic Biol Med,* 27, 882–90.

KOSTNER, G. M., OETTL, K., JAUHIAINEN, M., EHNHOLM, C., ESTERBAUER, H. & DIEPLINGER, H. 1995. Human plasma phospholipid transfer protein accelerates exchange/transfer of alpha-tocopherol between lipoproteins and cells. *Biochem J,* 305, 659–67.

KOYA, D. & KING, G. L. 1998. Protein kinase C activation and the development of diabetic complications. *Diabetes,* 47, 859–66.

KOYA, D., LEE, I. K., ISHII, H., KANOH, H. & KING, G. L. 1997. Prevention of glomerular dysfunction in diabetic rats by treatment with d-alpha-tocopherol. *J Am Soc Nephrol,* 8, 426–35.

KOYAMA, H., RAINES, E. W., BORNFELDT, K. E., ROBERTS, J. M. & ROSS, R. 1996. Fibrillar collagen inhibits arterial smooth muscle proliferation through regulation of Cdk2 inhibitors. *Cell,* 87, 1069–78.

KUNSCH, C. & MEDFORD, R. M. 1999. Oxidative stress as a regulator of gene expression in the vasculature. *Circ Res,* 85, 753–66.

KURATA, S., SENOO, H. & HATA, R. 1993. Transcriptional activation of type I collagen genes by ascorbic acid 2-phosphate in human skin fibroblasts and its failure in cells from a patient with alpha 2(I)-chain-defective Ehlers-Danlos syndrome. *Exp Cell Res,* 206, 63–71.

LANDES, N., PFLUGER, P., KLUTH, D., BIRRINGER, M., RUHL, R., BOL, G. F., GLATT, H. & BRIGELIUS-FLOHE, R. 2003. Vitamin E activates gene expression via the pregnane X receptor. *Biochem Pharmacol,* 65, 269–73.

LECOMPTE, S., SZABO DE EDELENYI, F., GOUMIDI, L., MAIANI, G., MOSCHONIS, G., WIDHALM, K., MOLNAR, D., KAFATOS, A., SPINNEKER, A., BREIDENASSEL, C., DALLONGEVILLE, J., MEIRHAEGHE, A. & BOREL, P. 2011. Polymorphisms in the CD36/FAT gene are associated with plasma vitamin E concentrations in humans. *Am J Clin Nutr,* 93, 644–51.

LEHR, H. A., FREI, B. & ARFORS, K. E. 1994. Vitamin C prevents cigarette smoke-induced leukocyte aggregation and adhesion to endothelium in vivo. *Proc Natl Acad Sci U S A,* 91, 7688–92.

LEHR, H. A., FREI, B., OLOFSSON, A. M., CAREW, T. E. & ARFORS, K. E. 1995. Protection from oxidized LDL-induced leukocyte adhesion to microvascular and macrovascular endothelium in vivo by vitamin C but not by vitamin E. *Circulation,* 91, 1525–32.

LEHR, H. A., WEYRICH, A. S., SAETZLER, R. K., JUREK, A., ARFORS, K. E., ZIMMERMAN, G. A., PRESCOTT, S. M. & MCINTYRE, T. M. 1997. Vitamin C blocks inflammatory platelet-activating factor mimetics created by cigarette smoking. *J Clin Invest,* 99, 2358–64.

LEMAIRE-EWING, S., DESRUMAUX, C., NEEL, D. & LAGROST, L. 2010. Vitamin E transport, membrane incorporation and cell metabolism: Is alpha-tocopherol in lipid rafts an oar in the lifeboat? *Mol Nutr Food Res,* 54, 631–40.

LENER, T., BURGSTALLER, G., CRIMALDI, L., LACH, S. & GIMONA, M. 2006. Matrix-degrading podosomes in smooth muscle cells. *Eur J Cell Biol,* 85, 183–9.

LEVINE, M. 1986. New concepts in the biology and biochemistry of ascorbic acid. *N Engl J Med,* 314, 892–902.

LI, L., MAMPUTU, J. C., WIERNSPERGER, N. & RENIER, G. 2005. Signaling pathways involved in human vascular smooth muscle cell proliferation and matrix metallopro-teinase-2 expression induced by leptin: inhibitory effect of metformin. *Diabetes,* 54, 2227–34.

LI, W., HELLSTEN, A., JACOBSSON, L. S., BLOMQVIST, H. M., OLSSON, A. G. & YUAN, X. M. 2004. Alpha-tocopherol and astaxanthin decrease macrophage infiltra-tion, apoptosis and vulnerability in atheroma of hyperlipidaemic rabbits. *J Mol Cell Cardiol,* 37, 969–78.

LIU, L. & MEYDANI, M. 2002. Combined vitamin C and E supplementation retards early progression of arteriosclerosis in heart transplant patients. *Nutr Rev,* 60, 368–71.

LOVE-GREGORY, L., SHERVA, R., SCHAPPE, T., QI, J. S., MCCREA, J., KLEIN, S., CONNELLY, M. A. & ABUMRAD, N. A. 2011. Common CD36 SNPs reduce protein expression and may contribute to a protective atherogenic profile. *Hum Mol Genet,* 20, 193–201.

LOVE-GREGORY, L., SHERVA, R., SUN, L., WASSON, J., SCHAPPE, T., DORIA, A., RAO, D. C., HUNT, S. C., KLEIN, S., NEUMAN, R. J., PERMUTT, M. A. & ABUMRAD, N. A. 2008. Variants in the CD36 gene associate with the metabolic syndrome and high-density lipoprotein cholesterol. *Hum Mol Genet,* 17, 1695–704.

LUNA, J., MASAMUNT, M. C., LLACH, J., DELGADO, S. & SANS, M. 2011a. Palm oil tocotrienol rich fraction reduces extracellular matrix production by inhibiting trans-forming growth factor-beta1 in human intestinal fibroblasts. *Clin Nutr,* 30, 858–64.

LUNA, J., MASAMUNT, M. C., RICKMANN, M., MORA, R., ESPANA, C., DELGADO, S., LLACH, J., VAQUERO, E. & SANS, M. 2011b. Tocotrienols have potent antifibro-genic effects in human intestinal fibroblasts. *Inflamm Bowel Dis,* 17, 732–41.

LYNCH, S. M., GAZIANO, J. M. & FREI, B. 1996. Ascorbic acid and atherosclerotic cardio-vascular disease. *Subcell Biochem,* 25, 331–67.

MA, X., BACCI, S., MLYNARSKI, W., GOTTARDO, L., SOCCIO, T., MENZAGHI, C., IORI, E., LAGER, R. A., SHROFF, A. R., GERVINO, E. V., NESTO, R. W., JOHNSTONE, M. T., ABUMRAD, N. A., AVOGARO, A., TRISCHITTA, V. & DORIA, A. 2004. A common haplotype at the CD36 locus is associated with high free fatty acid levels and increased cardiovascular risk in Caucasians. *Hum Mol Genet,* 13, 2197–205.

MAEDA, N., HAGIHARA, H., NAKATA, Y., HILLER, S., WILDER, J. & REDDICK, R. 2000. Aortic wall damage in mice unable to synthesize ascorbic acid. *Proc Natl Acad Sci U S A,* 97, 841–6.

MARCHIOLI, R. 1999. Antioxidant vitamins and prevention of cardiovascular disease: Laboratory, epidemiological and clinical trial data [In Process Citation]. *Pharmacol Res,* 40, 227–38.

MARDONES, P. & RIGOTTI, A. 2004. Cellular mechanisms of vitamin E uptake: relevance in alpha-tocopherol metabolism and potential implications for disease. *J Nutr Biochem,* 15, 252–60.

MARDONES, P., STROBEL, P., MIRANDA, S., LEIGHTON, F., QUINONES, V., AMIGO, L., ROZOWSKI, J., KRIEGER, M. & RIGOTTI, A. 2002a. Alpha-tocopherol metabolism is abnormal in scavenger receptor class B type I (SR-BI)-deficient mice. *J Nutr,* 132, 443–9.

MARTIN, A. & FREI, B. 1997. Both intracellular and extracellular vitamin C inhibit athero-genic modification of LDL by human vascular endothelial cells. *Arterioscler Thromb Vasc Biol,* 17, 1583–90.

MAUCH, S., KOLB, C., KOLB, B., SADOWSKI, T. & SEDLACEK, R. 2002. Matrix metal-loproteinase-19 is expressed in myeloid cells in an adhesion-dependent manner and associates with the cell surface. *J Immunol,* 168, 1244–51.

MAY, J. M. 1999. Is ascorbic acid an antioxidant for the plasma membrane? *FASEB J,* 13, 995–1006.

MAY, J. M., COBB, C. E., MENDIRATTA, S., HILL, K. E. & BURK, R. F. 1998. Reduction of the ascorbyl free radical to ascorbate by thioredoxin reductase. *J Biol Chem*, 273, 23039–45.

MAY, J. M., QU, Z. C. & COBB, C. E. 2004. Human erythrocyte recycling of ascorbic acid: Relative contributions from the ascorbate free radical and dehydroascorbic acid. *J Biol Chem*, 279, 14975–82.

MCCARY, C. A., YOON, Y., PANAGABKO, C., CHO, W., ATKINSON, J. & COOK-MILLS, J. M. 2012. Vitamin E isoforms directly bind PKCalpha and differentially regulate activation of PKCalpha. *Biochem J*, 441, 189–98.

MILMAN, U., BLUM, S., SHAPIRA, C., ARONSON, D., MILLER-LOTAN, R., ANBINDER, Y., ALSHIEK, J., BENNETT, L., KOSTENKO, M., LANDAU, M., KEIDAR, S., LEVY, Y., KHEMLIN, A., RADAN, A. & LEVY, A. P. 2008. Vitamin E supplementation reduces cardiovascular events in a subgroup of middle-aged individuals with both type 2 diabetes mellitus and the haptoglobin 2-2 genotype: A prospective double-blinded clinical trial. *Arterioscler Thromb Vasc Biol*, 28, 341–7.

MORI, Y., HATAMOCHI, A., TAKEDA, K. & UEKI, H. 1994. Effects of tretinoin tocoferil on gene expression of the extracellular matrix components in human dermal fibroblasts in vitro. *J Dermatol Sci*, 8, 233–8.

MORLEY, S., THAKUR, V., DANIELPOUR, D., PARKER, R., ARAI, H., ATKINSON, J., BARNHOLTZ-SLOAN, J., KLEIN, E. & MANOR, D. 2010. Tocopherol transfer protein sensitizes prostate cancer cells to vitamin E. *J Biol Chem*, 285, 35578–89.

MULVIHILL, E. R., JAEGER, J., SENGUPTA, R., RUZZO, W. L., REIMER, C., LUKITO, S. & SCHWARTZ, S. M. 2004. Atherosclerotic plaque smooth muscle cells have a distinct phenotype. *Arterioscler Thromb Vasc Biol*, 24, 1283–9.

MUNTEANU, A., TADDEI, M., TAMBURINI, I., BERGAMINI, E., AZZI, A. & ZINGG, J. M. 2006. Antagonistic effects of oxidized low density lipoprotein and alpha-tocopherol on CD36 scavenger receptor expression in monocytes: Involvement of protein kinase B and peroxisome proliferator-activated receptor-gamma. *J Biol Chem*, 281, 6489–97.

MUNTEANU, A. & ZINGG, J. M. 2007. Cellular, molecular and clinical aspects of vitamin E on atherosclerosis prevention. *Mol Aspects Med*, 28, 538–590.

MUNTEANU, A., ZINGG, J. M. & AZZI, A. 2004. Anti-atherosclerotic effects of vitamin E: Myth or reality? *J Cell Mol Med*, 8, 59–76.

MUSTACICH, D. J., VO, A. T., ELIAS, V. D., PAYNE, K., SULLIVAN, L., LEONARD, S. W. & TRABER, M. G. 2007. Regulatory mechanisms to control tissue alpha-tocopherol. *Free Radic Biol Med*, 43, 610–8.

NA, N., DELANGHE, J. R., TAES, Y. E., TORCK, M., BAEYENS, W. R. & OUYANG, J. 2006. Serum vitamin C concentration is influenced by haptoglobin polymorphism and iron status in Chinese. *Clin Chim Acta*, 365, 319–24.

NAIDU, K. A. 2003. Vitamin C in human health and disease is still a mystery? An overview. *Nutr J*, 2, 7.

NAKATA, Y. & MAEDA, N. 2002. Vulnerable atherosclerotic plaque morphology in apolipoprotein E-deficient mice unable to make ascorbic acid. *Circulation*, 105, 1485–90.

NAPOLI, C., WILLIAMS-IGNARRO, S., DE NIGRIS, F., LERMAN, L. O., ROSSI, L., GUARINO, C., MANSUETO, G., DI TUORO, F., PIGNALOSA, O., DE ROSA, G., SICA, V. & IGNARRO, L. J. 2004. Long-term combined beneficial effects of physical training and metabolic treatment on atherosclerosis in hypercholesterolemic mice. *Proc Natl Acad Sci U S A*, 101, 8797–802.

NARUSHIMA, K., TAKADA, T., YAMANASHI, Y. & SUZUKI, H. 2008. Niemann-Pick C1-like 1 mediates alpha-tocopherol transport. *Mol Pharmacol*, 74, 42–9.

NESPEREIRA, B., PEREZ-ILZARBE, M., FERNANDEZ, P., FUENTES, A. M., PARAMO, J. A. & RODRIGUEZ, J. A. 2003. Vitamins C and E downregulate vascular VEGF and VEGFR-2 expression in apolipoprotein-E-deficient mice. *Atherosclerosis*, 171, 67–73.

NESS, A., EGGER, M. & SMITH, G. D. 1999. Role of antioxidant vitamins in prevention of cardiovascular diseases. Meta-analysis seems to exclude benefit of vitamin C supplementation. *BMJ,* 319, 577.

NEUVILLE, P., GEINOZ, A., BENZONANA, G., REDARD, M., GABBIANI, F., ROPRAZ, P. & GABBIANI, G. 1997. Cellular retinol-binding protein-1 is expressed by distinct subsets of rat arterial smooth muscle cells in vitro and in vivo. *Am J Pathol,* 150, 509–21.

NEUZIL, J., DONG, L. F., WANG, X. F. & ZINGG, J. M. 2006. Tocopherol-associated protein-1 accelerates apoptosis induced by alpha-tocopheryl succinate in mesothelioma cells. *Biochem Biophys Res Commun,* 343, 1113–7.

NI, J., PANG, S. T. & YEH, S. 2007. Differential retention of alpha-vitamin E is correlated with its transporter gene expression and growth inhibition efficacy in prostate cancer cells. *Prostate,* 67, 463–71.

NI, J., WEN, X., YAO, J., CHANG, H. C., YIN, Y., ZHANG, M., XIE, S., CHEN, M., SIMONS, B., CHANG, P., DI SANT'AGNESE, A., MESSING, E. M. & YEH, S. 2005. Tocopherol-associated protein suppresses prostate cancer cell growth by inhibition of the phosphoinositide 3-kinase pathway. *Cancer Res,* 65, 9807–16.

NIER, B., WEINBERG, P. D., RIMBACH, G., STOCKLIN, E. & BARELLA, L. 2006. Differential gene expression in skeletal muscle of rats with vitamin E deficiency. *IUBMB Life,* 58, 540–8.

NOEL, S. E., LAI, C. Q., MATTEI, J., PARNELL, L. D., ORDOVAS, J. M. & TUCKER, K. L. 2010. Variants of the CD36 gene and metabolic syndrome in Boston Puerto Rican adults. *Atherosclerosis,* 211, 210–5.

NUMAKAWA, Y., NUMAKAWA, T., MATSUMOTO, T., YAGASAKI, Y., KUMAMARU, E., KUNUGI, H., TAGUCHI, T. & NIKI, E. 2006. Vitamin E protected cultured cortical neurons from oxidative stress-induced cell death through the activation of mitogen-activated protein kinase and phosphatidylinositol 3-kinase. *J Neurochem,* 97, 1191–202.

OEMAR, B. S. & LUSCHER, T. F. 1997. Connective tissue growth factor. Friend or foe? *Arterioscler Thromb Vasc Biol,* 17, 1483–9.

OEMAR, B. S., WERNER, A., GARNIER, J. M., DO, D. D., GODOY, N., NAUCK, M., MARZ, W., RUPP, J., PECH, M. & LUSCHER, T. F. 1997. Human connective tissue growth factor is expressed in advanced atherosclerotic lesions. *Circulation,* 95, 831–9.

OGIER, N., KLEIN, A., DECKERT, V., ATHIAS, A., BESSEDE, G., LE GUERN, N., LAGROST, L. & DESRUMAUX, C. 2007. Cholesterol accumulation is increased in macrophages of phospholipid transfer protein-deficient mice: Normalization by dietary alpha-tocopherol supplementation. *Arterioscler Thromb Vasc Biol,* 27, 2407–12.

OOMMEN, S., VASU, V. T., LEONARD, S. W., TRABER, M. G., CROSS, C. E. & GOHIL, K. 2007. Genome wide responses of murine lungs to dietary alpha-tocopherol. *Free Radic Res,* 41, 98–109.

ORAM, J. F., VAUGHAN, A. M. & STOCKER, R. 2001. ATP-binding cassette transporter A1 mediates cellular secretion of alpha-tocopherol. *J Biol Chem,* 276, 39898–902.

ORBE, J., RODRIGUEZ, J. A., ARIAS, R., BELZUNCE, M., NESPEREIRA, B., PEREZ-ILZARBE, M., RONCAL, C. & PARAMO, J. A. 2003. Antioxidant vitamins increase the collagen content and reduce MMP-1 in a porcine model of atherosclerosis: Implications for plaque stabilization. *Atherosclerosis,* 167, 45–53.

ORLANDI, A., EHRLICH, H. P., ROPRAZ, P., SPAGNOLI, L. G. & GABBIANI, G. 1994. Rat aortic smooth muscle cells isolated from different layers and at different times after endothelial denudation show distinct biological features in vitro. *Arterioscler Thromb,* 14, 982–9.

OZER, N. K., BOSCOBOINIK, D. & AZZI, A. 1995. New roles of low density lipoproteins and vitamin E in the pathogenesis of atherosclerosis. *Biochem Mol Biol Int,* 35, 117–24.

PADAYATTY, S. J., KATZ, A., WANG, Y., ECK, P., KWON, O., LEE, J. H., CHEN, S., CORPE, C., DUTTA, A., DUTTA, S. K. & LEVINE, M. 2003. Vitamin C as an antioxidant: Evaluation of its role in disease prevention. *J Am Coll Nutr,* 22, 18–35.

PADH, H. 1991. Vitamin C: Newer insights into its biochemical functions. *Nutr Rev,* 49, 65–70.

PARK, J. Y., HA, S. W. & KING, G. L. 1999. The role of protein kinase C activation in the pathogenesis of diabetic vascular complications. *Perit Dial Int,* 19 Suppl 2, S222–7.

PATEL, J., MATNOR, N. A., IYER, A. & BROWN, L. 2011. A regenerative antioxidant protocol of vitamin E and alpha-lipoic acid ameliorates cardiovascular and metabolic changes in fructose-red rats. *Evid Based Complement Alternat Med,* 2011, 120801.

PETERSZEGI, G., DAGONET, F. B., LABAT-ROBERT, J. & ROBERT, L. 2002. Inhibition of cell proliferation and fibronectin biosynthesis by Na ascorbate. *Eur J Clin Invest,* 32, 372–80.

PINNEL, S. R., MURAD, S. & DARR, D. 1987. Induction of collagen synthesis by ascorbic acid. A possible mechanism. *Arch Dermatol,* 123, 1684–6.

PORTER, T. D. 2003. Supernatant protein factor and tocopherol-associated protein: An unexpected link between cholesterol synthesis and vitamin E (review). *J Nutr Biochem,* 14, 3–6.

QUAGLINO, D., FORNIERI, C., BOTTI, B., DAVIDSON, J. M. & PASQUALI-RONCHETTI, I. 1991. Opposing effects of ascorbate on collagen and elastin deposition in the neonatal rat aorta. *Eur J Cell Biol,* 54, 18–26.

RAINES, E. W. 2000. The extracellular matrix can regulate vascular cell migration, proliferation, and survival: Relationships to vascular disease. *Int J Exp Pathol,* 81, 173–82.

RASK-MADSEN, C. & KING, G. L. 2005. Proatherosclerotic mechanisms involving protein kinase C in diabetes and insulin resistance. *Arterioscler Thromb Vasc Biol,* 25, 487–96.

REBOUL, E., KLEIN, A., BIETRIX, F., GLEIZE, B., MALEZET-DESMOULINS, C., SCHNEIDER, M., MARGOTAT, A., LAGROST, L., COLLET, X. & BOREL, P. 2006. Scavenger receptor class B type I (SR-BI) is involved in vitamin E transport across the enterocyte. *J Biol Chem,* 281, 4739–4745.

REDDANNA, P., RAO, M. K. & REDDY, C. C. 1985. Inhibition of 5-lipoxygenase by vitamin E. *FEBS Lett,* 193, 39–43.

REN, G., CRAMPTON, M. S. & YAP, A. S. 2009. Cortactin: Coordinating adhesion and the actin cytoskeleton at cellular protrusions. *Cell Motil Cytoskeleton,* 66, 865–73.

RICCIARELLI, R., MARONI, P., OZER, N., ZINGG, J. M. & AZZI, A. 1999. Age-dependent increase of collagenase expression can be reduced by alpha-tocopherol via protein kinase C inhibition. *Free Radic Biol Med,* 27, 729–37.

RICCIARELLI, R., TASINATO, A., CLEMENT, S., OZER, N. K., BOSCOBOINIK, D. & AZZI, A. 1998. Alpha-tocopherol specifically inactivates cellular protein kinase C alpha by changing its phosphorylation state. *Biochem J,* 334, 243–9.

RICCIARELLI, R., ZINGG, J. M. & AZZI, A. 2000. Vitamin E reduces the uptake of oxidized LDL by inhibiting CD36 scavenger receptor expression in cultured aortic smooth muscle cells. *Circulation,* 102, 82–7.

RICCIARELLI, R., ZINGG, J. M. & AZZI, A. 2001. Vitamin E: Protective role of a Janus molecule. *FASEB J,* 15, 2314–25.

RIGOTTI, A. 2007. Absorption, transport, and tissue delivery of vitamin E. *Mol Aspects Med,* 28, 423–36.

ROSENFELD, L. 1997. Vitamine—vitamin. The early years of discovery. *Clin Chem,* 43, 680–5.

RUHRBERG, C. 2003. Growing and shaping the vascular tree: Multiple roles for VEGF. *Bioessays,* 25, 1052–60.

SAITOH, Y., YUMOTO, A. & MIWA, N. 2009. Alpha-tocopheryl phosphate suppresses tumor invasion concurrently with dynamic morphological changes and delocalization of cortactin from invadopodia. *Int J Oncol,* 35, 1277–88.

SALONEN, J. T., NYYSSONEN, K., SALONEN, R., LAKKA, H. M., KAIKKONEN, J., PORKKALA-SARATAHO, E., VOUTILAINEN, S., LAKKA, T. A., RISSANEN, T., LESKINEN, L., TUOMAINEN, T. P., VALKONEN, V. P., RISTONMAA, U. & POULSEN, H. E. 2000. Antioxidant Supplementation in Atherosclerosis Prevention (ASAP) study: A randomized trial of the effect of vitamins E and C on 3-year progression of carotid atherosclerosis. *J Intern Med,* 248, 377–86.

SAMANDARI, E., VISARIUS, T., ZINGG, J. M. & AZZI, A. 2006. The effect of gamma-tocopherol on proliferation, integrin expression, adhesion, and migration of human glioma cells. *Biochem Biophys Res Commun,* 342, 1329–33.

SANYAL, A. J., CHALASANI, N., KOWDLEY, K. V., MCCULLOUGH, A., DIEHL, A. M., BASS, N. M., NEUSCHWANDER-TETRI, B. A., LAVINE, J. E., TONASCIA, J., UNALP, A., VAN NATTA, M., CLARK, J., BRUNT, E. M., KLEINER, D. E., HOOFNAGLE, J. H. & ROBUCK, P. R. 2010. Pioglitazone, vitamin E, or placebo for nonalcoholic steatohepatitis. *N Engl J Med,* 362, 1675–85.

SATO, K., NIKI, E. & SHIMASAKI, H. 1990. Free radical-mediated chain oxidation of low density lipoprotein and its synergistic inhibition by vitamin E and vitamin C. *Arch Biochem Biophys,* 279, 402–5.

SATO, Y., ARAI, H., MIYATA, A., TOKITA, S., YAMAMOTO, K., TANABE, T. & INOUE, K. 1993. Primary structure of alpha-tocopherol transfer protein from rat liver. Homology with cellular retinaldehyde-binding protein. *J Biol Chem,* 268, 17705–10.

SATO, Y., HAGIWARA, K., ARAI, H. & INOUE, K. 1991. Purification and characterization of the alpha-tocopherol transfer protein from rat liver. *FEBS Lett,* 288, 41–5.

SAYED-AHMED, M. M., KHATTAB, M. M., GAD, M. Z. & MOSTAFA, N. 2001. L-carnitine prevents the progression of atherosclerotic lesions in hypercholesterolaemic rabbits. *Pharmacol Res,* 44, 235–42.

SCHGOER, W., MUELLER, T., JAUHIAINEN, M., WEHINGER, A., GANDER, R., TANCEVSKI, I., SALZMANN, K., ELLER, P., RITSCH, A., HALTMAYER, M., EHNHOLM, C., PATSCH, J. R. & FOEGER, B. 2008. Low phospholipid transfer protein (PLTP) is a risk factor for peripheral atherosclerosis. *Atherosclerosis,* 196, 219–26.

SCHINDLER, R. & MENTLEIN, R. 2006. Flavonoids and vitamin E reduce the release of the angiogenic peptide vascular endothelial growth factor from human tumor cells. *J Nutr,* 136, 1477–82.

SCHNEIDER, M., VERGES, B., KLEIN, A., MILLER, E. R., DECKERT, V., DESRUMAUX, C., MASSON, D., GAMBERT, P., BRUN, J. M., FRUCHART-NAJIB, J., BLACHE, D., WITZTUM, J. L. & LAGROST, L. 2004. Alterations in plasma vitamin E distribution in type 2 diabetic patients with elevated plasma phospholipid transfer protein activity. *Diabetes,* 53, 2633–9.

SEN, C. K. & PACKER, L. 1996. Antioxidant and redox regulation of gene transcription. *FASEB J,* 10, 709–20.

SENO, T., INOUE, N., MATSUI, K., EJIRI, J., HIRATA, K., KAWASHIMA, S. & YOKOYAMA, M. 2004. Functional expression of sodium-dependent vitamin C transporter 2 in human endothelial cells. *J Vasc Res,* 41, 345–51.

SHIBAYAMA, H., HISAMA, M., MATSUDA, S., OHTSUKI, M. & IWAKI, M. 2008. Effect of a novel ascorbic derivative, disodium isostearyl 2-O-L-ascorbyl phosphate on human dermal fibroblasts: Increased collagen synthesis and inhibition of MMP-1. *Biol Pharm Bull,* 31, 563–8.

SHINOMIYA, K., FUKUNAGA, M., KIYOMOTO, H., MIZUSHIGE, K., TSUJI, T., NOMA, T., OHMORI, K., KOHNO, M. & SENDA, S. 2002. A role of oxidative stress-generated eicosanoid in the progression of arteriosclerosis in type 2 diabetes mellitus model rats. *Hypertens Res,* 25, 91–8.

SIOW, R. C., RICHARDS, J. P., PEDLEY, K. C., LEAKE, D. S. & MANN, G. E. 1999a. Vitamin C protects human vascular smooth muscle cells against apoptosis induced by moderately oxidized LDL containing high levels of lipid hydroperoxides. *Arterioscler Thromb Vasc Biol,* 19, 2387–94.

SIOW, R. C., SATO, H., LEAKE, D. S., ISHII, T., BANNAI, S. & MANN, G. E. 1999b. Induction of antioxidant stress proteins in vascular endothelial and smooth muscle cells: Protective action of vitamin C against atherogenic lipoproteins. *Free Radic Res,* 31, 309–18.

SIRIKCI, O., OZER, N. K. & AZZI, A. 1996. Dietary cholesterol-induced changes of protein kinase C and the effect of vitamin E in rabbit aortic smooth muscle cells. *Atherosclerosis,* 126, 253–63.

STAUBLE, B., BOSCOBOINIK, D., TASINATO, A. & AZZI, A. 1994. Modulation of activator protein-1 (AP-1) transcription factor and protein kinase C by hydrogen peroxide and D-alpha-tocopherol in vascular smooth muscle cells. *Eur J Biochem,* 226, 393–402.

STUART, L. M., DENG, J., SILVER, J. M., TAKAHASHI, K., TSENG, A. A., HENNESSY, E. J., EZEKOWITZ, R. A. & MOORE, K. J. 2005. Response to *Staphylococcus aureus* requires CD36-mediated phagocytosis triggered by the COOH-terminal cytoplasmic domain. *J Cell Biol,* 170, 477–85.

SUARNA, C., WU, B. J., CHOY, K., MORI, T., CROFT, K., CYNSHI, O. & STOCKER, R. 2006. Protective effect of vitamin E supplements on experimental atherosclerosis is modest and depends on preexisting vitamin E deficiency. *Free Radic Biol Med,* 41, 722–30.

SUN, Y. & OBERLEY, L. W. 1996. Redox regulation of transcriptional activators. *Free Radic Biol Med,* 21, 335–48.

SUNG, B. H. & WEAVER, A. M. 2011. Regulation of lysosomal secretion by cortactin drives fibronectin deposition and cell motility. *Bioarchitecture,* 1, 257–60.

SUNG, B. H., ZHU, X., KAVERINA, I. & WEAVER, A. M. 2011. Cortactin controls cell motility and lamellipodial dynamics by regulating ECM secretion. *Curr Biol,* 21, 1460–9.

TAKAHASHI, T., LORD, B., SCHULZE, P. C., FRYER, R. M., SARANG, S. S., GULLANS, S. R. & LEE, R. T. 2003. Ascorbic acid enhances differentiation of embryonic stem cells into cardiac myocytes. *Circulation,* 107, 1912–6.

TAKANAGA, H., MACKENZIE, B. & HEDIGER, M. A. 2004. Sodium-dependent ascorbic acid transporter family SLC23. *Pflugers Arch,* 447, 677–82.

TANG, F. Y. & MEYDANI, M. 2001. Green tea catechins and vitamin E inhibit angiogenesis of human microvascular endothelial cells through suppression of IL-8 production. *Nutr Cancer,* 41, 119–25.

TAPPEINER, C., MEYENBERG, A., GOLDBLUM, D., MOJON, D., ZINGG, J. M., NESARETNAM, K., KILCHENMANN, M. & FRUEH, B. E. 2010. Antifibrotic effects of tocotrienols on human Tenon's fibroblasts. *Graefes Arch Clin Exp Ophthalmol,* 248(1), 65–71.

TASINATO, A., BOSCOBOINIK, D., BARTOLI, G. M., MARONI, P. & AZZI, A. 1995. D-alpha-tocopherol inhibition of vascular smooth muscle cell proliferation occurs at physiological concentrations, correlates with protein kinase C inhibition, and is independent of its antioxidant properties. *Proc Natl Acad Sci U S A,* 92, 12190–4.

TERASAWA, Y., LADHA, Z., LEONARD, S. W., MORROW, J. D., NEWLAND, D., SANAN, D., PACKER, L., TRABER, M. G. & FARESE, R. V., JR. 2000. Increased atherosclerosis in hyperlipidemic mice deficient in alpha-tocopherol transfer protein and vitamin E. *Proc Natl Acad Sci U S A,* 97, 13830–4.

TINKER, D. & RUCKER, R. B. 1985. Role of selected nutrients in synthesis, accumulation, and chemical modification of connective tissue proteins. *Physiol Rev,* 65, 607–57.

TRIANTAFILOU, M., GAMPER, F. G., HASTON, R. M., MOURATIS, M. A., MORATH, S., HARTUNG, T. & TRIANTAFILOU, K. 2006. Membrane sorting of toll-like receptor (TLR)-2/6 and TLR2/1 heterodimers at the cell surface determines heterotypic associations with CD36 and intracellular targeting. *J Biol Chem,* 281, 31002–11.

ULRICH-MERZENICH, G., METZNER, C., SCHIERMEYER, B. & VETTER, H. 2002. Vitamin C and vitamin E antagonistically modulate human vascular endothelial and smooth muscle cell DNA synthesis and proliferation. *Eur J Nutr,* 41, 27–34.

UPSTON, J. M., KRITHARIDES, L. & STOCKER, R. 2003. The role of vitamin E in athero-sclerosis. *Prog Lipid Res,* 42, 405–22.

VAN GOETHEM, E., GUIET, R., BALOR, S., CHARRIERE, G. M., POINCLOUX, R., LABROUSSE, A., MARIDONNEAU-PARINI, I. & LE CABEC, V. 2011. Macrophage podosomes go 3D. *Eur J Cell Biol,* 90, 224–36.

VASU, V. T., HOBSON, B., GOHIL, K. & CROSS, C. E. 2007. Genome-wide screening of alpha-tocopherol sensitive genes in heart tissue from alpha-tocopherol transfer protein null mice (ATTP(-/-)). *FEBS Lett,* 581, 1572–8.

VILLACORTA, L., AZZI, A. & ZINGG, J. M. 2007. Regulatory role of vitamin E and C on extracel-lular matrix components of the vascular system. *Mol Aspects Med,* 28, 507–37.

VILLACORTA, L., GRACA-SOUZA, A. V., RICCIARELLI, R., ZINGG, J. M. & AZZI, A. 2003. Alpha-tocopherol induces expression of connective tissue growth factor and antagonizes tumor necrosis factor-alpha–mediated downregulation in human smooth muscle cells. *Circ Res,* 92, 104–10.

VOEGELE, A. F., JERKOVIC, L., WELLENZOHN, B., ELLER, P., KRONENBERG, F., LIEDL, K. R. & DIEPLINGER, H. 2002. Characterization of the vitamin E-binding properties of human plasma afamin. *Biochemistry,* 41, 14532–8.

WANG, H. & KEISER, J. A. 1998. Vascular endothelial growth factor upregulates the expres-sion of matrix metalloproteinases in vascular smooth muscle cells: Role of flt-1. *Circ Res,* 83, 832–40.

WANG, X., LEMAIRE, S. A., CHEN, L., SHEN, Y. H., GAN, Y., BARTSCH, H., CARTER, S. A., UTAMA, B., OU, H., COSELLI, J. S. & WANG, X. L. 2006. Increased collagen deposition and elevated expression of connective tissue growth factor in human thoracic aortic dissection. *Circulation,* 114, I200–5.

WASHBURN, M. P. & WELLS, W. W. 1999. The catalytic mechanism of the glutathione-dependent dehydroascorbate reductase activity of thioltransferase (glutaredoxin). *Biochemistry,* 38, 268–74.

WAY, K. J., KATAI, N. & KING, G. L. 2001. Protein kinase C and the development of diabetic vascular complications. *Diabet Med,* 18, 945–59.

WHITBREAD, A. K., MASOUMI, A., TETLOW, N., SCHMUCK, E., COGGAN, M. & BOARD, P. G. 2005. Characterization of the omega class of glutathione transferases. *Methods Enzymol,* 401, 78–99.

WIEGMAN, P. J., BARRY, W. L., MCPHERSON, J. A., MCNAMARA, C. A., GIMPLE, L. W., SANDERS, J. M., BISHOP, G. G., POWERS, E. R., RAGOSTA, M., OWENS, G. K. & SAREMBOCK, I. J. 2000. All-trans-retinoic acid limits restenosis after balloon angioplasty in the focally atherosclerotic rabbit: A favorable effect on vessel remodel-ing. *Arterioscler Thromb Vasc Biol,* 20, 89–95.

WITZTUM, J. L. & STEINBERG, D. 1991. Role of oxidized low density lipoprotein in ath-erogenesis. *J Clin Invest,* 88, 1785–92.

WODZINSKA, J. M. 2005. Transglutaminases as targets for pharmacological inhibition. *Mini Rev Med Chem,* 5, 279–92.

WONG, B. W., RAHMANI, M., LUO, Z., YANAGAWA, B., WONG, D., LUO, H. & MCMANUS, B. M. 2009. Vascular endothelial growth factor increases human cardiac microvascular endothelial cell permeability to low-density lipoproteins. *J Heart Lung Transplant,* 28, 950–7.

WRIGHT, M. E., PETERS, U., GUNTER, M. J., MOORE, S. C., LAWSON, K. A., YEAGER, M., WEINSTEIN, S. J., SNYDER, K., VIRTAMO, J. & ALBANES, D. 2009. Association of variants in two vitamin E transport genes with circulating vitamin E concentrations and prostate cancer risk. *Cancer Res,* 69, 1429–38.

YOU, H., YU, W., SANDERS, B. G. & KLINE, K. 2001. RRR-alpha-tocopheryl succinate induces MDA-MB-435 and MCF-7 human breast cancer cells to undergo differentiation. *Cell Growth Differ,* 12, 471–80.

YUASA-KAWASE, M., MASUDA, D., YAMASHITA, T., KAWASE, R., NAKAOKA, H., INAGAKI, M., NAKATANI, K., TSUBAKIO-YAMAMOTO, K., OHAMA, T., MATSUYAMA, A., NISHIDA, M., ISHIGAMI, M., KAWAMOTO, T., KOMURO, I. & YAMASHITA, S. 2012. Patients with CD36 deficiency are associated with enhanced atherosclerotic cardiovascular diseases. *J Atheroscler Thromb,* 19(3), 263–75.

ZHANG, B., TANAKA, J., YANG, L., SAKANAKA, M., HATA, R., MAEDA, N. & MITSUDA, N. 2004a. Protective effect of vitamin E against focal brain ischemia and neuronal death through induction of target genes of hypoxia-inducible factor-1. *Neuroscience,* 126, 433–40.

ZHANG, M., ALTUWAIJRI, S. & YEH, S. 2004b. RRR-alpha-tocopheryl succinate inhibits human prostate cancer cell invasiveness. *Oncogene,* 23, 3080–8.

ZHU, X. Y., RODRIGUEZ-PORCEL, M., BENTLEY, M. D., CHADE, A. R., SICA, V., NAPOLI, C., CAPLICE, N., RITMAN, E. L., LERMAN, A. & LERMAN, L. O. 2004. Antioxidant intervention attenuates myocardial neovascularization in hypercholesterolemia. *Circulation,* 109, 2109–15.

ZIMMER, S., STOCKER, A., SARBOLOUKI, M. N., SPYCHER, S. E., SASSOON, J. & AZZI, A. 2000. A novel human tocopherol-associated protein: cloning, in vitro expression, and characterization. *J Biol Chem,* 275, 25672–80.

ZINGG, J. M. 2007a. Modulation of signal transduction by vitamin E. *Mol Aspects Med,* 28, 481–506.

ZINGG, J. M. 2007b. Molecular and cellular activities of vitamin E analogues. *Mini Rev Med Chem,* 7, 543–58.

ZINGG, J. M. 2012. Vitamin E and disease risk: Research focus turns on genetic polymorphisms and molecular mechanisms. *Vitam Trace Elem,* 1, e110.

ZINGG, J. M. & AZZI, A. 2004. Non-antioxidant activities of vitamin E. *Cur Med Chem,* 11, 1113–133.

ZINGG, J. M. & AZZI, A. 2007. *Smooth Muscle Cells and Alpha-Tocopherol,* Wallingford, Oxfordshire, UK: CABI International.

ZINGG, J. M. & AZZI, A. 2009. Comment re: Vitamin E transport gene variants and prostate cancer. *Cancer Res,* 69, 6756; author reply 6756.

ZINGG, J. M., AZZI, A. & MEYDANI, M. 2008a. Genetic polymorphisms as determinants for disease-preventive effects of vitamin E. *Nutr Rev,* 66, 406–14.

ZINGG, J. M., KEMPNA, P., PARIS, M., REITER, E., VILLACORTA, L., CIPOLLONE, R., MUNTEANU, A., DE PASCALE, C., MENINI, S., CUEFF, A., AROCK, M., AZZI, A. & RICCIARELLI, R. 2008b. Characterization of three human sec14p-like proteins: Alpha-tocopherol transport activity and expression pattern in tissues. *Biochimie,* 90, 1703–15.

ZINGG, J. M., LIBINAKI, R., LAI, C. Q., MEYDANI, M., GIANELLO, R., OGRU, E. & AZZI, A. 2010. Modulation of gene expression by alpha-tocopherol and alpha-tocopheryl phosphate in THP-1 monocytes. *FRBM,* 49, 1989–2000.

ZINGG, J. M., MEYDANI, M. & AZZI, A. 2012. Alpha-tocopheryl phosphate: An activated form of vitamin E important for angiogenesis and vasculogenesis? *Biofactors,* 38, 24–33.

ZINGG, J. M., RICCIARELLI, R., ANDORNO, E. & AZZI, A. 2002. Novel 5' exon of scavenger receptor CD36 is expressed in cultured human vascular smooth muscle cells and atherosclerotic plaques. *Arterioscler Thromb Vasc Biol,* 22, 412–7.

9 Vitamin K and Vascular Calcification

*Chandrasekar Palaniswamy, Wilbert S. Aronow,
Jagadish Khanagavi, and Arunabh Sekhri*

CONTENTS

9.1 INTRODUCTION

Vitamin K has garnered much attention in recent years due to the flux of newer oral anticoagulant medications that could be substituted for warfarin, a vitamin K antagonist (VKA). Concurrently, research in the last few years has also helped in better understanding of the various other physiological roles of this vital trace element. Danish biochemist Henrik Dam is credited with the discovery of vitamin K. He observed that chickens fed with low-fat and sterol-free diets developed a syndrome of spontaneous subcutaneous and intramuscular hemorrhage.[1] The quest for isolating the ingredient that caused this deficiency syndrome led him to discover that it was fat soluble. Doisy delineated the biochemistry of vitamin K: vitamin K1 (phylloquinone) present in green plants, and vitamin K2 (menaquinones) produced by bacterial action.[2] In 1943, Dam and Doisy shared the Nobel Prize in Physiology or Medicine for the discovery of vitamin K and its chemical nature. Naphthoquinone is the functional group of all compounds with the vitamin K cofactor activity, so that the mechanism of action is similar for all K vitamins. The different lipophilicity of the various side chains leads to differences in intestinal absorption, transport, tissue distribution, and bioavailability. Synthetic derivatives called vitamin K3 are water soluble and bypass the natural utilization in the body.

9.2   VITAMIN K–BINDING PROTEINS

There are many vitamin K–dependent proteins that need γ-carboxylation for their physiologic activity (Table 9.1). Vitamin K exerts its physiological action by acting as a cofactor in the process of gamma carboxylation of its various dependent proteins. This process is illustrated in Figure 9.1. Dietary vitamin K is reduced by the NAD(P)H-dependent quinone reductase(s) to vitamin K quinol. This is utilized by γ-glutamyl carboxylase to modify glutamyl residues to γ-carboxylated glutamyl residues in vitamin K–dependent proteins resulting in their functionally active form. The clotting factors II, VII, IX, and X are γ-carboxylated in the liver to be active. Anticoagulant factors—protein C, protein S, and protein Z—are γ-carboxylated predominantly in the liver and to some extent in extrahepatic tissues. The other proteins are osteocalcin, the matrix Gla- protein (MGP), growth arrest-specific gene 6 (GAS6) protein, and transmembrane Gla proteins. This carboxylation is linked to oxidation of vitamin K to form vitamin K epoxide, which is in turn recycled back to the reduced form by the enzyme vitamin K epoxide reductase (VKOR). Warfarin inhibits VKOR, thereby diminishing available vitamin K stores and inhibiting production of functioning coagulation factors.[3]

9.3   MGP AND VASCULAR CALCIFICATION

Vascular calcification is a process that involves accumulation of calcium deposits in vessel walls, resulting in increased arterial wall stiffness. Extra-osseous calcification occurs in areas of chronic inflammation, as in atherosclerotic lesions, in which oxidized lipids are the inflammatory stimulus.[4] However, nonatherosclerotic calcification can occur in metabolic disorders like end-stage renal disease, diabetes, hyperparathyroidism, and vitamin K deficiency.[5,6,7] This involves a complex interplay of various factors such as serum calcium and phosphate levels, activity of calcification promoters and inhibitors. Vascular calcification is associated with increased morbidity and mortality, predominantly from cardiovascular causes.[8]

MGP, a small secretory protein that was originally described in 1983, is a potent inhibitor of tissue calcification.[9] It is expressed in several tissues including bone, cartilage, vascular smooth muscle cells (VSMCs), kidney, lung, and heart. Rats treated with warfarin developed massive cartilage calcification, notably in the

---

TABLE 9.1
Vitamin K–Dependent Proteins

1. Coagulation factors II, VII, IX, X
2. Protein C, protein S, protein Z
3. Matrix Gla-protein
4. Osteocalcin
5. Growth arrest–specific protein 6
6. Proline-rich Gla-proteins
7. Transforming growth factor-beta–inducible protein
8. Periostin

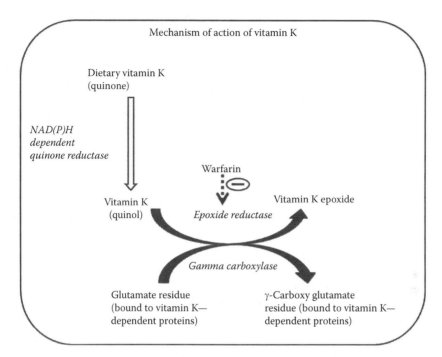

FIGURE 9.1 Dietary vitamin K is reduced by the NAD(P)H-dependent quinone reductase(s) to vitamin K quinol. This is utilized by γ-glutamyl carboxylase to modify glutamyl residues to γ-carboxylated glutamyl residues in vitamin K–dependent proteins, resulting in their functionally active form. This carboxylation is linked to oxidation of vitamin K to form vitamin K epoxide, which is in turn recycled back to the reduced form by the enzyme vitamin K epoxide reductase (VKOR). Warfarin inhibits VKOR, thereby diminishing available vitamin K stores and inhibiting production of functioning coagulation factors.

epiphyses and facial bones, leading to growth retardation and maxillonasal hypoplasia.[10] Subsequently, MGP was identified in cartilage, and loss of MGP function resulting in cartilage calcification was recognized.[11] At that point, it was considered that the role of MGP was limited to prevention of cartilage calcification. An initial study done on mice lacking MGP showed that they developed spontaneous calcification of arteries and cartilage and died later of vascular rupture. This identified MGP as the first recognized inhibitor of vascular calcification in vivo.[12] Spronk et al. found that MGP accumulated at the borders of vascular calcification in human tissue specimens.[13] These investigators suggested that undercarboxylated MGP is biologically inactive, and that poor vascular vitamin K status may be a risk factor for vascular calcification.[7]

MGP maintains the contractile phenotype of VSMCs, inhibiting their differentiation into chondrocyte- and osteoblast-like cells (see Figure 9.2). Bone morphogenetic protein-2 (BMP-2), a member of the transforming growth factor-beta superfamily, is an osteogenic growth factor that induces VSMCs toward the osteoblastic phenotype. Binding of BMP-2/4 to its receptor results in up-regulation of key osteogenic transcription factors: core-binding factor α-1 (Cbfα1), Msx2, and osterix. MGP

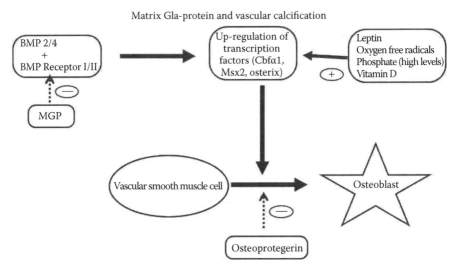

FIGURE 9.2 Vascular smooth muscle cells are induced toward the osteoblastic phenotype by Bone Morphogenetic Proteins (BMP) type 2 and 4. Binding of BMP-2/4 to its receptor results in up-regulation of key osteogenic transcription factors: core binding factor-$\alpha1$ (Cbf$\alpha1$), Msx2, and osterix. Matrix Gla-protein (MGP) prevents interaction of BMP-2 with its receptor, thereby inhibiting calcification. Expression of the osteogenic transcription factor, Cbf$\alpha1$, is also up-regulated by oxidative stress, leptins, vitamin D, and high phosphate levels. Inhibitors of intracellular BMP signaling include osteoprotegerin, which inhibits osteoblastic phenotype in vascular smooth muscle cells, and osteopontin, which inhibits hydroxyapatite formation.

prevents interaction of BMP-2 with its receptor, thereby inhibiting calcification. The interaction of MGP with BMP-2 is dependent on the presence of calcium ions and the gamma carboxylation Gla domain of MGP. Undercarboxylated forms of MGP are ineffective in inhibiting the BMP-2 actions. Figure 9.1 illustrates the currently accepted mechanism of prevention of vascular calcification by MGP, and the role of various promoters and inhibitors of vascular calcification.[3] Sweatt et al. elucidated the interaction of MGP and BMP-2 in the process of vascular calcification.[14] Using immunohistochemistry, these investigators showed that calcified lesions in the aortic wall of aging rats contained elevated concentrations of MGP that was poorly γ-carboxylated and did not bind BMP-2. They demonstrated the existence of a BMP-2/MGP complex in vivo, consistent with a role for MGP as a BMP-2 inhibitor. They postulated that age-related arterial calcification may be a consequence of increased levels of undercarboxylated MGP, allowing unopposed BMP-2 activity. MGP is also known to interact with vitronectin, an extracellular matrix component.[15] This binding is not calcium dependent and involves C-terminal binding and, hence, is independent of gamma carboxylation. This interaction is postulated to influence the effect of MGP on BMP-2. Other inhibitors of BMP include osteoprotegerin and osteopontin.[3] The presence of MGP at the site of elastin fibers in healthy human arteries suggests an interaction of MGP with elastin that might prevent calcification.[16] Elastin is a substrate for initiation of calcium crystal formation and works as a nidus. Apoptosis of VSMCs may also serve as a nidus for arterial calcification. MGP may also prevent apoptosis of VSMCs, thereby protecting from vascular calcification.[17] In MGP-deficient mice, local expression of MGP in VSMCs prevents vascular calcification. However, hepatic expression of MGP resulting in high serum levels does not prevent vascular calcification. This suggests that MGP prevents arterial calcification by acting locally and not systemically.

A number of factors affect expression of MGP in tissues. The promoter region for MGP contains regulatory sequences including possible binding sites for retinoic acid and vitamin D. Retinoic acid and transforming growth factor–beta decrease MGP expression in VSMCs.[18,19] Vitamin D3 at physiological levels increases expression of MGP. In areas of high extracellular calcium, MGP expression is up-regulated in VSMCs, likely through the calcium sensing receptors.[20] This helps to limit pathological calcification. The principal determinant that affects the functional activity of MGP in tissues is gamma carboxylation. This is dependent on the availability of the reduced form of vitamin K recycled from the epoxide form by VKOR. Drugs like warfarin inhibit this enzyme, leading to decreased MGP activity.

In humans, mutations in the gene encoding for MGP cause the Keutel syndrome, a rare disorder characterized by abnormal cartilage calcification, peripheral pulmonary stenosis, and extensive arterial calcification.[21,22] Interestingly, unlike mouse models of MGP knockout, patients with Keutel syndrome survive into adulthood, signifying the role of other factors in humans that inhibit vascular calcification.

9.4   WARFARIN AND VASCULAR CALCIFICATION

Warfarin has been the mainstay of oral anticoagulant therapy for many years, principally for atrial fibrillation, mechanical heart valves, or venous thromboembolism. Adverse-effects profile of patients on warfarin therapy parallel what would be seen in vitamin K deficiency. Due to its inhibitory effect on VKOR in the liver, warfarin affects the function of the MGP and has been associated with vascular calcification in various animal and human studies.

9.4.1   ANIMAL STUDIES

Howe and Webster designed a study to cause extra-hepatic vitamin K deficiency with a warfarin treatment regimen.[23] Rats were treated with daily doses of warfarin and concurrent vitamin K1 from birth for 5–12 weeks. This treatment was presumed to cause extra-hepatic vitamin K deficiency without affecting the vitamin K–dependent coagulation factors. At the end of treatment, the examination of the vascular system of these rats revealed extensive arterial calcification. A study by Price et al. showed that warfarin causes focal calcification of the elastic lamellae in the tunica media of major arteries and in aortic valves in mice.[24] These investigators found that the calcification of arteries induced by warfarin was similar to that seen in MGP-deficient mice and suggested that warfarin induces arterial calcification by inhibiting carboxylation of MGP, thereby inactivating the putative calcification-inhibitory activity of the protein. Later, they also showed that concurrent warfarin administration increased the extent of calcification in the media of vitamin D–treated mice.[25] They hypothesized that inactivation of MGP with warfarin causes living arteries to calcify, and that addition of MGP to medium containing warfarin prevents this calcification. In a subsequent study, they demonstrated that addition of warfarin to culture medium caused extensive Alizarin red staining for calcification in the living carotid artery segment, whereas no staining could be detected in living carotid arteries incubated in the same medium without warfarin.[16] No calcification could be detected if the living arteries were incubated in a culture medium containing warfarin with no serum, which confirms the role of serum in arterial calcification in this system. Purified bovine MGP also prevented warfarin-induced calcification of devitalized arteries in the same medium.

9.4.2 HUMAN STUDIES ON VASCULAR CALCIFICATION

A series of sixteen patients with cutaneous necrosis from calcific uremic arteriolopathy identified warfarin as an independent risk factor.[26] In two of these patients, cessation of warfarin therapy and substitution with low-molecular-weight heparin resulted in clinical improvement. Schori and Stungis reported a case of arterial calcification in a patient with long-term warfarin use.[27] In a retrospective analysis of eight patients who developed calciphylaxis, six patients were on anticoagulation therapy with warfarin.[28] All patients were obese women, with metabolic syndrome and poorly controlled hypertension. These patients did not have significant alterations of calcium metabolism to explain calciphylaxis. Histopathologic examination of these lesions revealed calcium deposits in arterioles with vascular thrombosis. These authors concluded that anticoagulant therapy was one of the risk factors to develop calciphylaxis, in the absence of severe disorders of calcium metabolism.

In a cross-sectional analysis of seventy patients (46 men, mean age 68 ± 13 years) on warfarin therapy without known coronary artery disease, after adjustment for cardiovascular risk factors, no correlation was observed between duration of warfarin use and coronary artery calcification score on multivariate analysis ($r = 0.075$, $p = 0.537$).[29]

9.4.3 HUMAN STUDIES ON VALVULAR CALCIFICATION

Schurgers et al. measured the grade of aortic valve calcification in valves removed from patients undergoing surgical valve replacement.[30] Calcifications in valves from patients receiving preoperative oral anticoagulant treatment were significantly larger than in patients not receiving preoperative oral anticoagulants. In a study of eighty-six patients (53 men, mean age 71 ± 8 years) with calcific aortic valve disease, patients on warfarin therapy (n = 23) had increased coronary artery calcification (coronary Agatston score 1561 ± 1141 vs. 738 ± 978; $p = 0.024$), and aortic valvular calcification (valvular Agatston score 2410 ± 1759 vs. 1070 ± 1085; $p = 0.002$) compared to patients not on warfarin therapy.[31] Subsequently, they showed that patients with calcific aortic valve disease had significantly lower levels of circulating inactive MGP as compared to a reference population free of coronary and valvular calcifications. They concluded that warfarin treatment may decrease local expression of MGP, resulting in decreased circulating MGP levels and subsequently increased aortic valve calcifications as an adverse side effect.[32]

In a retrospective cohort study among 108 hemodialysis patients, subjects with long-term warfarin exposure (36.7 ± 19.7 months) were more likely to have severe aortic valve calcification (p = 0.04). This was independent of dialysis use, calcium, and calcitriol intake. The odds ratio of severe calcification following eighteen months of warfarin use was 3.77 (95% CI, 0.97–14.70; p = 0.055).[33] Coronary-artery calcification is common and progressive in young adults with end-stage renal disease who are undergoing dialysis.[34] Hemodialysis patients also have a suboptimal vitamin K status, as shown by low serum vitamin K1 level and a high percentage of uncarboxyated osteocalcin.[35]

In the Japanese Aortic Stenosis Study, a retrospective observational study of 556 subjects aged 50 years or older, and with calcification in any aortic valve leaflet or peak aortic jet velocity ≥2 m/sec, use of warfarin was identified to be a prognostic factor in early stage disease (peak aortic jet velocity of ≥2 m/sec), and not for late-stage disease (peak aortic jet velocity of ≥3 m/sec).[36] These investigators concluded that we should be vigilant about progression of calcific aortic valve disease in patients treated with warfarin. In a study of 1155 patients (mean age 74 years) with nonvalvular atrial fibrillation, 725 (63%) were treated with warfarin and 430 (37%) without warfarin. Mitral valve calcification, mitral annular calcification, or aortic valve calcification with two-dimensional echocardiograms was present in 473 of 725 patients (65%) on warfarin versus 225 of 430 patients (52%) not on warfarin (p < 0.0001).[37] On stepwise logistic regression analysis, there was a significant association between the use of warfarin and the risk of calcification (odds ratio = 1.71, 95% CI = [1.34–2.18]) after adjustment for confounding risk factors. The authors concluded that use of warfarin in patients with nonvalvular atrial fibrillation is associated with an increased prevalence of mitral valve calcification, mitral annular calcification, or aortic valve calcification.[37]

9.5   VITAMIN K IN PREVENTION AND TREATMENT OF VASCULAR CALCIFICATION

Until recently, calcification was thought to be an irreversible phenomenon. Current understanding, however, dictates that calcification is an active process with inhibitors and stimulators of calcification. Warfarin-induced vascular calcification in rats was prevented, and in some cases reversed, by high vitamin K intake.[38] This was associated with significantly less VSMC apoptosis and with significant regression of arterial calcification. Calcium deposits were removed by phagocytosis carried out by the surrounding VSMCs.[39] Vitamin K is present in different forms. Extrahepatic tissues such as vessel walls specifically accumulate vitamin K2 (menaquinones), even when the diet exclusively contains vitamin K1 (phylloquinone). The present recommended dietary allowance for vitamin K (90–120 micrograms) is based on the hepatic K1 requirement for coagulation factor synthesis. In a study on warfarin-treated rats that were fed diets containing K1 or K2 (MK-4), despite their similar in vitro cofactor activity, vitamin K2 and not K1 inhibited warfarin-induced arterial calcification. The total hepatic vitamin K1 accumulation was threefold higher than that of vitamin K2, whereas aortic vitamin K2 was three times that of K1.[40] Vitamin K2 could also down-regulate osteoprotegerin, suggesting that it acts as an anticalcification component in the vessel wall.[41] However, Schurgers et al. proposed that in the high-K1 group, K1 had been converted to K2 to such an extent that in the high-K1 group, arterial K2 had comparable tissue concentrations as in the K2-treated group. They concluded that at very high intakes of K1 (200-fold the daily requirement), both these forms of vitamin K might help decrease arterial calcification.[38] In a population-based study of 4807 subjects with no history of myocardial infarction at baseline that examined the association of dietary intake of vitamin K1 and K2 with aortic calcification and coronary heart disease, intake of vitamin K2 inversely related to all-cause mortality (relative risk = 0.91 [0.75, 1.09] in mid tertile, and 0.74 [0.59, 0.92] in upper tertile, compared to lower tertile) and severe aortic calcification (odds ratio = 0.71 [0.50, 1.00] in mid tertile, and 0.48 [0.32, 0.71] in upper tertile, compared to lower tertile). Vitamin K1 intake was not related to any of the above outcomes in this study.[42] These data suggest that a high vitamin K2 intake may be an effective interventional strategy to decrease the vascular calcification risk, and thus cardiovascular mortality in the general population. In a randomized placebo-controlled intervention study, 181 postmenopausal women were given either a placebo or a supplement containing minerals and vitamin D (MD-group), or the same supplement with vitamin K1 (MDK-group).[43] A total of 150 participants completed the study, and analysis was performed on 108 participants. Elastic properties of the common carotid artery (compliance coefficient, distensibility coefficient) in the MDK-group remained unchanged over the three-year period, but decreased in the MD- and placebo-group. No significant differences were observed in the change of intima-media thickness among the three groups. The authors concluded that supplementation of vitamins K1 and D has a beneficial effect on the elastic properties of the arterial vessel wall. However, there is no randomized prospective data in humans to confirm that high vitamin K intake may protect from adverse cardiovascular events in the general population or in patients with prior coronary artery disease.

9.6 OTHER THERAPIES FOR VASCULAR CALCIFICATION

In patients with warfarin-induced calcification, stopping warfarin and using an alternative anticoagulant can help. This has been shown in patients with cutaneous necrosis from calcific uremic arteriolopathy.[26] In another case report, a patient with biopsy-proven calciphylaxis thought to be attributable to warfarin was treated with a therapeutic substitution of anticoagulant and hyperbaric oxygen therapy leading to resolution of cutaneous lesions.[44] The mechanism by which hyperbaric therapy might benefit patients with calciphylaxis is unclear. Statins have numerous pleiotropic effects, resulting from their ability to block synthesis of isoprenoid intermediates and inhibit prenylation of Rho family guanosine triphosphatases.[45] Cerivastatin and atorvastatin inhibit in vitro calcification of human vascular smooth muscle cells induced by inflammatory mediators in a dose-dependent manner. Inhibition of Rho and its downstream target, Rho kinase may mediate this inhibitory effect of statins.[46] In vivo administration of atorvastatin reduced vitamin D3 and warfarin-induced arterial calcification, plasma calcium concentration, and alkaline phosphatase levels in rats.[47] The protective effects of atorvastatin on vascular remodeling in renovascular hypertensive rats is attributed to decreased expression of osteopontin.[48]

Etidronate, clodronate, and several other first-generation bisphosphonates had been shown to inhibit vitamin D–induced artery calcification in rats.[49] This was thought to be due to the inhibitory effect of bisphosphonates on formation of hydroxyapatite crystals. Later, alendronate and ibandronate were demonstrated to inhibit warfarin-induced calcification of arteries and heart valves in rats at doses comparable to the doses that inhibit bone resorption.[50] However, in a randomized trial of fifty patients with stage 3-4 chronic kidney disease comparing weekly alendronate to placebo, there was no difference in vascular calcification progression assessed by computed tomography with alendronate compared with placebo.[51]

Osteoprotegerin, a secreted protein of the tumor necrosis factor family that inhibits osteoclast differentiation and activation, has been shown to protect against warfarin-induced calcification in rats.[52]

9.7   CONCLUSION

Although animal studies and small studies in humans have shown this strong association of vitamin K deficiency with vascular calcification, this has not been utilized as a standard therapy for vascular calcification. It is still unclear if vitamin K replenishment can affect the morbidity and mortality associated with vascular and valvular calcification in humans. Further studies are needed to determine the appropriate formulation of vitamin K (K1 vs. K2), dose, and duration of vitamin K therapy that would lead to beneficial effects in prevention of vascular calcification. Because of their very low toxicity and potentially beneficial effects on attenuation of arterial calcification, vitamin K supplementation can be considered in susceptible patients, such as patients with end-stage renal disease who are on hemodialysis.

REFERENCES

1. Dam H, Schonheyder F. A deficiency disease in chicks resembling scurvy, *Biochem J* 1934;28:1355–1359.
2. Doisy EA, Binkley SB, Thayer SA, McKee RW. Vitamin K, *Science* 1940;91:58–62.
3. Palaniswamy C, Sekhri A, Aronow WS, Kalra A, Peterson SJ. Association of warfarin use with valvular and vascular calcification: A review, *Clin Cardiol* 2011;34:74–81.
4. Rifkin RD, Parisi AF, Folland E. Coronary calcification in the diagnosis of coronary artery disease, *Am J Cardiol* 1979;44:141–147.
5. Friedman SA, Novack S, Thomson GE. Arterial calcification and gangrene in uremia, *N Engl J Med* 1969;280:1392–1394.
6. Massry SG, Gordon A, Coburn JW, et al. Vascular calcification and peripheral necrosis in a renal transplant recipient. Reversal of lesions following subtotal parathyroidectomy, *Am J Med* 1970;49:416–422.
7. Schurgers LJ, Dissel PE, Spronk HM, et al. Role of vitamin K and vitamin K–dependent proteins in vascular calcification, *Z Kardiol* 2001;90 Suppl 3:57–63.
8. Lehto S, Niskanen L, Suhonen M, Ronnemaa T, Laakso M. Medial artery calcification. A neglected harbinger of cardiovascular complications in non-insulin-dependent diabetes mellitus, *Arterioscler Thromb Vasc Biol* 1996;16:978–983.

9. Price PA, Urist MR, Otawara Y. Matrix Gla protein, a new gamma-carboxyglutamic acid-containing protein which is associated with the organic matrix of bone, *Biochem Biophys Res Commun* 1983;117:765–771.

10. Howe AM, Webster WS. The warfarin embryopathy: A rat model showing maxillonasal hypoplasia and other skeletal disturbances, *Teratology* 1992;46:379–390.

11. Hale JE, Fraser JD, Price PA. The identification of matrix Gla protein in cartilage, *J Biol Chem* 1988;263:5820–5824.

12. Luo G, Ducy P, McKee MD, et al. Spontaneous calcification of arteries and cartilage in mice lacking matrix GLA protein, *Nature* 1997;386:78–81.

13. Spronk HM, Soute BA, Schurgers LJ, et al. Matrix Gla protein accumulates at the border of regions of calcification and normal tissue in the media of the arterial vessel wall, *Biochem Biophys Res Commun* 2001;289:485–490.

14. Sweatt A, Sane DC, Hutson SM, Wallin R. Matrix Gla protein (MGP) and bone morphogenetic protein-2 in aortic calcified lesions of aging rats, *J Thromb Haemost* 2003;1:178–185.

15. Nishimoto SK, Nishimoto M. Matrix Gla protein C-terminal region binds to vitronectin. Co-localization suggests binding occurs during tissue development, *Matrix Biol* 2005;24:353–361.

16. Price PA, Chan WS, Jolson DM, Williamson MK. The elastic lamellae of devitalized arteries calcify when incubated in serum: Evidence for a serum calcification factor, *Arterioscler Thromb Vasc Biol* 2006;26:1079–1085.

17. Reynolds JL, Joannides AJ, Skepper JN, et al. Human vascular smooth muscle cells undergo vesicle-mediated calcification in response to changes in extracellular calcium and phosphate concentrations: A potential mechanism for accelerated vascular calcification in ESRD, *J Am Soc Nephrol* 2004;15:2857–2867.

18. Strebel RF, Girerd RJ, Wagner BM. Cardiovascular calcification in rats with hypervitaminosis A, *Arch Pathol* 1969;87:290–297.

19. Cancela ML, Price PA. Retinoic acid induces matrix Gla protein gene expression in human cells, *Endocrinology* 1992;130:102–108.

20. Farzaneh-Far A, Proudfoot D, Weissberg PL, Shanahan CM. Matrix Gla protein is regulated by a mechanism functionally related to the calcium-sensing receptor, *Biochem Biophys Res Commun* 2000;277:736–740.

21. Munroe PB, Olgunturk RO, Fryns JP, et al. Mutations in the gene encoding the human matrix Gla protein cause Keutel syndrome, *Nat Genet* 1999;21:142–144.

22. Hur DJ, Raymond GV, Kahler SG, Riegert-Johnson DL, Cohen BA, Boyadjiev SA. A novel MGP mutation in a consanguineous family: Review of the clinical and molecular characteristics of Keutel syndrome, *Am J Med Genet A* 2005;135: 36–40.

23. Howe AM, Webster WS. Warfarin exposure and calcification of the arterial system in the rat, *Int J Exp Pathol* 2000;81:51–56.

24. Price PA, Faus SA, Williamson MK. Warfarin causes rapid calcification of the elastic lamellae in rat arteries and heart valves, *Arterioscler Thromb Vasc Biol* 1998;18:1400–1407.

25. Price PA, Faus SA, Williamson MK. Warfarin-induced artery calcification is accelerated by growth and vitamin D, *Arterioscler Thromb Vasc Biol* 2000;20:317–327.

26. Coates T, Kirkland GS, Dymock RB, et al. Cutaneous necrosis from calcific uremic arteriolopathy, *Am J Kidney Dis* 1998;32:384–391.

27. Schori TR, Stungis GE. Long-term warfarin treatment may induce arterial calcification in humans: Case report, *Clin Invest Med* 2004;27:107–109.

28. Verdalles Guzman U, de la Cueva P, Verde E, et al. Calciphylaxis: Fatal complication of cardiometabolic syndrome in patients with end stage kidney disease, *Nefrologia* 2008;28:32–36.

29. Villines TC, O'Malley PG, Feuerstein IM, Thomas S, Taylor AJ. Does prolonged warfarin exposure potentiate coronary calcification in humans? Results of the warfarin and coronary calcification study, *Calcif Tissue Int* 2009;85:494–500.

30. Schurgers LJ, Aebert H, Vermeer C, Bultmann B, Janzen J. Oral anticoagulant treatment: Friend or foe in cardiovascular disease? *Blood* 2004;104:3231–3232.

31. Koos R, Mahnken AH, Muhlenbruch G, et al. Relation of oral anticoagulation to cardiac valvular and coronary calcium assessed by multislice spiral computed tomography, *Am J Cardiol* 2005;96:747–749.

32. Koos R, Krueger T, Westenfeld R, et al. Relation of circulating Matrix Gla-Protein and anticoagulation status in patients with aortic valve calcification, *Thromb Haemost* 2009;101:706–713.

33. Holden RM, Sanfilippo AS, Hopman WM, Zimmerman D, Garland JS, Morton AR. Warfarin and aortic valve calcification in hemodialysis patients, *J Nephrol* 2007;20:417–422.

34. Goodman WG, Goldin J, Kuizon BD, et al. Coronary-artery calcification in young adults with end-stage renal disease who are undergoing dialysis, *N Engl J Med* 2000;342:1478–1483.

35. Pilkey RM, Morton AR, Boffa MB, et al. Subclinical vitamin K deficiency in hemodialysis patients, *Am J Kidney Dis* 2007;49:432–439.

36. Yamamoto K, Yamamoto H, Yoshida K, et al. Prognostic factors for progression of early- and late-stage calcific aortic valve disease in Japanese: The Japanese Aortic Stenosis Study (JASS) Retrospective Analysis, *Hypertens Res* 2010;33:269–274.

37. Lerner RG, Aronow WS, Sekhri A, et al. Warfarin use and the risk of valvular calcification, *J Thromb Haemost* 2009;7:2023–2027.

38. Schurgers LJ, Spronk HM, Soute BA, Schiffers PM, DeMey JG, Vermeer C. Regression of warfarin-induced medial elastocalcinosis by high intake of vitamin K in rats, *Blood* 2007;109:2823–2831.

39. Proudfoot D, Davies JD, Skepper JN, Weissberg PL, Shanahan CM. Acetylated low-density lipoprotein stimulates human vascular smooth muscle cell calcification by promoting osteoblastic differentiation and inhibiting phagocytosis, *Circulation* 2002;106:3044–3050.

40. Spronk HM, Soute BA, Schurgers LJ, Thijssen HH, De Mey JG, Vermeer C. Tissue-specific utilization of menaquinone-4 results in the prevention of arterial calcification in warfarin-treated rats, *J Vasc Res* 2003;40:531–537.

41. Wallin R, Schurgers L, Wajih N. Effects of the blood coagulation vitamin K as an inhibitor of arterial calcification, *Thromb Res* 2008;122:411–417.

42. Geleijnse JM, Vermeer C, Grobbee DE, et al. Dietary intake of menaquinone is associated with a reduced risk of coronary heart disease: The Rotterdam Study, *J Nutr* 2004;134:3100–3105.

43. Braam LA, Hoeks AP, Brouns F, Hamulyak K, Gerichhausen MJ, Vermeer C. Beneficial effects of vitamins D and K on the elastic properties of the vessel wall in postmenopausal women: A follow-up study, *Thromb Haemost* 2004;91:373–380.

44. Banerjee C, Woller SC, Holm JR, Stevens SM, Lahey MJ. Atypical calciphylaxis in a patient receiving warfarin then resolving with cessation of warfarin and application of hyperbaric oxygen therapy, *Clin Appl Thromb Hemost* 2010;16:345–350.

45. Palaniswamy C, Selvaraj DR, Selvaraj T, Sukhija R. Mechanisms underlying pleiotropic effects of statins, *Am J Ther* 2010;17:75–78.

46. Kizu A, Shioi A, Jono S, Koyama H, Okuno Y, Nishizawa Y. Statins inhibit in vitro calcification of human vascular smooth muscle cells induced by inflammatory mediators, *J Cell Biochem* 2004;93:1011–1019.

47. Li H, Tao HR, Hu T, et al. Atorvastatin reduces calcification in rat arteries and vascular smooth muscle cells, *Basic Clin Pharmacol Toxicol* 2010;107:798–802.

48. Yang S, Zhou JZ, Chen YH. Effects of atorvastatin calcium on osteopontin in renovascular hypertensive rats, *Beijing Da Xue Xue Bao* 2010;42:147–150.
49. Francis MD, Russell RG, Fleisch H. Diphosphonates inhibit formation of calcium phosphate crystals in vitro and pathological calcification in vivo, *Science* 1969;165:1264–1266.
50. Price PA, Faus SA, Williamson MK. Bisphosphonates alendronate and ibandronate inhibit artery calcification at doses comparable to those that inhibit bone resorption, *Arterioscler Thromb Vasc Biol* 2001;21:817–824.
51. Toussaint ND, Lau KK, Strauss BJ, Polkinghorne KR, Kerr PG. Effect of alendronate on vascular calcification in CKD stages 3 and 4: A pilot randomized controlled trial, *Am J Kidney Dis* 2010;56:57–68.
52. Price PA, June HH, Buckley JR, Williamson MK. Osteoprotegerin inhibits artery calcification induced by warfarin and by vitamin D, *Arterioscler Thromb Vasc Biol* 2001;21:1610–1616.

10 Therapeutic Effects of Pyridoxamine and Benfotiamine

Shyamala Dakshinamurti and
Krishnamurti Dakshinamurti

CONTENTS

10.1 MECHANISMS OF VASCULAR DYSFUNCTION

The implications of advanced glycation end products (AGEs) and advanced lipoxidation end products (ALEs) on the pathogenesis of diabetes-mediated uremic complications as well as on aging and atherosclerosis have been recognized for some time. In addition, the roles of reactive carbonyl compounds and of reactive oxygen species (ROS) in pathologies related to diabetes, atherosclerosis, Alzheimer's disease, as well as the aging process have received much consideration. Any chemical that would have inhibitory action on any of these individual steps leading to pathologies affecting the micro- and macrovascular systems would have therapeutic potential. Many such chemicals have been identified of which pyridoxamine, a vitamin B_6 vitamer, and benfotiamine, a lipid-soluble form of thiamine (vitamin B_1), have a very significant place in the therapy of micro- and macrovascular defects associated with many chronic diseases.

10.2   CARDIOVASCULAR COMPLICATIONS OF DIABETES MELLITUS: MECHANISMS

The reducing sugar glucose, which is elevated in the diabetic condition, can modify proteins through condensation of its aldehyde group with the ε-amino group of lysine residues or the N-terminal amino group forming a Schiff base. This reversible process is referred to as the Maillard reaction and is dependent on the concentration of glucose and, thus, is exacerbated in the diabetic condition. The Schiff bases isomerize to intermediate ketamine Amadori products including glycated hemoglobin (HbA$_{1c}$). In the later phase, over a period of weeks and months, glucose-independent reactions including rearrangements, condensation, dehydration, polymerization, and fragmentation lead to a host of metabolites referred to as advanced glycation end products. Lipid peroxidation of polyunsaturated fatty acids similarly gives rise to advanced lipoxidation end products. The preponderant glycation adduct is N$_e$-fructosyl-lysine 1. Endogenously formed dicarbonyl compounds, referred to as the reactive carbonyl species (RCS), such as glyoxal, methyl glyoxal, and 3-deoxyglucosone, derived from protein-bound glycation intermediates or from glucose itself, are also potent glycating agents. Dicarbonyls can react directly with proteins to form AGEs. Hydroimidazolones derived from arginine residues modified by glyoxal, methyl glyoxal, and 3-deoxyglucosone are the predominant AGEs, although other AGEs such as Nε-carboxymethyl lysine (CML) and Nε-carboxyethyl lysine (CEL) and the protein cross links such as pentosidine and glucosepane are also formed (1). ALEs comprise a range of molecular species including amino acids (carboxy methyl lysine [CML]), glyoxal derived lysine dimers (GOLD), methylglyoxal-derived lysine dimers (MOLD), carbohydrates (pentosidine), and lipids (malondialdehyd-lysine) (2). Histopathological evidence attests to the accumulation of AGEs in a variety of tissues such as the renal cortex, glomerular mesangium, and basement membrane. AGEs have been associated with oxidative and nitrosative stresses in both in vitro and in vivo studies.

Evidence for the role of AGEs in vascular complications has been presented (1). There is increased concentration of AGE residues in tissue sites such as the renal glomeruli, retina, and peripheral nerve where vascular complications arise. Experimentally, the injection of AGE precursor, methyl glyoxal, or AGE-modified proteins, induces vascular damage similar to that seen in the diabetic condition.

Cell surface receptors that bind AGE-modified proteins (RAGE) have been identified. These receptors, which are multiligand members of the immunoglobulin super family of cell surface receptors, are expressed on microvascular pericytes and endothelial cells. The signal transduction receptor RAGE activates multiple intracellular signaling pathways that operate through phosphatidyl inositol-3 kinase (PI-3K), K$_i$-Ras, and mitogen-activated protein kinase (MAPK) pathways leading to nuclear translocation of nuclear factor-κB (NF-κB) involved in the expression of several genes including the pro-inflammatory cytokines (3, 4). Activation of NF-κB results in a positive feedback up-regulation of RAGE. RAGE, through stimulation of NAD(P)H oxidase, produces reactive oxygen species (ROS), which also alter proteins, lipids, and DNA (5). In the diabetic state there is an overexpression of RAGE. The diabetic homozygous RAGE null rodent models were resistant to the development of diabetic nephropathy, confirming a role for RAGE in the development of inflammatory and thrombogenic reactions that contribute to the vascular complications in the diabetic condition (1, 6, 7).

Reactive oxygen species (ROS) are generated through autoxidation of glucose or through post-Amadori or oxidative degradation reactions. The toxicity of ROS is ascribed to their high reactivity and ability to damage proteins and nucleic acids as well as activation of pathogenic signaling pathways (8). Superoxide anion, hydrogen peroxide, hydroxyl, peroxyl, and alkoxyl radicals are the major ROS species generated. The oxidation of various amino acid side chains of proteins such as the oxidation of cysteine to sulfonate, methionine to its sulfoxide, and hydroxylation of tryptophan by ROS contributes to the structural and functional damage of proteins (2). ROS are implicated in the complications of diabetes as well as other chronic diseases.

Mitochondria are the energy powerhouses of the cell, producing ATP required for all biosynthetic processes. Glycation of mitochondrial proteins and reactive dicarbonyl damage to mitochondria result in complications of diabetes (9). The enzyme glyoxalase I (GLOI) is the key and rate-limiting enzyme of the glyoxalase system involved in the detoxification of dicarbonyls and prevention of the formation of AGEs. Overexpression of GLOI in endothelial cells, under hyperglycemic conditions, reduces reactive dicarbonyls under various pathologic conditions (10, 11). Some naturally occurring flavones increase the expression of GLOI (12) and have therapeutic potential in decreasing the formation of AGEs and, thus, in decreasing the diabetic complications including neuropathy and nephropathy.

The activation of peroxisome proliferator-activated receptor-γ (PPARγ) by rosiglitazone or pioglitazone blocks the harmful effect of AGE-induced NOS expression and its sequelae (13, 14). PPARα agonists decreased TNFα-induced RAGE expression through suppression of transcriptional factor NF-κB activation. Rosiglitazone treatment reduces extracellular matrix accumulation and proteinuria in AGE-injected rats (13). The suppression of RAGE expression seems to be the target of action of PPARγ agonists (15, 16).

Various studies indicate that hyperglycemia is the most significant factor in the onset and development of the vascular complications of diabetes of which chronic kidney disease (CKD) is a major component. Inhibition of the renin-angiotensin system (RAS) with angiotensin-converting enzyme inhibitors (ACEIs) and/or angiotension II receptor blocker (ARB) is the main pharmacotherapy for CKD. In addition to their hemodynamic action, ARBs and ACEIs also block the formation of reactive carbonyl precursors of AGE as well as AGE-RAGE–induced ROS generation (17).

Of the various compounds that inhibit or correct individual steps in the pathophysiological sequence leading to vascular complications of diabetes, two stand out: pyridoxamine and benfotiamine (Figure 10.1). Pyridoxamine is a member of the vitamin B_6 vitamer group. Benfotiamine is a lipid-soluble vitamin B_1 analogue. Their use as therapeutic agents in the treatment of diabetic vascular complications has much potential in view of their lack of toxicity.

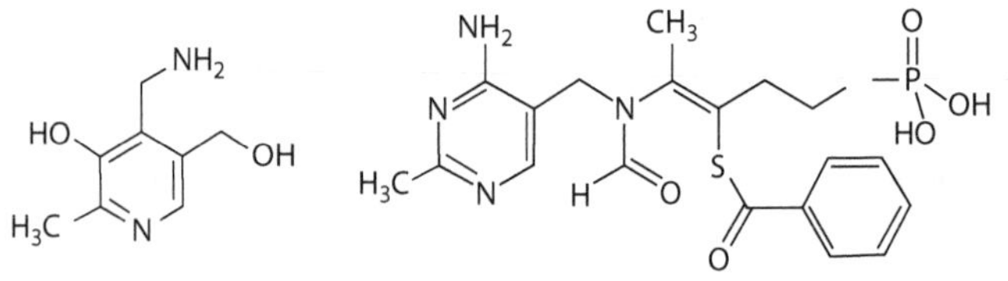

Pyridoxamine   S-benzoylthiamine-O-monophosphate

FIGURE 10.1   Chemical structure of pyridoxamine and benfotiamine.

10.3   PYRIDOXAMINE

Pyridoxamine (PM) is a transient intermediate in the enzymatic transamination reactions. The phenolic hydroxyl group at position 3 and the aminomethyl group at position 4 of the pyridinium ring are the crucial structural features of pyridoxamine. The nonenzymatic transamination reaction requires binding and coordination of metal ion by pyridoxamine. In view of the reactivity of PM with carbonyl groups in transamination reactions, it was hypothesized that PM would be a potent inhibitor of the Maillard reaction. PM was shown to inhibit this reaction and was significantly more effective than aminoguanidine in the formation of antigenic AGEs with model proteins when incubated with glucose (2).

AGEs, because of their low molecular weight, are cleared by the kidney. However, under conditions of renal impairment, serum and tissue concentrations of AGEs build up. Tissue accumulation of AGE can directly damage extracellular matrix through AGE-specific receptors (RAGE), which are expressed on the surface of various cells such as monocytes, macrophages, neurons, endothelial cells, smooth muscle cells, and fibroblasts. Following the interaction between AGEs and RAGEs, there is a stimulation of genes for cytokines, growth factors, and adhesion molecules. Stimulation of cell proliferation, increase in vascular permeability, induction of migration of macrophages, stimulation of endothelium formation, increased synthesis of collagen, fibronectin, and proteoglycans, as well as procoagulation tissue factors ensue. AGE receptors contribute to a number of chronic diseases such as diabetes, amyloidosis, inflammatory conditions, and tumors (18).

Many compounds with AGE inhibitor activity have been identified, such as aminoguanidine, phenacyl thiazolium bromide, 2-isopropylidenehydrazano 4-oxo-thiazolidin-5-yl-acetanilide, 2,3-diaminophenazine, vitamin C, vitamin E, and pyridoxamine (19). The KK-A^y/Ta mouse has been developed as a very useful experimental model for type 2 diabetes as the associated hyperglycemia, glucose intolerance, hyperinsulinemia, obesity, and microalbuminuria closely resemble the human condition in diabetic nephropathy. Tanimoto et al. (20) have shown that treatment with high dose of pyridoxamine improved the urinary albumin/creatine ratio and fasting serum triglyceride and 3-deoxyglucosome. Pyridoxamine also prevented accumulations of Nε-(carboxymethyl) lysine, nitrotyrosine, TGF-β1, and laminin-β1 in kidney. The therapeutic potential of inhibitors of AGE production such as angiotensin-converting enzyme (ACE) inhibitors, angiotensin-receptor antagonists, metal chelators, and antioxidants has been recorded (18). The benefits of

ACE inhibitors go beyond their hemodynamic effects. The therapeutic effect of these compounds has been attested in clinical trials (21, 22).

Low molecular weight carbonyl compounds are formed by autoxidation of glucose or Schiff base intermediates through reaction of glucose with the amino residues of proteins. Glyoxal, methylglyoxal, and glycoaldehyde are the major carbonyl compounds. These low molecular weight compounds diffuse from the sites of their formation to sites of pathological protein modification. The modification of collagen IV by methylglyoxal inhibited its adhesion to endothelial cells, and when the collagen IV modification was performed in the presence of pyridoxamine, adhesion to endothelial cells was much less affected, indicating the role of PM in trapping methylglyoxal (2). Under cell culture conditions, pyridoxamine suppressed the oxidative DNA damage induced by high glucose concentration (23). Studies using various animal models indicate that pyridoxamine is effective in the treatment of diabetic nephropathy, retinopathy, and neuropathy (24, 25).

10.4 ROLE OF REACTIVE OXYGEN SPECIES AND LOSS OF PROTEIN FUNCTIONALITY

Pyridoxamine was also shown to inhibit the post-Amadori reactions by inhibiting the oxidative reactions through binding to catalytic redox metal ions (26). Thus, PM affects both the pre- and post-Amadori AGE formation. The autoxidation of glucose and the oxidative degradation of Amadori intermediates generate reactive oxygen species (ROS) such as superoxide anion, hydrogen peroxide, hydroxyl, and methylglyoxal radicals (27). Reactive oxygen species initiate pathogenic signaling, and they are implicated in complications of diabetes as well as in other diseases such as atherosclerosis and cancer (28). Pyridoxamine decreases the hydroxyl radical generation by Fenton reaction. PM also reacts with hypochlorous acid, an ROS generated by activated neutrophils and monocytes at sites of inflammation (29). PM reduced oxidative stress in the streptozotocin-induced diabetic hamster (30). These studies indicate a protective role for PM in ROS-related toxicity (2). The effect of high dose pyridoxamine is seen to be related to its antioxidant action. AGEs and oxidative stress activate angiotension II signaling, leading to induction of TGF-β1-Smad signaling in mesangial cells (20).

The formation of ROS leads to ROS-mediated protein damage as the hydroxyl radical reacts with a number of amino acid side chains of proteins (31). The potential of pathogenic protein damage is high as the generation of ROS could be in proximity to potential targets. Hydroxyl radical can oxidize amino acid side chains of proteins, leading to protein backbone fragmentation that could be beyond the site of initial ROS production (32). This leads to loss of protein functionality. Hence, prevention of protein backbone fragmentation is a useful strategy for the treatment of diabetic complications such as nephropathy. In in vitro studies, using ribonuclease A, lysozyme, and serum albumin as models, pyridoxamine was demonstrated to inhibit the fragmentation of protein backbone induced by different mechanisms. The anti-protein fragmentation effect of pyridoxamine was due to its scavenging of hydroxyl radical. The therapeutic effect of pyridoxamine has been confirmed in clinical trials (33).

10.5   ROLE OF CHELATION

In recent studies, a new class of AGE inhibitors, referred to as Lalezari-Rahbar (LR) compounds, has been identified. They are aromatic acids with ureido and carboxamide functional groups. They have been shown to protect against diabetic complications in animal models (34, 35). In terms of their effect in experimental studies, their actions are similar to those of aminoguanidine and pyridoxamine. However, LR compounds lack nucleophilic groups and hence do not trap carbonyl compounds under physiological conditions. They, along with another class of compounds, the edaravone derivatives (36), are inhibitors of metal-catalyzed oxidation reactions because of their chelating activity. As both these groups of compounds protect against diabetes complications in animals without being carbonyl trapping agents, it was proposed (35) that metal chelation alone is adequate to prevent formation of AGEs and their sequelae. Chelators inhibit autoxidative glycosylation and glyoxidation and also prevent enzymatic and metal-catalyzed ROS production. The LR compounds inhibit inflammatory response in monocytes stimulated with RAGE ligands and block the increase in monocyte endothelial cell adhesion. Similarly, anti-hypertensive agents are renoprotective, independent of their blood pressure–lowering action, through their chelating and antioxidant activity (37).

Pyridoxamine protects against the complications of diabetes through multiple mechanisms of action such as (1) blocking oxidation of Amadori intermediate; (2) trapping of reactive carbonyls; (3) scavenging of ROS, and (4) being a profound metal chelator (35, 38). Pyridoxamine has no toxicity and is well tolerated. As it acts through multiple mechanisms, each with moderate activity, the net clinical effect could be significant. A literature search on clinical trials of drugs to treat diabetic kidney disease (DKD) has indicated that pyridoxamine is among ten drugs that have shown evidence of a beneficial effect in treating DKD patients, as indicated by improvements in glomerular filtration rate, albumin to creatinine ratio, protein-uria, or serum creatinine concentration (39, 40). Although no significant change in renal function was seen in their clinical trial (41), these authors as well as Chen and Francis in their editorial (42) suggest that the effect of advanced glycation inhibition may be seen before the onset of significant pathologic change.

10.6   BENFOTIAMINE

Benfotiamine (S-[(z)-2-[([?]4-amino-2 methylpyrimidin-5-yl) methyl formylamino]-5-phosphonoxypent-2-en-3yl] benzene carbothiolate) has an open thiazole ring. After absorption, the thiazole ring is closed. The ecto-alkaline phosphatase on the brush border of intestinal mucosal cells dephosphorylates it to S-benzoyl thiamine. Being lipophilic it diffuses through the membrane and mucosal cells and distributes in circulation. Compared with an equivalent dose of thiamine, the plasma concentration of thiamine is five-fold more after administration of benfotiamine (43). It is converted into thiamine diphosphate, which is a coenzyme of transketolase, pyruvate dehydrogenase, and α-ketoglutarate complexes. The expression of transketolase in renal glomeruli is increased by high-dose thiamine or benfotiamine (44). Vascular cells are sensitive to high glucose concentrations. Glucose-induced apoptosis has been shown in podocytes, mesangial cells, dorsal root ganglion cells, cardiac myocytes, and renal tubular epithelial cells associated with increased expression of proapoptotic Bax proteins and activation of NF-κB (45). Constant and intermittent high glucose enhances endothelial apoptosis through mitochondrial superoxide overproduction (46).

10.7 ROLE IN AGE AND ROS PRODUCTION

Among the mechanisms leading to toxicity in the vasculature are the increased flux through the polyol pathway, increased formation of AGEs, the activation of PKC, and increased flux through the hexosamine pathway (44). AGEs accumulate at sites of microvascular complications, leading to impairment of protein function and dysfunction of the vasculature. Aldose reductase, a key enzyme of the polyol pathway, reduces toxic aldehydes to inactive alcohols. When overwhelmed by excess glucose, it is reduced to sorbitol using NADPH. This results in a pseudohypoxia and susceptibility to oxidative stress (47). Treatment of diabetic dogs with Sorbinil, an aldose reductase inhibitor, prevented defective nerve conduction in long-term diabetic dogs. The role of persistent hyperglycemia in AGE formation and the role of AGE in protein modification and alteration of gene transcription leading to vascular dysfunction have been referred to in earlier sections. Also, the role of hyperglycemia in the synthesis of lipid second messenger diacylglycerol and enhancement of PKC synthesis with the resulting decreased synthesis of endothelial nitric oxide synthase has been referred to (44). Hyperglycemia induces increase in fructose-6-phosphate, which is converted to glucosamine-6-phosphate and UDP N-acetylglucosamine, which act through transcription factors to alter gene expression. The common feature of the above listed biochemical pathways is the overproduction of ROS. Overexpression of superoxide dismutase in db/db mice attenuates the renal injury (48). Benfotiamine, by increasing transketolase activity, reduces the accumulation of triosephosphates and decreases PKC activation, oxidative stress, and the renal complications in diabetes (44).

10.8 VASCULAR ENDOTHELIAL DYSFUNCTION

Benfotiamine restores vasodilation in diabetic mice through PKB/Akt-mediated increase of angiogenesis and inhibition of apoptosis (49). Hyperglycemia impairs endothelial progenitor cell (EPC) number and their differentiation. The administration of benfotiamine improves some aspects of EPC differentiation and phenotype, restores eNOS activity, and increases angiogenesis (50).

The vascular endothelium in its healthy state is antiatherogenic, antiproliferative, regulates vascular tone, and maintains blood flow. Diseases such as diabetes, atherosclerosis, hypertension, coronary artery disease, and stroke are associated with vascular endothelial dysfunction (VED). Under experimental conditions, administration of uric acid or nicotine or sodium arsenite for extended periods to rats produced VED by impairing the integrity of vascular endothelium and decreasing serum and aortic concentrations of nitrite/nitrate. The acetylcholine-induced endothelium-dependent relaxation was attenuated. Concurrent treatment of these rats with benfotiamine prevented the nicotine, sodium arsenite, or uric acid–induced VED and improved the integrity of the vascular endothelium by reducing oxidative stress, acting through mechanisms unrelated to AGE formation (51). Activation of Akt/PKB by benfotiamine results in activation of eNOS and improvement in vascular endothelial function. The action of benfotiamine is similar to those of antioxidants such as N-acetylcysteine, resveratrol, curcumin, and vitamins C and E in resolving the inflammation associated with bacterial endotoxin (52–54). Benfotiamine prevents high glucose–induced DNA fragmentation and caspase-3 activity and hence, endothelial cell damage.

Macrophages have a key role in inflammatory and immune reactions. The binding of the lipopolysaccharide (LPS) component of the bacterial cell wall to receptors on macrophage cell membrane induces the release of pro-inflammatory markers such as the cytokines, chemokines, growth factors, iNOS, and COX-2, which control the growth and dissemination of the pathogens (55). Excessive inflammatory responses could lead to severe septic shock. LPS-induced inflammatory signals lead to cell death of macrophages. The release of inflammatory markers promotes oxidative stress followed by programmed cell death by the Bcl-2 family of proteins. Benfotiamine has been shown to prevent the arachidonic acid pathway–generated inflammatory lipid mediators and the expression of enzymes such as COX-2 and LOX-5. This inhibition results in decreased synthesis of pro-inflammatory prostaglandins. Benfotiamine is a more potent anti-inflammatory compound than specific COX and LOX inhibitors (56).

10.9 POST-MI SURVIVAL AND ANGIOGENESIS

The pentosephosphate pathway plays a crucial role in the maintenance of cardiomyocyte contractility during ischemia. Under conditions of oxidative stress there is an increase in the activity of glucose-6-phosphate dehydrogenase (G6PD), the rate-limiting enzyme of the pentose phosphate pathway to modulate cytosolic redox state. The ensuing increase in vascular endothelial growth factor (VEGF) receptor 2, B PKB/Akt, and endothelial nitric oxide synthase (eNOS) promotes angiogenesis. In diabetes mellitus, due to the impaired activities of G6PD and transketolase, there is a depletion of reducing agents and accumulation of glycolysis end products with harmful effects on the cardiovascular system. In experimental studies Katare et al. (57) have shown that benfotiamine improved post myocardial infarct survival, functional recovery, and neovascularization in both diabetic and nondiabetic mice. They have shown that benfotiamine significantly increased G6PD and transketolase activity, reduced oxidative stress, and preserved cardiac viability and vascularization. The blunted response of G6PD-dependent VBGFR 2/Akt/eNos pathway in the diabetic condition is reversed by benfotiamine (58). Increased expression of Akt inhibits apoptosis of cardiomyocytes. In the diabetic condition there is a decrease in phospho-inositide-3-kinase (P13K)/Akt/Proiviral integration site for Moloney murine leukemia virus-1 (Pim-1) expression and the Pim-1 signaling pathway. Pim-1 induces Bcl-2 and Bcl-xL proteins and the phosphorylation/inactivation of transcription factor Bad. Benfotiamine restores the level of Pim-1 in diabetic hearts with its beneficial function (57, 58).

10.10   OTHER CHRONIC CONDITIONS

Impairment of glucose metabolism is a major feature of Alzheimer's disease. Pan et al. (59) have examined the effect of benfotiamine on cognitive impairment and pathology in the amyloid precursor proteins/presenilin-1 transgenic mice, a mouse model of Alzheimer's disease. Chronic treatment with benfotiamine enhanced the spatial memory of the mice and also reduced both amyloid plaque numbers and phosphorylated tau levels in the cortical areas. Benfotiamine increased the phosphorylation level of glycogen synthase kinase-3α and 3β and reduced their enzyme activities, a newer aspect of the effect of benfotiamine.

A variety of mechanisms are brought into play to produce the biological effects of benfotiamine. These include (1) control of redox status, (2) decrease in AGE formation, (3) stimulation of prosurvival G6PD/Akt/Pim-1 pathway, as well as (4) inhibition of glycogen synthase kinase 3. These augment the therapeutic potential of benfotiamine. In clinical studies, treatment with benfotiamine resulted in significant improvement in diabetic polyneuropathy (60). Benfotiamine was also demonstrated to prevent macro- and microvascular endothelial dysfunction in diabetic patients (61).

Both pyridoxamine and benfotiamine have been shown to be effective in preventing or treating the vascular dysfunction syndrome associated with both experimental and clinical diabetes mellitus. This is accomplished through their participation in multiple metabolic pathways. Although in many instances pyridoxamine and benfotiamine affect the same steps in different metabolic pathways, each one has, in addition, specific sites of activity. Whether this would lead to synergistic effects of these compounds is yet to be investigated.

REFERENCES

1. Ahmed N and Thornalley PJ. Advanced glycation end products: What is their relevance to diabetic complications? *Diabetes, Obesity and Metabolism*, 9: 233–245 (2007).
2. Voziyan PA and Hudson BG. Pyridoxamine as a multifunctional pharmaceutical: Targeting pathogenic glycation and oxidative damage, *Cellular and Molecular Life Sciences*, 62: 1671–1681 (2005).
3. Lukic IK, Humpert PM, Nawrath PP, and Bierhaus A. The RAGE pathway: Activation and perpetuation in the pathogenesis of diabetic neuropathy. *Ann NY Acad Sci*, 1126: 76–80 (2008).
4. Vincent AM, Perrone L, Sullivan KA, Backus C, Sastry AM, Lastostie C, and Feldman EL. Receptor for advanced glycation end products activation injures primary sensory neurons via oxidative stress. *Endocrinology*, 48: 548–558 (2007).
5. Jack M and Wright D. Role of advanced glycation end products and glyoxalase I in diabetic peripheral sensory neuropathy. *Translational Research*, 159: 355–365 (2012).
6. Yamagishi S, Nakamura K, Matsui T, Ueda S, Fukami K, and Okuda S. Agents that block advanced glycation end product (AGE)—RAGE (receptor for AGEs)—oxidative stress system: A novel therapeutic strategy for diabetic vascular complications. *Expert Opin Investig Drugs*, 17: 983–996 (2008).
7. Yamagishi S, Maeda S, Matsui T, Ueda S, Fukami K, and Okuda S. Role of advanced glycation end products (AGEs) and oxidative stress in vascular complications in diabetes. *Biochemica et Biophysica Acta,* 1820: 663–671 (2012).

8. Lee HB, Yu MR, Yang Y, Jiang Z, and Ha H. Reactive oxygen species–regulated signaling pathways in diabetic nephropathy. *J Am Sec Nephrol,* 14: S241–245 (2003).

9. Rabbani N and Thornalley PJ. Dicarbonyls linked to damage in the powerhouse: Glycation of mitochondrial proteins and oxidative stress. *Biochem Soc Trans,* 36: 1045–1050 (2008).

10. Brouwers O, Niessen PM, Ferreira I, Miyata T, Scheffer PG, Teerlink T, Schrauwen P, Brownlee M, Stehouwer CD, and Schalkwijk CG. Overexpression of glyoxalase–I reduces hyperglycemia-induced levels of advanced glycation end products and oxidative stress in diabetic rats. *J Biol Chem,* 286: 1374–1380 (2011).

11. Ahmed U, Dobler D, Larkin SJ, Rabbani N, and Thornally PJ. Reversal of hyperglycemia-induced angiogenesis deficit of human endothelial cells by over expression of glyoxalase I in vitro. *Ann NY Acad Sci,* 1126: 262–264 (2008).

12. Maher P, Dargusch R, and Ehren JL. Fisetin lowers methylglyoxal-dependent protein glycation and limits the complication of diabetes. *PLOS One,* 6: e21226 (2011).

13. Tang SC, Leung JC, Chan LY, Tsang AW, and Lai KN. Activation of tubular epithelial cells in diabetic nephropathy and the role of the peroxisome proliferator-activated receptor-γ agonist. *J Amer Soc Nephrol,* 17: 1633–1643 (2006).

14. Yu X, Li C, Li X, and Cai L. Rosiglitazone prevents advanced glycation end products–induced renal toxicity likely through suppression of plasminogen activator inhibitor–1. *Toxicol Sci,* 96: 346–356 (2007).

15. Yamagishi S, Matsui T, Nakmurak, Takenchi M, and Inoye H. Telmisartan inhibits advanced glycation end products (AGEs)-elicited endothelial cell injury by suppressing AGE receptor (RAGE) expression via peroxisome proliferator-activated receptor-gamma activation. *Protein Pept Lett,* 15: 850–853 (2008).

16. Yoshida T, Yamagishi S., Matsui T, Nakamura K, Ueno T, Takeuchi M, and Sata M. Telmisartan, an angiotensin II type 1 receptor blocker, inhibits advanced glycation end product (AGE)–elicited hepatic insulin resistance via peroxisome proliferator–activated receptor–gamma activation. *J Int Med Res,* 36: 237–243 (2008).

17. Thomas MC, Tikellis C, Burns WM, Bailkowski K, Cao Z, Coughlan MT, Jandelit-Dahm K, Cooper ME, and Forbes JM. Interactions between renin-angiotensin system and advanced glycation in the kidney. *J Am Sec Nephrol,* 16: 2976–2987 (2005).

18. Mendez JD, Jianling X, Aguilar-Hernandez M, and Mendez-Valenzuela N. Trends in advanced glycation end products research in diabetes mellitus and its complications. *Mol Cell Biochem,* 341: 33–41 (2010).

19. Nakamura S, Makita Z, Ishikawa S, Yasumura K, Fujii W, and Yanagisawa K. Progression of nephropathy in spontaneous diabetic rats is prevented by OPB–9195, a novel inhibitor of advanced glycation. *Diabetes,* 46: 895–899 (1997).

20. Taniomoto M, Gohda T, Kaneko S, Hagiwara S, Murakoshi M, Aoki T, Yamada K, Ito T, Matsumoto M, Horikoshi S, and Tomino Y. Effect of pyridoxamine (K-163), an inhibitor of advanced glycation end products, on type 2 diabetic nephropathy in KK-A^y/Ta mice. *Metabolism,* 56: 160–167 (2007).

21. Heart Outcomes Prevention Evaluation Study Investigators. Effects of ramipril on cardiovascular and microvascular outcomes in people with diabetes mellitus: Results of the HOPE study and MICRO-HOPE substudy. *Lancet,* 355: 253–259 (2000).

22. Brenner BM, Cooper ME, deZeeuw D, Keane WF, Mitch WE, Parving H-H, Remuzzi G, Snapinn SM, Zhang Z, and Shainfar S. Effects of losartan on renal and cardiovascular outcomes in patients with type 2 diabetes and nephropathy. *N Eng J Med,* 345: 861–869 (2001).

23. Shimoi K, Okitsu A, Green MH, Lowe JE, Ohta T, Kaji K, Terato H, Ide H, and Kinae N. Oxidative DNA damage induced by high glucose and its suppression in human umbilical vein endothelial cells. *Mutat Res,* 480–481: 371–378 (2001).

24. Cameron N, Cotter M, Alderson NL, Thorpe SR, and Baynes JW. Pyridoxamine treatment improves nerve function in diabetic rats. *Diabetes,* 52: A192 (2003).

25. Metz TO, Alderson NL, Thorpe SR, and Baynes JW. Pyridoxamine, an inhibitor of advanced glycation and lipoxidation reactions: A novel therapy for treatment of diabetic complications. *Arch Biochem Biophys,* 419: 41–49 (2003).

26. Voziyan PA, Khalifah RG, Thibaudeu C, Yildiz A, Jacob J, and Hudson BG. Modification of proteins in vitro by physiological levels of glucose: Pyridoxamine inhibits conversion of Amadori intermediates to advanced glycation end-products through binding of redox metal ions. *J Biol Chem,* 278: 46616–46624 (2003).

27. Yim MB, Yim HS, Lee C, Kang SO, and Chock PB. Protein glycation: Creation of catalytic sites for free radical generation, *Ann NY Acad Sci,* 928: 48–53 (2001).

28. Brownlee M. Biochemistry and molecular cell biology of diabetic complications *Nature,* 414: 813–820 (2001).

29. Daumer KM, Khan AV, and Steinbeck MJ. Chlorination of pyridinium compounds. Possible role of hypochlorite, N-chloramines and chlorine in the oxidation of pyridinoline cross-links of articular cartilage collagen type II during acute inflammation. *J Biol Chem,* 275: 3468–34692 (2000).

30. Takatori A, Ishi Y, Itagaki S, Kyuwa S, and Yoshikawa Y. Amelioration of the beta-cell dysfunction in diabetic APA hamsters by antioxidants and AGE inhibitor treatment. *Diabetes Metab Res Rev,* 20: 311–218 (2004).

31. Chetyrkin SV, Mathis ME, Ham A-JL, Hachey DL, Hudson BG, and Voziyan PA. Propagation of protein glycation damage involves modification of tryptophan residues via reactive oxygen species: Inhibition by pyridoxamine. *Free Radical Biology & Medicine,* 44: 1276–1285 (2008).

32. Chetyrkin S, Mathis M, McDonald WH, Shackelford X, Hudson B, and Voziyan P. Pyridoxamine protects protein backbone from oxidative fragmentation. *Biochem Biophys Res Communs,* 411: 574–579 (2011).

33. Williams ME, Bolton WK, Khalifah RG, Degenhardt TP, Schotzinger RJ, and McGill JB. Effects of pyridoxamine in combined phase 2 studies of patients with type 1 and type 2 diabetes and overt nephropathy. *Am J Nephrol,* 27: 605–614 (2007).

34. Figarola JL, Loera S, Weng Y, Shanmugam N, Natarajan R, and Rabbar S. LR-90 prevents dyslipidemic and diabetic nephropathy in the Zucker diabetic fatty rat. *Diabetologia,* 51: 882–891 (2008).

35. Nagai R, Murray DB, Metz TO, and Baynes JW. Chelation: A fundamental mechanism of action of AGE inhibitors, AGE breakers and other inhibitors of diabetic complications. *Diabetes,* 61: 549–559 (2012).

36. Izuahar Y, Nangaku M, Takizawa S, Takahashi S, Shao J, Oishi H, Kobayashi H, Van Ypersele de Strihou C, and Miyata T. A novel class of advanced glycation inhibitors ameliorates renal and cardiovascular damage in experimental rat models. *Nephrol Dial Transplant,* 23: 497–509 (2008).

37. Miyata T and van Ypersele de Strihou C. Renoprotection of angiotensin receptor blockers: Beyond blood pressure lowering. *Nephrol Dial Transplant,* 21: 846–849 (2006).

38. Dakshinamurti S and Dakshinamurti K. Vitamin B_6 in *Handbook of Vitamins.* 5th ed., Zemplani J, Rucker RB, Suttie JW, eds. CRC Press, Boca Raton, Florida. (2012).

39. Shepher B, Nash C, Smith C, DiMarco A, Petty J, and Szeweiw S. Update on potential drugs for the treatment of diabetic kidney disease. *Clinical Therapeutics,* 34: 1237–1246 (2012).

40. Williams ME, Bolton WK, Khalifah RG, Degenhardt TP, Schotzinger RJ, and McGill JB. Effects of pyridoxamine in combined phase 2 studies of patients with type 1 and type 2 diabetes and overt nephropathy. *Am J Nephrol,* 27: 605–614 (2007).

41. Lewis EJ, Green T, Spitalewiz S, Blumenthal S, Bert T, Hunsicker LG, Pohl MA, Rhode RD, Raz I, Yerushalmy Y, Yagil Y, Herskovitis T, Atkins RC, Reutens AT, Packham DK, and Lewis JB. Pyridorin in type 2 diabetic nephropathy. *J Am Soc Nephrol,* 23: 131–136 (2012).

42. Chen JLT and Francis J. Pyridoxamine, advanced glycation inhibition and diabetic nephropathy. *J Am Soc Nephrol*, 23: 3–12 (2012).

43. Balakumar P, Rohilla A, Krishnan P, Solairaj P, and Thanga Thirupathi A. The multifaceted therapeutic potential of benfotiamine. *Pharmacol Res*, 61: 482–488 (2010).

44. Beltramo E, Berrone E, Tarallo S, and Porta M. Effects of thiamine and benfotiamine on intracellular glucose metabolism and relevance in the prevention of diabetic complications. *Acta Diabetol*, 45: 131–141 (2008).

45. Sustzak K, Raff AC, Schiffer M, and Bottinger EP. Glucose-induced reactive oxygen species cause apoptosis of podocytes and podocyte depletion at the onset of diabetic nephropathy. *Diabetes*, 55: 225–231 (2006).

46. Piconi L, Quagliaro L, Assaloni R, Da Ros R, Maier A, Zuodar G, and Ceriello A. Constant and intermittent high glucose enhances endothelial cell apoptosis through mitochondrial superoxide overproduction. *Diab Met Res Rev*, 22: 198–203 (2006).

47. Du XL, Edelstein D, Rosetti L, Fantus IG, Goldberg H, Ziyadeh F, Wu J, and Brownlee M. Hyperglycemia-induced mitochondrial superoxide overproduction activates the hexosamine pathway and induces plasminogen activator inhibitor-1 expression by increasing Sp1 glycosylation. *Proc Natl Acad Sci USA*, 97: 12222–12226 (2000).

48. De Rubertis FR, Craven PA, Mehlman MF, and Salah EM. Attenuation of renal injury in db/db mice overexpressing superoxide dismutase: Evidence for reduced superoxide-nitric oxide interaction. *Diabetes*, 53: 762–768 (2004).

49. Gadau S, Emanuelli C, Van Linthout S, Graiani G, Todaro M, Meloni M, Campesi I, Invernici G, Spillman F, Ward K, and Madeddu P. Benfotiamine accelerates the healing of ischemic diabetic limbs in mice through protein kinase B/Akt–mediated potentiation of angiogenesis and inhibition of apoptosis. *Diabetologia*, 49: 405–420 (2006).

50. Marchetti V, Menghini R, Rizza S, Vivanti A, Feccia T, Lauro D, Fukamizu A, Lauro R, and Federici M. Benfotiamine counteracts glucose toxicity effects on endothelial progenitor cell differentiation via Akt/FoxO signaling. *Diabetes*, 55: 2231–2237 (2006).

51. Balakumar P, Sharma R, and Singh M. Benfotiamine attenuates nicotine and uric acid-induced vascular endothelial dysfunction in the rat. *Pharmacol Res*, 58: 356–363 (2008).

52. Verma S, Reddy K, and Balakumar P. The defensive effect of benfotiamine in sodium arsenite-induced experimental vascular endothelial dysfunction. *Biol Trace Elem Res*, 137: 96–109 (2010).

53. Sompamit K, Kukongviriyapan U, Nakmareong S, Pannangpetch P, and Kukongviriyapan V. Curcumin improves vascular function and alleviates oxidative stress in non-lethal lipopolysaccharide-induced endotoxemia in mice. *Eur J Pharmacol*, 616: 192–199 (2009).

54. Wilson JX. Mechanism of action of vitamin C in sepsis: Ascorbate modulates redox signaling in endothelium. *Biofactors*, 35: 5–13 (2009).

55. Yadav UCS, Kalariya NM, Srivastava SK, and Ramana KV. Protective role of benfotiamine, a fat-soluable vitamin B_1 analogue, in lipopolysaccharide-induced cytotoxic signals in murine macrophages. *Free Radical Biol & Med*, 48: 1423–1434 (2010).

56. Shoeb M and Ramana KV. Anti-inflammatory effects of benfotiamine are mediated through the regulation of the arachidonic acid pathway in macrophages. *Free Radical Biol & Med*, 52: 182–190 (2012).

57. Katare R, Caporali A, Emanuelli C, and Madeddu P. Benfotiamine improves functional recovery of the infarcted heart via activation of pro-survival G6PD/Akt signaling pathway and modulation of neurohormonal response. *J Mol Cell Cardiol*, 49: 625–638 (2010).

58. Katare RG, Caporali A, Oikawa A, Meloni M, Emianneli C, and Madeddu P. Vitamin B analog benfotiamine prevents diabetes-induced diastolic dysfunction and heart failure through Akt/Pim-1-mediated survival pathway. *Circ Heart Fail*, 3: 294–305 (2010).

59. Pan X, Gong N, Zhao J, Yu Z, Gu F, Chen J, Sun X, Zhao L, Yu M, Xu Z, Dong W, Quin Y, Fei G, Zhong C, and Xu T-L. Powerful beneficial effects of benfotiamine on cognitive impairment and β-amyloid deposition in amyloid precursor protein/presenilin-1 transgenic mice. *Brain,* 133: 1342–1351 (2010).

60. Haupt E, Ledermann H, and Kopeke W. Benfotiamine in the treatment of diabetic polyneuropathy—a three-week randomized, controlled pilot study (BEDIP study). *Int J Clin Pharmacol Ther,* 43: 71–77 (2005).

61. Stirban A, Negrean M, Stratmann B, Gawlowski T, Horstmann T, Gotting C, Kleesiak K, Mueller-Roesel M, Kochinsky T, Uribari J, Vlassara H, and Tschope D. Benfotiamine prevents macro-and micro-vascular endothelial dysfunction and oxidative stress following a meal rich in advanced glycation end products in individuals with type 2 diabetes. *Diabetes Care,* 29: 2064–2071 (2006).

11 Multifaceted Therapeutic Potential of Vitamin B$_6$

Shyamala Dakshinamurti and
Krishnamurti Dakshinamurti

CONTENTS

The term *vitamin B$_6$* refers to a group of naturally occurring pyridine derivatives represented by pyridoxine (pyridoxol, PN), pyridoxal (PL), and pyridoxamine (PM), and their phosphorylated derivatives. They are collectively referred to as *vitamin B$_6$ vitamers*. The natural free forms of the vitamers could be converted to the key coenzymatic form, pyridoxal phosphate (PLP), by the action of two enzymes, a kinase and an oxidase. There are more than 140 PLP-dependent enzymatic reactions, and they are distributed in all organisms. These enzymes comprise diverse groups such as the oxidoreductases, transferases, hydrolases, lyases, and isomerases. About 1.5% of the genes of free-living prokaryotes encode PLP enzymes. These enzymes participate in the metabolism of amino acids, carbohydrates, and lipids, indicating the versatility of PLP-dependent enzymes.

11.1   NEUROBIOLOGY OF VITAMIN B$_6$

The crucial role played by vitamin B$_6$ in the nervous system is evident from the fact that the putative neurotransmitters, dopamine (DA), norepinephrine (NE), serotonin (5-HT), and γ-amino butyric acid (GABA), as well as taurine, sphingolipids, and polyamines, are synthesized by PLP-dependent enzymes. Of the PLP enzymes, those involved in the decarboxylations respectively of glutamic acid, 5-hydroxytryptophan, and ornithine are of considerable significance and can explain most of the neurological defects of vitamin B_6 deficiency in all species studied.

The endogenous production of hydrogen sulfide (H_2S) depends upon PLP enzymes such as cystathionine beta synthase (CBS), cystathionine gamma lyase (CSE), and 3 mercaptopyruvate sulfotransferase (3MST). CBS is highly expressed in the hippocampus and cerebellum. The astrocytes are the main cells in the brain to produce H_2S. Hydrogen sulfide is involved in the regulation of intracellular signaling molecules such as protein kinase A, receptor tyrosine kinase, mitogen kinase, and oxidative stress signaling. Ion channels such as the L and T type calcium channels, and potassium and chloride channels, are regulated by H_2S. Also, the release and function of neurotransmitters such as gamma aminobutyric acid (GABA), N-methyl-D-aspartate (NMDA), glutamate, and catecholamines require H_2S as a signal. Thus, H_2S has a number of neuronal functions and is important in neuroprotection (1). H_2S does not circulate in plasma at a measurable concentration and, hence, acts in various organs and tissues in a paracrine fashion. Recurrent febrile seizures result in hippocampal damage through loss of receptor $GABA_B$ R1 and $GABA_B$ R2. H_2S, through its effect on calcium channels, increases Ca^{2+}-dependent transcription of these receptor proteins and, thus, protects the hippocampus. H_2S promotes cell survival of both the neurons and glia. As the molecule has a short half life, its effects are transmitted through its action on messenger molecules or ion channels. Hence, H_2S operates as a "switch" molecule whose in vivo synthesis is regulated by PLP.

Vitamin B_6 in its various forms has antioxidant properties that compare favorably with those of the well-established antioxidant vitamins such as vitamin C and the tocopherols. Pyridoxine and pyridoxamine inhibit superoxide radicals and prevent lipid peroxidation, protein glycosylation, and Na+,K+-ATPase activity in high glucose–treated erythrocytes and hydrogen peroxide–treated monocytes (2) and endothelial cells (3). In bovine endothelial cells, treatment with homocysteine and copper increased extracellular hydrogen peroxide levels. Treatment with pyridoxal or EDTA prevented such increases and enhanced the viability of the cells by supporting apoptosis.

Hyperglycemia-induced oxidative stress has an important role in the pathogenesis of diabetic complications. Human monocytes exposed to 2-deoxy-D-ribose exhibited loss of cell viability, overproduction of ROS, depletion of glutathione, and apoptosis. Treatment with PLP inhibited these as well as lipid peroxidation and protein oxidation (4). Pyridoxine has a very high level of quenching of hydroxyl radicals (5). Mono- and bicyclic amino pyridinols have been synthesized from pyridoxine hydrochloride and have been shown to have antioxidant properties (6).

11.2 NEUROTRANSMITTERS AND VITAMIN B_6

The putative neurotransmitters, dopamine (DA), norepinephrine (NE), serotonin (5-HT), and γ-amino butyric acid (GABA), as well as taurine, sphingolipids, and polyamines, are synthesized by PLP-dependent enzymes. There is considerable variation in the affinities of the various apoenzymes for PLP. This explains the observed differential susceptibility of various PLP enzymes to decrease during vitamin B$_6$ depletion in animals and humans. The decarboxylations, respectively, of glutamic acid, 5-hydroxytryptophan, and ornithine are considerably decreased in vitamin B$_6$ deficiency in all species studied.

The enzyme L-aromatic amino acid decarboxylase (AADC, EC 4.1.1.28) lacks substrate specificity and has been considered to be involved in the formation of the catecholamines and serotonin. There are many differences in the optimal conditions for enzyme activity, including kinetics, affinity for PLP, activation and inhibition by specific chemicals, and regional differences in the distribution of DOPA and 5-HTP decarboxylation activities. Nonparallel changes in brain monoamines in the vitamin B$_6$–deficient rat have been reported (7–9). Brain content of dopamine and norepinephrine were not decreased during deficiency, whereas serotonin was significantly decreased.

γ-aminobutyric acid (GABA) is present almost exclusively in the nervous system of invertebrates and vertebrates. It is formed from glutamic acid through the action of glutamic acid decarboxylase (GAD) and is catabolized by transamination catalyzed by GABA transaminase (GABA-T) to yield succinic semialdehyde (SSA). Both GAD and GABA-T are PLP enzymes. GABA is an inhibitory neurotransmitter, whereas glutamic acid is an excitatory neurotransmitter. GABA, GAD, and GABA-T are localized predominantly in the regions of the brain that are inhibitory in function. The concentration of GABA in the cerebellum is particularly high in the Purkinje cells. The neurophysiological action of GABA studied by iontophoretic application resembles that observed in postsynaptic inhibition produced by electrical stimulation. Apart from the involvement of GABA in the etiology of certain convulsive seizures, abnormalities in GABA-ergic neuronal pathways contributing to other CNS disorders such as depression, anxiety, and panic disorders have been recognized. In the moderately vitamin B$_6$–deficient rat, there are biologically significant decreases in the activities of GAD65 and AADC (5 HTP-DC) acting on 5-hydroxytryptophan leading to decreases in neurotransmitters GABA and serotonin (5-HT). DOPA decarboxylase activity is not affected during PLP depletion, resulting in no change or even in an increase in catecholamine levels in the nervous system. Decreased brain serotonin in the vitamin B$_6$–deficient rat is implicated in physiological changes such as decreased deep-body temperature and altered sleep pattern with shortening of deep slow-wave sleep and REM sleep. The effects of vitamin B$_6$ depletion on sleep parallel the effects of experimental serotonergic deficit (10).

11.3 NEUROENDOCRINOLOGY OF VITAMIN B$_6$

The hypothalamus is one of the areas of the brain of vitamin B$_6$–deficient rats with significant decreases in PLP and serotonin compared with vitamin B$_6$–replete controls. There is no decrease in the contents of dopamine and norepinephrine. The concept of the regulatory role of the hypothalamus through the neurotransmitters is generally accepted. Regulation of the release of stimulatory or inhibitory factors by the hypothalamus involves complex neural circuitry in which the serotonergic and dopaminergic neurons represent links in the control mechanisms (11). The hypothalamus of the normal animal has high concentrations of both dopamine and serotonin, which are essentially antagonistic in their effects on pituitary hormone regulation.

In determining the locus of the biochemical lesion leading to the hypothyroid state in vitamin B_6 deficiency, various possibilities such as primary with a defective thyroid gland, secondary with a defective pituitary thyrotroph, or tertiary with a defective hypothalamus were considered. The chronic deficiency of TRH in the deficient rat is indicated by an increase in the number of TRH receptors with no change in receptor affinity. Results reported (12) are consistent with a hypothalamic type of hypothyroidism in the vitamin B_6–deficient rat caused by the specific decrease in hypothalamic serotonin level.

The pineal gland transduces photoperiodic information and, hence, has a crucial role in the temporal organization of various metabolic, physiological, and behavioral processes. Melatonin is the major secretory product of the pineal gland. Tryptophan is hydroxylated in the pinealocyte to 5-hydroxy-tryptophan (5 HTP) and decarboxylated to yield serotonin. Serotonin is converted to N-acetyl serotonin (NAS) by the enzyme N-acetyl-transferase (NAT). NAS is converted to melatonin by hydroxyindole-O-methyltransferase. Melatonin synthesis is stimulated by β-adrenergic post-ganglionic sympathetic fibers from the superior cervical ganglion, which are stimulated in the dark. Melatonin levels in tissues and body fluids show both circadian and seasonal rhythms. Pineal levels of 5-HT and 5-HIAA were significantly lower in the deficient rats. Treatment of deficient rats with pyridoxine restored the levels of 5-HT, NAS, and melatonin to levels seen in vitamin B_6–replete controls. Such reversal was evident both during day- and nighttime periods. There was no difference in pineal NAT between deficient and control animals. However, pineal 5-HTP decarboxylase activity was significantly decreased in vitamin B_6–deficient rats (13). Tryptophan hydroxylation is considered to be the rate-limiting step in the syntheses of serotonin. Several studies indicate that a decrease in pineal 5-HT can reduce melatonin synthesis. In vivo administration of aromatic amino acid decarboxylase inhibitors such as benzerazide or monofluoromethyl dopa result in a reduction in the synthesis of pineal 5-HT and melatonin levels without altering pineal NAT activity. Thus, 5-HT availability, in addition to other known factors, could be important in the regulation of the synthesis of melatonin.

The secretion of prolactin (PRL) is controlled by both stimulatory and inhibitory factors of hypothalamic origin. The inhibitory control is exerted primarily by dopamine, which is released from the tuberoinfundibular dopaminergic (TIDA) neurons into the pituitary portal circulation. Evidence based on peripheral administration of serotonin precursors, agonists or antagonists, intraventricular injection of serotonin, and electrical stimulation of the raphe nucleus indicates that central serotonergic projections to the hypothalamus are involved in the stimulation of PRL. Administration of pyridoxine to deficient rats resulted in a significant increase in plasma PRL (14).

Vitamin B_6 status has significant effects on the central production of serotonin and GABA, neurotransmitters that control pain perception, anxiety, and depression. Various reports suggest that high-dose pyridoxine, through its effects on neurotransmitters, might have favorable impacts on dysphoric mental states (15, 16).

11.4 PYRIDOXINE-DEPENDENCY SEIZURES

For about half a century, pyridoxine dependency has been recognized as an inborn abnormality. Infants present, generally soon after birth, with seizures that are resistant to the commonly used antiepileptic drugs and respond only to pharmacologic doses of pyridoxine. It is a rare autosomal recessive genetic disorder. In view of the prevalence of atypical variants of this disorder, it is thought to be under-recognized. A pyridoxine-dependent condition has to be considered in all children with intractable epilepsy up to three years of age (17). A variety of seizure types such as myoclonic seizures, atonic seizures, partial and generalized seizures, as well as infantile spasms are all associated with this condition. The unusual rhythmic in utero movements reported retrospectively by some mothers might represent fetal seizures (18). At present there is no biochemical test to confirm pyridoxine-dependency seizures, and clinical diagnosis is the only mode of recognition. Response to pyridoxine monotherapy and recurrence of seizures following withdrawal of treatment is the only confirmatory test. Such testing is fraught with difficulty due to ethical considerations. Pyridoxine dependency is to be distinguished from vitamin B$_6$ deficiency, first reported in infants fed on commercial milk formula where autoclaving destroyed the vitamin B$_6$ content (19).

The intravenous administration of 50–100 mg of pyridoxine generally results in a dramatic cessation of seizures. In some cases, doses up to 500 mg as well as repeated dosing might be needed. This additional requirement of pyridoxine is for life. Once the initial seizures are controlled, the minimum dose necessary for maintenance of a seizure-free state can be titrated. Delay in achieving milestones, developmental defects, as well as permanent brain damage are the concomitants of untreated pyridoxine-dependency condition. Most patients have some degree of cognitive impairment, particularly in language expression. Brain imaging studies indicate gray and white matter atrophy in these patients. The progressive MRI (magnetic resonance imaging) changes are suggestive of selective neuronal loss (20). Magnetic resonance spectroscopy is complementary to MRI and shows the presence of various metabolites in the samples. Such study revealed a decrease in N-acetyl aspartate to creatine ratio in the frontal and parieto-occipital cortices indicative of neuronal loss (21). Recent studies (22) indicate that increasing the dose of pyridoxine in pyridoxine-dependent children without seizures could improve their IQ, indicating a role for pyridoxine in normal brain development and in functions other than in controlling the excitable state.

Autopsy studies on pyridoxine-dependent seizure patients showed elevated glutamate and decreased GABA levels in the frontal and occipital cortices (21). Similar observations in the cerebrospinal fluid of affected patients point to a defect in the conversion of glutamate to GABA (23). Pyridoxal phosphate is the coenzyme of GAD. An abnormality of this enzyme was presumed to be responsible for the impairment in the synthesis of GABA from glutamate. Of the two isoforms of GAD, GAD-65 is PLP dependent, and the defective binding of PLP to this enzyme was suggested to be the cause of the decreased synthesis of GABA in vitamin B$_6$–dependent seizures. Studies using cultured fibroblasts obtained from affected patients showed a reduction in the PLP-dependent GABA synthesis. More recent studies, however, have excluded a mutation in either of the two isoforms of GAD as the molecular defect responsible for pyridoxine-dependent seizures (24). The report of elevated blood and CSF levels of pipecolic acid, a precursor to a PLP-dependent reaction in the pathway of the metabolic conversion of lysine to glutaryl-CoA, is significant as it has been shown that a high concentration of pipecolic acid inhibits GABA uptake (25).

Four conditions of inborn errors of metabolism decrease brain concentrations of vitamin B_6 and result in seizures in the newborn. Hyperprolinemia type 2, antiquitin deficiency, pyridoxine phosphate oxidase deficiency, and hypophosphatasia are all associated with early onset seizures that do not respond to conventional antiepileptic medication (26). Mutations of the antiquitin gene leading to antiquitin (ATQ) deficiency are recognized to be the major cause of early onset seizures. ATQ functions as an aldehyde dehydrogenase in the pathway of lysine catabolism, and a deficiency leads to accumulation of α-aminoadipic semialdehyde, piperidine-6-carboxylate, and pipecolic acid (27, 28). Seizures in infants are treated with pyridoxine given intravenously. Seizures due to hyperprolinemia, hypophosphatasia, and ATQ deficiency respond to any form of vitamin B_6. However, infantile seizures caused by pyridoxine phosphate oxidase deficiency respond only to pyridoxal phosphate (29). Hence, prompt treatment of all infants and neonates having epileptic seizures with pyridoxal phosphate permits the normal development of afflicted patients due to any of the conditions (30, 31).

There are other seizure conditions in which pyridoxine therapy finds a place. Infantile spasm (spastic convulsions), in combination with diffuse electroencephalographic abnormalities (hypsarrhythmia), is referred to as "West syndrome." Mental retardation is associated with this condition (32). ACTH is effective for the short-term treatment of infantile spasms. In view of the elevated therapy-associated morbidity, valproic acid and vigabatrin have been used (33). Following reports of beneficial effects of high doses of pyridoxine, initial treatment for one to two weeks with high doses of pyridoxine is the established therapy in some European countries and in Japan (34). Combined therapy with high-dose pyridoxine in association with low-dose corticotrophin has also been reported as a promising therapy for seizure control, normalized EEG, and intellectual outcome.

11.5 SEIZURES AND VITAMIN B_6

In view of the clinical and biochemical manifestations of vitamin B_6 deficiency in young and adult animals, it was of interest to produce and characterize vitamin B_6 deficiency in the very young rat. The report of the existence of a critical period in the development of the central nervous system indicated the importance of inducing deficiency during or prior to this period. Deficient pups had a significantly lower content of PLP in their brains. Related to this observation was the occasional finding, among the vitamin B_6–deficient group, of pups with spontaneous convulsions that became noticeable at about three to four days of postnatal age. These fits were characterized by a high-pitched scream followed by generalized convulsions of a few seconds' duration and repeated many times within a one- to three-minute time period. The motility, perception, and alertness of the deficient neonates were inferior to that of the controls. This was the first report of the production of congenital pyridoxine deficiency (35). In view of the high mortality of the deficient pups, they could not be used in studies on the development of the central nervous system, which in the rat extends from the tenth to the twenty-first day after birth. In a further study, female rats were fed a vitamin B$_6$–deficient diet from the first postpartum day, and the pups were fed the deficient diet from the time they were weaned until they were five to six weeks of age. The lower levels of GABA in the brains of deficient pups are directly related to the decrease in the activity of the GAD holoenzyme. The GAD apoenzyme levels were quite significantly increased, possibly due to stimulation of the apoprotein synthesis by the low concentration of GABA. The effects of deficiency on various electrophysiological parameters were examined. The bursts of high-voltage spikes during spontaneous EEG activity, as well as the spontaneous convulsions observed, reflect the decrease in cerebral GABA concentration in deficient rats. The more complicated changes in cortical auditory evoked potentials in the vitamin B$_6$–deficient rats are the result of the retardation of the normal ontogenetic development of the CNS of these rats. The inability of the cerebral cortex of the deficient pups to follow the increasing frequency of any kind of repetitive stimulus as well as a marked decrease in the amplitude of evoked potentials were quite apparent and correlated with decreases in both PLP and GABA levels in various brain areas.

The thalamus acts as a relay station for various peripheral and central inputs to the cerebral cortex. Hence, the electroresponsiveness of thalamic VPL (ventro-posterior lateral) neurons in normal control and vitamin B$_6$–deficient adult rats in response to local administration of convulsants such as picrotoxin or pentylene tetrazole were investigated. The extent of neuronal recovery following intrathalamic administration of either GABA or pyridoxine or systemic administration of pyridoxine was assessed using computerized EEG analysis. The results demonstrated an antiepileptic effect of exogenously applied GABA and pyridoxine on thalamic VPL neurons, with pyridoxine having a much slower effect than GABA. Brain levels of PLP, GAD, and GABA responded to the systemic administration of pyridoxine. Even in normal rats, GAD is unsaturated with respect to the cofactor PLP (36, 37). Brain levels of glutamate were significantly increased following administration of picrotoxin or pentylene tetrazole. Excitatory neurotransmitters such as glutamate are linked to the initiation and spread of seizure discharge, and GABA is responsible for the termination of seizure activity. Neuronal recovery following pyridoxine is related to the synthesis of GABA through activation of GAD (30, 37–39).

11.6 NEUROTOXINS AND NEUROPROTECTION BY VITAMIN B$_6$

Domoic acid, a rigid structural analog of glutamate, is a neuroexcitant. It was identified as the toxic contaminant of cultivated mussels responsible for the outbreaks of acute food poisoning characterized by gastrointestinal and neurologic symptoms (40). The hippocampal CA-3 region was chosen for the study of seizure activity as this region has minimum seizure threshold compared to other cerebral areas. Acute intrahippocampal administration of picomole amounts of domoic acid led to EEG epileptiform seizure discharge activity. Domoic acid was 125 times more potent than kainic acid, a well-known neuroexcitant. Local administration of GABA or pyridoxine attenuated the seizure activity (37, 41). Following domoic acid injection, GABA levels decreased significantly in various brain regions. Domoic acid inhibited GAD activity. As tissue levels of glutamate do not represent the neuronal pool of glutamate, we studied the effect of domoic acid on the in vitro release of glutamate in tissue superfusion experiments. The KCl-induced depolarization leads to release of neurotransmitter glutamate. This was augmented by domoic acid. The direct application of GABA to the hippocampus of rats exhibiting domoic acid–induced seizure activity resulted in suppression of spike discharges. The slower effect of pyridoxine is related to the augmented pyridoxal phosphate–dependent formation of GABA from glutamate. Similar observations have been reported by others (42).

It has been reported that serotonin functions in the stabilization of brain regional GABA-ergic neurons and in seizure control. Serotonin has been shown to have a stimulatory effect on brain GABA-ergic neurotransmission. The neuroprotective action of pyridoxine flows from its effect on the synthesis of both GABA and serotonin. Hippocampal changes in developing mice at postnatal age ten to thirty days following intrauterine exposure to a single dose of domoic acid at day thirteen of gestation were studied (43). The mice exhibited age-related developmental neurotoxicity. Brain regional GABA levels were significantly reduced, and glutamate levels increased in these mice. Neuronal death was apparent in the offspring at thirty days of chronological age. This delayed neurotoxicity cannot be attributed to the acute effect of domoic acid and might be related to the increased sensitivity of hippocampal cells to the high concentration of endogenous glutamate.

In other experiments, electroencephalographic recordings in the cerebral cortex of adult mice given a single subconvulsive dose of domoic acid exhibited typical spike and wave discharges. Administration of drugs such as sodium valproate or nimodipine or pyridoxine simultaneously with or after domoic aid treatment resulted in significantly less spike and wave activity. Administration of these same drugs forty-five minutes prior to the administration of domoic acid also significantly decreased domoic acid–induced EEG background (44). Mechanistically, sodium valproate and pyridoxine significantly attenuated the domoic acid–induced increase in the levels of glutamate, increase in calcium influx, C-fos, jun B, and jun D. Pyridoxine, acting through pyridoxal phosphate, appears to decrease intracellular levels of glutamate by increasing influx, to decrease levels of GABA and increase levels of proto-oncogenes and glutamic acid decarboxylase activity, and to decrease calcium influx through its action on cell-surface calcium channels.

In a subsequent study, primary cultures of hippocampal neurons were exposed to domoic acid. This resulted in apoptotic changes. ^{1}H-NMR proton spectra of domoic acid–treated neuronal cultures exhibited significantly increased glutamate and reduced GABA levels. Pretreatment of cells with pyridoxal phosphate prior to domoic acid exposure reduced glutamate and increased GABA levels. Exposure of cells to domoic acid or Bay-K 8644 increased the calcium influx into neurons. This was again significantly decreased by preincubating hippocampal cells with either nimodipine or pyridoxal phosphate. The domoic acid–induced proto-oncogene expression was also significantly suppressed by preincubation of the cell culture with either nimodipine or pyridoxal phosphate (44). These results would suggest that domoic acid–induced apoptosis might occur through increased calcium influx, proto-oncogene induction, and intranucleosomal DNA fragmentation. Similar results were obtained by exposure of cultures of neuroblastoma/glioma 108/15 cells to domoic acid and to pyridoxine, respectively.

Aluminum has been suggested to be a risk factor in the pathogenesis of Alzheimer disease. In experimental work, aluminum significantly impairs dendritic connectivity within the rat hippocampus. When rats were fed pyridoxine for thirty days following aluminum injections the branch points in all areas of the hippocampus increased significantly (45). This effect might be due to the effect of pyridoxine on neurotransmitter and polyamine syntheses.

Ischemia/reperfusion injury to brain is a major cause of neurological abnormalities. The injection of pyridoxal phosphate into the lateral cerebral ventricle after ischemic insult significantly protected CA1 pyramidal neurons of the hippocampus from ischemic damage (46) through increased synthesis of GABA. Pyridoxine has also been shown to inhibit the release of glutamate that was induced by exposing synaptasomes to the K$^+$ channel blocker 4-aminopyridine (47). This inhibition was shown to be through a suppression of voltage-dependent Ca++ channels (44).

The effects of pyridoxine administration on cell death, cell proliferation, neuroblast differentiation, and the GABA-ergic system in the mouse dendrite gyrus was examined (48). GFAP-positive cells, a marker for cell proliferation, and double-cortin, a marker for neuroblast differentiation, were significantly increased in the pyridoxine-treated mice. The levels of glutamic acid decarboxylase GAD 67 and pyridoxal phosphate oxidase (PNPO) were significantly increased in the dentate gyrus, indicating that pyridoxine had a significant effect on cell proliferation and neuroblast differentiation.

In further work, Yoo et al. (49) have shown that pyridoxine treatment, in association with the histone deacetylase inhibitor sodium butyrate, significantly restored the age-related reductions in memory function, cell proliferation, neuroblast differentiation, and pCREB immunoreactivity in the dentate gyrus of a mouse model of aging induced by D-galactose. In subsequent work, these investigators (50) have shown the restorative potential of pyridoxine on ischemic damage in the hippocampal CA1 region of Mongolian gerbils. Chronic administration of pyridoxine enhanced neuroblast differentiation in the dentate gyrus and neurogenesis in the hippocampal CA1 region by up-regulating BDNF expression in the hippocampus.

11.7 CARDIOVASCULAR FUNCTION AND VITAMIN B$_6$

Hypertension is one of the major causes of chronic illness in Western societies, where about 20%–30% of the adult population has some degree of blood pressure elevation. The moderately vitamin B$_6$–deficient male rat has been introduced as an additional animal model for the study of experimental hypertension (51–53).

Male Sprague-Dawley rats were fed a vitamin B$_6$–deficient diet for up to twelve weeks (36) and compared with a group pair-fed a pyridoxine-supplemented diet. The systolic blood pressure changes in the vitamin B$_6$–deficient rat can be classified into three phases: prehypertensive (1–4 weeks), hypertensive (5–11 weeks), and posthypertensive (from week 12). During the hypertensive phase the rats were only moderately vitamin B$_6$ deficient and have been biochemically characterized in terms of tissue vitamin B$_6$ levels. They were functionally deficient in the neurotransmitters serotonin and GABA. Treatment of the hypertensive vitamin B_6–deficient rats with dietary pyridoxine corrected both the deficiency state and the hypertensive condition. The possibility that the reversible hypertension was related to sympathetic stimulation was examined. The concentration of norepinephrine (NE) in plasma is a valid reflection of sympathetic activity. Both epinephrine and norepinephrine levels in the plasma of hypertensive vitamin B_6–deficient rats were three-fold higher compared with controls. Significantly, NE turnover in the hearts of deficient hypertensive rats was three-fold higher as compared to controls. Treatment of the rats with pyridoxine returned both the blood pressure and catecholamine levels to normal within twenty-four hours. Pyridoxine administration to control rats had no significant effect on any of these parameters. The complete reversibility of hypertension in such a short time would preclude permanent structural damage to the vessel wall of the vitamin B_6–deficient rat. The lesion could be at the level of neurotransmitter regulation.

Serotonergic neurotransmission in the central nervous system controls a wide variety of functions such as blood pressure, emotional behavior, endocrine secretion, perception of pain, and sleep. It is possible that the decrease in neuronal 5-HT and the consequent changes in its receptors, particularly 5-HT_{1A}, may cause hypertension in vitamin B_6 deficiency. This was investigated in vitamin B_6–deficient rats after they had developed peak hypertension by examining the effects of various $5HT_{1A}$ agonists. They all had an acute hypotensive effect in these rats.

11.7.1   PYRIDOXAL PHOSPHATE AND CALCIUM CHANNELS

The end result of centrally mediated sympathetic stimulation is an increase in peripheral resistance. This is reflected in elevations of both resting and stimulated vascular tone in the resistance arteries of the moderately vitamin B_6–deficient hypertensive rats. Elevated peripheral resistance is the hallmark of hypertension as seen in other models of hypertension. The increase in tone of caudal artery segments from the hypertensive vitamin B_6–deficient rat is calcium dependent. The decrease in tone following the addition to the medium of the calcium channel antagonist, nifedipine, indicates that the increased peripheral resistance resulting from increased permeability of smooth muscle plasma membrane to Ca^{2+} might be central to the development of hypertension in the vitamin B_6–deficient rat.

Calcium influx occurs through plasma membrane Ca^{2+} channels that are voltage-operated or receptor mediated. Voltage-sensitive calcium channels open upon depolarization of the cell membrane, resulting in an inward movement of calcium ions. ATP is an important extracellular nucleotide that mediates its effect via plasma membrane–bound P2 receptors.

The slow channel (L-type) is the major pathway by which Ca^{2+} enters the cell during excitation for initiation and regulation of the force of contraction of cardiac and skeletal muscle. Vascular smooth muscle also contains the L-type channel. The possibility that in the vitamin B_6–deficient hypertensive rat a higher concentration of cytosolic free Ca^{2+} might be responsible for the higher tension in the vascular smooth muscle was evaluated. In the deficient hypertensive rats the $[^{45}Ca^{2+}]$ influx into the vascular smooth muscle was significantly increased to twice that of the control (54).

The alterations in [Ca^{+2}]$_i$ induced by KCl in vitamin B$_6$–deficient hypertensive and control rats were investigated. [Ca^{2+}]$_i$ was measured in isolated cardiomyocytes using the Fura-2 fluorescent technique. The KCl-induced [Ca^{2+}]$_i$ increase was significantly higher in cardiomyocytes isolated from vitamin B$_6$–deficient hypertensive rats. A single injection of vitamin B$_6$ (10 mg/kg body weight) to the deficient animal completely reversed the KCl-induced changes in [Ca^{2+}]$_i$ due to vitamin B$_6$ deficiency.

Chronic ingestion of simple carbohydrates such as sucrose or fructose has been shown to increase the SBP in several rat strains. This was attenuated by the inclusion of a vitamin B$_6$ supplement (five times the normal intake) in the diet. The possibility that a dietary supplement of vitamin B$_6$ could attenuate the elevation of SBP in genetically hypertensive animal models such as the Zucker obese or the spontaneously hypertensive rat (SHR) was examined. Male Zucker obese rats (fa/fa) fed a commercial rat chow developed hypertension in three to four weeks. The inclusion of a dietary vitamin B$_6$ supplement (five times the normal intake) resulted in a complete attenuation of the hypertension in the obese strain. This was reversible. In contrast to the effect seen in the Zucker obese rats, there was no response to the inclusion of a dietary vitamin B$_6$ supplement in SHRs. The changes in SBP in the Zucker as well as in the sucrose- or fructose-fed rats correlated with the changes in the uptake of calcium by the caudal artery segments in these groups (55).

In view of the previous results, the possibility that pyridoxine or more particularly pyridoxal phosphate could directly modulate the cellular calcium uptake process was investigated. BAY K 8644, a DHP-sensitive calcium channel agonist, stimulated calcium [^{45}Ca^{2+}] entry into artery segments from control rats. Pyridoxal phosphate dose-dependently reduced the BAY K 8644–stimulated calcium uptake by control artery segments (56). The basal uptake of [^{45}Ca^{2+}] by caudal artery segments from vitamin B$_6$–deficient hypertensive rats was at least twice the uptake by artery segments from control normal rats. Pyridoxal phosphate or nifedipine added to the incubation medium significantly decreased the [^{45}Ca^{2+}] uptake by artery segments from the vitamin B$_6$–deficient hypertensive rats. However, in the presence of BAY K 8644 in the incubation medium, both pyridoxal phosphate and nifedipine were much less effective in attenuating the [^{45}Ca^{2+}] uptake by artery segments from the deficient hypertensive rats. These in vitro direct antagonisms indicate the possibility that the calcium channel agonist BAY K 8644, the calcium channel antagonist nifedipine, and pyridoxal phosphate might all act at the same site on the calcium channel. We examined the effect of pyridoxal phosphate on the binding of tritiated nitrendipine, a dihydropyridine calcium channel antagonist, to membrane preparation from the caudal artery of normal rats. Pyridoxal phosphate treatment of crude membranes of caudal artery resulted in a significant decrease in the number of [^{3}H] nitrendipine binding sites. 1,4-Dihydropyridine derivatives such as cilnidipine inhibit both L-type and N-type calcium channels and are used therapeutically in the treatment of hypertension (57).

The neuroprotective effect of pyridoxal phosphate has been demonstrated in ischemic brain injury using a focal embolic model of stroke (58). The mechanism of action could be through prevention of a calcium overload, through the action of pyridoxal phosphate on calcium channels. In a further study, the influence of pyridoxal phosphate on the ATP-induced contractile activity of the isolated rat heart and the ATP-mediated increase in $[Ca^{2+}]_i$ in freshly isolated adult rat cardiomyocytes, as well as on the specific binding of ATP to cardiac sarcolemmal membrane, was examined in order to determine if pyridoxal phosphate is an effective antagonist of ATP receptors in the myocardium. The contractile activity of the isolated perfused rat heart was monitored upon infusion with 50 μm ATP in the presence or absence of 50 μm pyridoxal phosphate. The infusion of ATP caused an immediate increase (within seconds) in LVDP, +dP/dt, and –dP/dt. This effect was completely blocked in the hearts pretreated with pyridoxal phosphate for ten minutes. The antagonistic effect of pyridoxal phosphate was concentration dependent. The specificity of the effect of pyridoxal phosphate was established as propranolol, which prevented the positive inotropic action of isoproterenol, showed no effect on the positive inotropic action of ATP. Conversely, the contractile activity of isoproterenol was unaffected by pyridoxal phosphate (59).

The ATP-induced increase in $[Ca^{2+}]_i$ was significantly decreased in cardiomyocytes following pretreatment with pyridoxal phosphate. The effects of pyridoxal phosphate on both the high- and low-affinity binding sites for ATP on cardiac sarcolemma were examined. Pyridoxal phosphate almost completely blocked the low-affinity binding whereas the high-affinity binding was decreased by about 60%.

Pyridoxal phosphate in vitro attenuates the influx of extracellular calcium. This effect is achieved through modulation of ligand binding. This is analogous to the effect of pyridoxal phosphate on steroid hormone activity. The action of drugs at the calcium channels would indicate that endogenous factors or ligands might serve as physiological regulators, a function that is mimicked by synthetic calcium channel agonists or antagonists. The KCl-induced increase in $[Ca^{2+}]_i$ was increased in cardiomyocytes from vitamin B_6–deficient rats. Administration of vitamin B_6 to the deficient rats abolished this. It is possible that the augmentation of the KCl-induced increase in $[Ca^{2+}]_i$ in the deficient rat is related to an increase in Ca^{2+} influx through the sarcolemmal calcium channels. An increase in Ca^{2+} influx in smooth muscle cells causes an increase in tone of the smooth muscle and, hence, hypertension in the deficient animal (51).

Studies in humans have identified an independent association between low plasma vitamin B_6 concentration and a higher risk of coronary artery disease (60). In addition to a role for vitamin B_6 in atherosclerosis, other potential explanations have been considered. These include the role of pyridoxal phosphate in platelet aggregation through inhibition of adenosine-5′-diphosphate receptors, down-regulation of glycoprotein IIb gene expression (61), and the association between low pyridoxal phosphate and inflammatory markers (62). Low plasma concentration of pyridoxal phosphate has been inversely related to C-reactive protein (CRP) in the Framingham Heart Study Cohort (63). The relation between plasma PLP and major markers of acute-phase reactions in affecting coronary artery disease (CAD) was evaluated in a cohort of subjects who were characterized by angiography for severe coronary atherosclerosis and a control group of CAD-free individuals (64). They determined plasma PLP, fibrinogen, high-sensitivity CRP, serum lipid concentrations, and all major biochemical CAD risk factors including total homocysteine. A significant, inverse-graded relation was observed between PLP and both hs-CRP and fibrinogen. The CAD risk as a result of low PLP was additive when considered in combination with elevated hs-CRP concentration or with an increased ratio of LDL to HDL. Low plasma PLP concentrations were inversely associated with major markers of inflammation and independently associated with increased CAD risk (63). The association of low PLP concentration with higher risk of CAD remained even after the inclusion, in a multivariate logistic regression model, of hs-CRP, fibrinogen, and variables related to homocysteine metabolism. It has been suggested that the association of low PLP with CAD risk is mainly due to the effect of inflammation on plasma PLP concentration. Friso et al. (64) contend that the additive effect of low PLP to that conferred by hs-CPR was seen in the progressive increase in the estimate of CAD risk across increasing hs-CRP quintiles. This is supported by the observation of Aybak et al. (65) that a low PLP status was associated with stroke and only partially mediated via inflammation, as expressed by the major inflammation marker hs-CRP.

Our observations on the role of PLP in both the major calcium channels for the influx of extracellular calcium might proffer a viable biochemical explanation for the association between low PLP concentration and the risk for CAD. The increase in $[Ca^{2+}]_i$ in cardiomyocytes in the vitamin B$_6$–deficient rat might contribute to heart dysfunction and increased susceptibility to myocardial infarction and explain the beneficial effect of vitamin B$_6$ in patients with hypertension (66) and myocardial infarction. Pyridoxal phosphate also antagonizes the cardiac action of ATP by blocking purinoceptors on the myocardium. This is consistent with pharmacological studies showing the antagonistic effect of pyridoxal phosphate on ATP-induced changes in rat vagus and vas deferens. It is significant that pyridoxal phosphate has an inhibitory effect on both the major channels, the L-type as well as the ATP-mediated, for the influx of extracellular calcium into the cell.

Platelets are anucleated cells that express a number of receptors that govern platelet activity. Platelets are activated by a range of agonists and adhesion proteins. Thus, platelet aggregation events are activated by agents such as ADP, collagen, and thrombin receptor–activating peptide agonists. Antiplatelet drugs such as aspirin and Clopidogrel are used therapeutically for the prevention of vascular events. Pyridoxal phosphate is a mild inhibitor of ADP-, epinephrine-, collagen-, arachidonic acid–, and thrombin-induced platelet aggregation (67). Using the pyridoxal phosphate scaffold, a class of cardio- and cerebroprotective agents have been synthesized (68).

11.8   VITAMIN B$_6$: GENE EXPRESSION AND ANTICANCER EFFECT

In rats fed a diet adequate in vitamin B$_6$, the fraction of total pyridoxal phosphate found in the nuclei of liver cells was 21%, and this increased to 39% in rats fed a vitamin B$_6$–deficient diet, indicating a conservation of the vitamin in the nuclear compartment during deficiency. Pyridoxal phosphate in the cell nucleus is protein bound, and this protein has an apparent molecular mass of 50 to 55 kDa. Cells grown in the presence of 5 mM pyridoxine have a decreased glucocorticoid–dependent induction of enzymes such as tyrosine aminotransferase. Vitamin B$_6$ regulates transcriptional activation of human glucocorticoid receptors in the HeLa cells. The modulatory role in transcription is not restricted to the glucocorticoid receptor but extends to other members of the steroid hormone super family. The intracellular concentration of PLP could have a profound influence on steroid hormone–induced gene expression, with increased PLP levels resulting in a decreased transcriptional response to various steroid hormones and vice versa (69). The possibility of an interaction between vitamin B_6 and steroid hormone has been known for some time. The induction of aspartate aminotransferase in rat liver cytosol (c Asp AT) of adrenalectomized vitamin B_6–deficient rats by hydrocortisone was suppressed by the administration of pyridoxine using an oligonucleotide probe of glucocorticoid-response elements (GRE), and the binding of nuclear extract to oligonucleotide was assessed. The binding of nuclear extracts prepared from livers of vitamin B_6–deficient rats was greater than that of control rats. The preincubation of nuclear extracts with PLP resulted in a significant decrease in the binding of the extract to GRE (70).

The level of steroid-induced gene expression from simple promoters containing only hormone response elements and a TATA sequence was not affected by changes in the vitamin B_6 status. However, the modulatory influence of vitamin B_6 status was restored when a binding site for a transcription factor, nuclear factor (NFI), was included within the hormone-responsive promoter, indicating that PLP modulates gene expression through its influence on a functional interaction between the steroid hormone receptors and transcription factor NFI (71).

A general increase in gene expression that includes housekeeping genes such as β actin is seen in livers of vitamin B_6–deficient rats. This has been ascribed to the activation of RNA polymerase I and II in the deficient liver. Among the vitamin B_6–independent proteins, the expression of the albumin gene is increased in vitamin B_6 deficiency. A 170-nucleotide region immediately upstream of the transcription initiation site of the albumin gene is sufficient for tissue-specific expression of this gene. This region contains various transcription factors such as HNF-1 and C/EBP. Oligonucleotides that interacted with HNF-1 and C/EBP were synthesized, and the binding activity of liver nuclear extracts to each of these oligonucleotides was assessed. Binding activity of extracts prepared from vitamin B_6–deficient rat livers was greater than that from vitamin B_6–replete rat livers. The lower binding activity of extracts from vitamin B_6–replete rat liver was suggested to be due to inactivation of tissue-specific factors by PLP.

The growth of B16 melanoma in vitro was inhibited by 5 mM of pyridoxine or pyridoxal. Treatment of mice with 0.5 g pyridoxal per kg body weight reduced the growth of both new and established B16 melanoma. Topical administration of pyridoxal exhibited selective toxicity for melanoma and also resulted in the necrosis and regression of murine B16-C3 melanomas in mice. Pyridoxal supplementation was shown to reduce cell proliferation and DNA synthesis in both estrogen-dependent and -independent mammary carcinoma cell lines (72). The growth of hepatoma HepG2 cells was inhibited by the addition of pyridoxine or pyridoxal to the culture medium. There was concurrent inhibition of protein synthesis. They also reported that the growth of MH-134 hepatoma cells transplanted into C3H/He mice was significantly reduced by the administration of large amounts of pyridoxine to mice. Other reports indicate that vitamin B_6 suppressed azoxymethane-induced colon cancer. High dietary intake of vitamin B_6 has also been shown to suppress herpes simplex virus type 2-transformed (H-238) cell-induced tumor growth in BALB/c mice (73).

Epidemiological studies including the "Seven Country Study" indicate a reduced risk of lung and colorectal cancer in older men ingesting high doses of vitamin B_6

(74). An inverse relationship between vitamin B$_6$ status and incidence of prostate cancer has been reported. This relationship is supported by experimental work in mice. It has been suggested that inhibition of angiogenesis might be responsible for the anticancer effect of high doses of vitamin B$_6$. Vitamin B$_6$ was shown to suppress angiogenesis in a rat aortic ring angiogenesis model (75, 76). Pyridoxal and PLP were shown to suppress human umbilical vein endothelial cell (HUVEC) proliferation without affecting HUVEC tube formation (77). Of the vitamin B$_6$ vitamers, PLP was reported to be a strong inhibitor of DNA polymerase α and ε from a phylogenetically wide range of organisms from protists, plants, insects, fish, to mammals. These polymerase classes are related to DNA replication (78). Treatment with pharmacological doses of vitamin B$_6$ suppressed the expression of cell proliferation–related genes, c-myc and c-fos, in colon epithelium of mice treated with azoxymethane. PLP has also been shown to inhibit DNA topoisomerase I and II (79). DNA topoisomerases are ubiquitous and needed for strand separation, replication, and recombination. Inhibition of DNA topoisomerase arrests cell cycle and induces apoptosis (77). PLP has been shown to be an effective inhibitor of many enzymes that have binding sites for phosphate-containing substrates or effectors including RNA polymerase, reverse transcriptase, and DNA polymerase (80). Based on a meta-analysis of prospective studies, an inverse association between blood PLP levels and the risk of colorectal cancer has been reported (81).

Oxyplatin in combination with 5-fluorouracil is standard treatment for metastatic colorectal carcinoma. Peripheral sensory neuropathy is the dose-limiting side reaction for this treatment. Pyridoxine administration reduces the oxyplatin-induced neurotoxicity, thus allowing for more effective and less toxic treatment of such cancers (82). The preventive effects of vitamin B$_6$ on tumorigenesis might also derive from the strong antioxidant effect of vitamin B$_6$ (83). The expressions of c-myc and c-fos are induced by oxidative stress and suppressed by vitamin B$_6$. Vitamin B$_6$ deficiency in animals on a high-fat diet lends itself to increased lipid peroxidation. Nitric oxide (NO) plays an important role in colon carcinogenesis by elevating cyclooxygenase-2 and angiogenesis. The production of NO and the expression on iNOS mRNA are increased by oxidative stress and suppressed by pharmacological doses of vitamin B$_6$. These observations highlight the potential use of pharmacological doses of vitamin B$_6$ in cancer therapy. The anticancer effect of vitamin B$_6$ might be related to reductions in cell proliferation, angiogenesis, oxidative stress, inflammation, and nitric oxide synthesis.

11.9   VITAMIN B$_6$ AND IMMUNITY

Over the years the evidence for the requirement of vitamin B$_6$ in antibody production has accumulated. The thymus of rats made vitamin B$_6$ deficient was depleted of lymphocytes and there was atrophy of lymph nodes. Vitamin B$_6$ deficiency induced dietarily with or without a supplement of the antipyridoxine drug, deoxypyridoxine, resulted in impaired antibody formation following exposure to various antigens, and this was associated with a reduction in the number of antibody-forming cells in the spleens of the deficient rats. Cell-mediated immunity is also impaired in vitamin B$_6$–deficient rats.

The effects of a supplement of vitamin B_6 on the development of tumors and on the in vitro response associated with cell-mediated immunity were assessed. It was found that peripheral blood and splenic lymphocyte proliferative responses to T cell mitogens such as phytohemagglutinin and concanavalin A were higher in mice fed the highest levels of pyridoxine. High intake of vitamin B_6 also suppressed tumor development (73).

A decline in immune response is almost a concomitant of the aging process in animals and in humans, with the most significant effect on cell-mediated immunity, through a decrease in the number of T lymphocytes and also changes in T cell surface receptors. The effects of pyridoxine supplements on lymphocyte responses in elderly persons were studied. Lymphocyte proliferative response to both T and B cell antigens were reduced, and lymphocytes subpopulations were augmented in the vitamin B_6–supplemented group and correlated with plasma PLP levels (84).

Disease states such as uremia and arthritis are associated with immunological abnormalities. Treatment of uremic patients with pharmacologic doses of pyridoxine resulted in a significant increase in lymphocyte reactivity in mixed lymphocyte cultures. Plasma PLP was found to be 50% lower in patients with rheumatoid arthritis compared to controls matched for age, gender, race, and weight (85). All groups had similar intakes of the B vitamins. The plasma levels of PLP inversely correlated with production of tumor necrosis factor by unstimulated peripheral blood mononuclear cells.

Vitamin B_6 deficiency is widely prevalent among HIV-infected persons (86) despite an adequate dietary intake and, quite often, supplementation with multivitamins. It is suggested that this might be related to the HIV-related enteropathy. The relationship of pyridoxal phosphate to activation of CD4 T cells by antigen-presenting cells has been investigated. CD4 T cells are mediators in the initiation and continuation of the immune response, causing autoimmune diseases and allogeneic transplant rejection.

The CD4 glycoprotein is the characteristic surface receptor of all helper T cells. The extracellular part of CD4 molecules comprises four domains, D1–D4. CD4 binds to MHC class II through the D1 and D2 domains. It has been shown that pyridoxal phosphate binds very tightly to the D1 domain on CD4 and thus interferes with the CD4-MHC II interaction. Nonincorporation of CD4 into the activation complex could lead to T cell apoptosis. Further, the tight association of D1 and PLP would prevent protein-protein interaction of CD4 itself, its dimerization, or the interaction of the dimer with other molecules on the T cell surface, leading to apoptosis. PLP has been shown to be an anion channel blocker in a variety of cells. It has been suggested that PLP might have a role in the treatment of autoimmunity and in transplant rejection. In addition, as the interaction of HIV gp 120 and CD4 occurs through the D1 domain, PLP may have an anti-HIV effect as well. High concentrations of PLP inhibit viral coat protein (envelope glycoprotein) binding and infection of CD4+ T cells by isolates of HIV-1 in vitro (87). Thus PLP might function not only as an immune stimulator by increasing CD4 T cell count but also may protect uninfected CD4 T cells from infection by HIV-1. These effects of PLP are seen at a concentration of 50–70 µM. At issue is the mechanism of increasing the tissue concentration of PLP to therapeutic levels. The rapid hydrolysis of PLP by tissue nonspecific alkaline phosphatase is a major problem. Levamisole (2,3,5,6,-tetrahydro-b-phenylimidazo [2,1-b]) thiazole, an antihelmintic drug, is an inhibitor of this enzyme. A combination of phosphatase inhibitor with high ingested doses of pyridoxine might maintain the required high tissue concentrations of PLP to afford protection of uninfected CD4 T cells against HIV-1. Further work along these lines is warranted.

11.10   PYRIDOXAL PHOSPHATE ENZYMES AS DRUG TARGETS

Pyridoxal phosphate enzymes have been targets of therapeutic interventions. Abnormal pulsatile stimulation of striatal dopamine receptors leads to dysregulation of genes and proteins in downstream neurons and, consequently, alterations in neuronal firing patterns. These result in the motor complications associated with Parkinson's disease. Treatment of Parkinson's disease aims to restore normal dopaminergic transmission at striatal synapses. Levodopa slows the progression of Parkinson's disease and has a prolonged effect on the symptoms of the disease.

Human ornithine decarboxylase (hODC) is another PLP-target enzyme of considerable therapeutic significance. Ornithine decarboxylase is the rate-limiting step in polyamine biosynthesis. Polyamines are DNA-binding agents controlling its conformation. Mammalian ODC is a downstream mediator of myc-regulated processes and is up-regulated in proliferating cells. It is implicated as an oncogene in multiple types of tumors (88, 89). Inhibition of ODC suppresses tumor development, and, hence, ODC is a promising anticancer target. Inhibitors of hODC such as DMFO are currently being tested as anticancer agents. DMFO in combination with other drugs has shown positive effect as a cancer chemotherapeutic agent.

Pyridoxal phosphate–dependent enzymes are potential targets for developing antiparasitic drugs (90). There are several enzymes with high prevalence among the protozoan species. The three enzymes, serine hydroxymethyl transferase, aspartate aminotransferase, and cysteine desulfurase, are the smallest set of PLP enzymes present in all piroplasmida. Enzymes of the sulfur-containing amino acid metabolic pathway are significant in terms of their distribution as between the hosts and their disease-causing parasites. The enzymes of the forward transsulfuration pathway are present in protozoan parasites such as those causing amoebiases and trichomoniasis, but are lacking in their hosts. Another PLP enzyme, methionine γ-lyase (MGL) is important for the production of propionic acid for energy metabolism and for degradation of sulfur-containing amino acids. Host mammalian cells lack MGL, and, hence, it is a target enzyme for drug development against infection by protozoa such as *Trichomonas vulgaris* and *Entamoeba histolytica*. The halogenated methionine analog S-trifluoromethyl-L-homocysteine (TFM) and the amide derivative of TFM are toxic to the parasite and have potential as antiparasitic drugs (91).

Of the parasitic diseases the most devastating one is malaria with a worldwide death toll of more than a million people. It is transmitted by the mosquito. To combat this, the life cycle and metabolism of the parasite *Plasmodium falciparum* is affected through a reduction in the synthesis of xanthurenic acid (92). Xanthurenic acid, required for gametogenesis and fertility of the parasite, is synthesized in the tryptophan degradation pathway through the PLP-dependent enzyme kynurenine aminotransferase. Hence, this enzyme is a target for the development of antimalarial drugs.

African sleeping sickness, another widespread epidemic, is caused by *Trypanosoma brucei*, which is transmitted by flies. Ornithine decarboxylase (ODC), a key enzyme in the pathway for the synthesis of polyamines, is a target enzyme for drug development. Alpha-difluoromethyl ornithine, an irreversible inhibitor of ODC, has been developed for the treatment of sleeping sickness (93).

As the pyridoxal phosphate–dependent enzymes are the prime targets for the treatment of a number of parasitic diseases, Fang-Wu et al. (88) have devised a novel strategy for synthesizing small-molecule cell permeant inhibitors of specific PLP-dependent enzymes. Toward this, the imine adduct of pyridoxal phosphate and the specific amino acid substrate, the first intermediate in all PLP-dependent reactions of amino acids, was reduced to a stable amine. This was further modified to make it membrane permeant and to increase its affinity for the apo form of the target enzyme. Such compounds are very effective inhibitors of the target enzyme. This could be a general strategy applicable to all PLP-dependent target enzymes.

11.11   CONCLUSIONS

The diversity of the chemical reactions involving vitamin B_6 is due to the participation of pyridoxal phosphate in determining the three-dimensional structure of PLP-dependent enzymes as well as in determining the sites of elimination and replacement of substituents. Apart from its role in phospholipid metabolism through the synthesis of sphingosine and in glycogen metabolism through its structural role in glycogen phosphorylase, pyridoxal phosphate participates in the metabolism of amino acids. The formation of monoamine neurotransmitters through PLP-dependent decarboxylation of the precursors highlights the role of vitamin B_6 in the function of the nervous system. There is a range in the affinity of PLP to the apodecarboxylases with the result that during mild or moderate vitamin B_6 deficiency, the formation of some monoamine neurotransmitters is impaired whereas that of others is not. Thus, the nonparallel effect on monoamine neurotransmitter syntheses leads to significant alterations in the function of neuro and neuroendocrine systems. A moderate depletion of vitamin B_6 is characterized by significant decreases in the synthesis and secretion of GABA and serotonin with no change in the catecholamines. In addition, even under conditions not associated with a depletion of vitamin B_6, administration of pyridoxine to animals results in the augmented synthesis of some neurotransmitters, with significant biological effects. Pyridoxal phosphate also has significant effects on calcium transport, both through the voltage-dependent and the ATP-dependent pathways. As calcium is at the center of much metabolic regulation, this results in wide-ranging effects in the functioning of the organism. Here again, the administration of pyridoxine to animals, not associated with a vitamin-depleted state, have significant calcium-mediated biological effects, indicating a continuum of effects of pyridoxine administration. It is to be noted that ingestion of very high doses of pyridoxine does not have irreversible toxic effects. The associated neuropathy is reversed upon withdrawal of the huge supplement. The beneficial effects of a supplement of pyridoxine (vitamin B_6) on the nervous and cardiovascular systems and related disease processes and the protection afforded by pyridoxine under these conditions need further study.

Apart from its cofactor role, the affinity of PLP to diverse proteins is at the center of its non-cofactor biological role. PLP modulates gene expression through its influence on the interaction between steroid hormone receptors and the corresponding transcription factors. PLP binding to tissue-specific transcription factors makes them less accessible to their binding site on the target gene. The binding of PLP to the D1 domain of CD4 glycoprotein surface receptors of helper T cells seems to regulate cell-mediated immunity. The binding of PLP to cell surface calcium transport systems such as the L-type and the ATP-mediated calcium channels is the basis of cellular calcium changes mediated by PLP. The study of the mechanism of these protein-PLP interactions should provide valuable information and newer approaches to the therapeutic potential of PLP-related compounds.

REFERENCES

1. B.H. Tan, P.T.H. Wang, and J.S. Bian. Hydrogen sulfide: A novel signaling molecule in the central nervous system. *Neurochem. Intl.* 56: 3–19 (2010).
2. K. Kannan and S.K. Tain. Effect of vitamin B$_6$ on oxygen radicals, mitochondrial membrane potential and lipid peroxidation in A$_2$O$_2$–treated Vu937 monocytes. *Free Radic. Biol. Med.* 36: 423–428 (2004).
3. M.M. Mahpouz, S.Q. Zhou, and F.A. Kummerow. Vitamin B$_6$ compounds are capable of reducing the superoxide radical and lipid peroxide levels induced by A$_2$O$_2$ in vascular endothelial cells in culture. *J. Vitam. Nutr. Res.* 79: 218–229 (2009).
4. A. Ardestani, R. Yazdanparast, and A.S. Nejad. 2-Deoxy-D-ribose-induced oxidative stress causes apoptosis in human monocyte cells: Prevention by pyridoxal-5′-phosphate. *Toxical. In Vitro* 22: 968–979 (2008).
5. J.H. Matxain, B. Padn, M. Ristila, A. Strid, and L.A. Erickson. Evidence of high OH radical quenching efficiency by vitamin B$_6$. *J. Phys Chem B.* 113: 9629–9632 (2009).
6. T.G. Nam, J.M. Ku, C.L. Rector, H. Chjopi, N.A. Porter, and B.S. Jeong. Pyridoxine-derived licyche aminopyridinal antioxidants: Synthesis and their antioxidant activities. *Org. Biomal. Chem.* 9: 8475–8452 (2011).
7. K. Dakshinamurti, W.D. Leblancq, R. Herchl, and V. Havlicek. Nonparallel changes in brain monoamines of pyridoxine-deficient growing rats. *Exp. Brain Res.* 26: 255–266 (1976).
8. Y.L. Siow and K. Dakshinamurti. Effect of pyridoxine deficiency on aromatic amino acid decarboxylase in adult rat brain. *Exp. Brain Res.* 59: 575–581 (1985).
9. Y. L. Siow and K. Dakshinamurti. Effect of 1-methyl-4-phenyl-1,2,3,6 etrahydrophyridine and 1-methyl-4-phenyl-pyridinium on aromatic L-amino acid decarboxylase in rat brain. *Biochem. Pharmacol.* 35: 2640–2641 (1986).
10. K. Dakshinamurti. Neurobiology of pyridoxine. In: *Advances in Nutrition Research,* Vol. 4 (H.H. Draper, ed.). pp. 143–179, Plenum Press, New York (1982).
11. K. Dakshinamurti, C.S. Paulose, M. Viswanathan, and Y.L. Siow. Neuroendocrinology of pyridoxine deficiency. *Neurosci, Biobehav. Rev.* 12: 189–193 (1988).
12. K. Dakshinamurti, C.S. Paulose, and J. Vriend. Hypothroidismof hypothalamic origin in pyridoxine-deficient rats. *J. Endocrinol.* 109: 345–349 (1986).
13. M. Viswanathan, Y.L. Siow, C.S. Paulose, and K. Dakshinamurti. Pineal indoleamine metabolism in pyridoxine-deficient rats. *Brain Res.* 473: 37–42 (1988).
14. S.K. Sharma and K. Dakshinamurti. Effects of serotonergic agents on plasma prolactin levels in pyridoxine-deficient adult male rats. *Neurochem. Res.* 19: 687–692 (1994).
15. M.F. McCarty. High dose pyridoxine as an anti-stress strategy. *Medical Hypothesis* 54: 803–807 (2000).

16. S. Russo, I.P. Kema, R.M. Fokkema, J.C. Boon, P.H.B. Willemse, E.G. de Vries, J. Den Boer, and J. Korf. Tryptophan as a link between psychopathology and somatic states. *Psychosom. Med.* 65: 665–671 (2003).

17. S.M. Gospe, Jr. Pyridoxine-dependent seizures: Findings from recent studies pose new questions. *Pediatric Neurol.* 26: 181–185 (2002).

18. P.T. Clayton. B6-responsive disorders: A model of vitamin dependency. *J. Inherit. Metab. Dis.* 29: 317–326 (2006).

19. D.B. Coursin. Convulsive seizures in infants with pyridoxine deficient diet. *JAMA* 154: 406–408 (1954).

20. A. Sigirci, I. Orkan, and C. Yakinci. Pyridoxine-dependent seizures: Magnetic resonance spectroscopy findings. *J. Child. Neurol.* 19: 75–78 (2004).

21. A. Alkan, K. Sarac, and R. Kutlu. Early and late state subacute sclerosing panencephalitis: Chemical shift imagining and single voxel MR spectroscopy. *Am. Neuroradiol.* 24: 501–506 (2003).

22. P. Baxter. Pyridoxine-dependent seizures: A clinical and biochemical conundrum. *Biochim. Biophys. Acta* 1647: 36–41 (2003).

23. A. Kelly and C.A. Stanley. Disorders of glutamate metabolism. *MRDD Res. Rev.* 7: 287–295 (2001).

24. G. Battaglioli, D.R. Rosen, S.M. Gospe, Jr., and D.L. Martin. Glutamate decarboxylase is not genetically linked to pyridoxine-dependent seizures. *Neurology* 55: 309–311 (2000).

25. B. Plecko, S. Stockler-Ipsiroglu, E. Paske, W. Erwa, E.A. Struys, and C. Jacobs. Pipecolic acid elevation in plasma and cerebrospinal fluid of two patients with pyridoxine-dependent epilepsy. *Ann. Neurol.* 48: 121–125 (2000).

26. H.S. Wang and M.E. Kuo. Vitamin B_6 related epilepsy during childhood. *Chang Gung Med J.* 30: 396–401 (2007).

27. S. Stockler, B. Plecko, S.M. Gospe Jr., M. Coulter-Mackie, M. Connolly, C. Van Karnebeck, S. Mercimek-Mahmutoglu, H. Hartman, G. Scharer, E. Struijs, I. Tein, C. Jacobs, P. Clayton, and J.L. Van Hove. Pyridoxine-dependent epilepsy and antiquitin deficiency: Clinical and molecular characteristics and recommendations for diagnosis, treatment and follow up. *Mol. Genet. Metab.* 104: 48–60 (2011).

28. C.L. Bennett, Y. Chen, S. Hahn, I.A. Glass, and S.M. Gospe Jr. Prevalence of ALDH 7A1 mutation in North American pyridoxine-dependent seizure (PDS) patients. *Epilepsia* 50: 1167–1175 (2009).

29. H. Yamamoto, Y. Sasamoto, Y. Miyamoto, H. Murakami, and N. Kamiyama. A successful treatment with pyridoxal phosphate for West syndrome in hypophosphatasia. *Pediatr. Neurol.* 30: 216–218 (2004).

30. H.S. Wang, M.E. Kuo, M.L. Chou, P.C. Hung, K.L. Lin, M.Y. Hsieh, and M.Y. Chang. Pyridoxal phosphate is better than pyridoxine for controlling idiopathic intractable epilepsy. *Arch. Dis. Child.* 90: 512–515 (2005).

31. S. Ohtahara, Y. Yamatogi, and Y. Ohtsuka, Vitamin B_6 treatment of intractable seizures. *Brain and Devel.* 33: 783–789 (2011).

32. M.A. Mikati, G.A. Lepejian, and G.L. Holmes. Medical treatment of patients with infantile spasms. *Clin. Neuropharmacol.* 25: 61–70 (2002).

33. N. Fejerman, R. Cers-Simo, and R. Caraballo. Vigabatrin as a first-choice drug in the treatment of West syndrome. *J. Child Neurol.* 15: 161–165 (2000).

34. J. Pietz, C. Benninger, H. Schafer, D. Sontheimer, G. Mittermaier, and D. Rating. Treatment of infantile spasms with high-dosage vitamin B_6. *Epilepsia* 34: 757–763 (1993).

35. K. Dakshinamurti and M.C. Stephens. Pyridoxine deficiency in the neonate rat. *J. Neurochem.* 16: 1515–1522 (1969).

36. M.C. Stephens, V. Havlicek, and K. Dakshinamurti. Pyridoxine deficiency and development of the central nervous system in the rat. *J. Neurochem.* 18: 2407–2416 (1971).

37. S.K. Sharma and K. Dakshinamurti. Seizure activity in pyridoxine-deficient adult rats. *Epilepsia* 33: 235–247 (1992).
38. S.K. Sharma, B. Bolster, and K. Dakshinamurti. Picrotoxin and pentylene tetrazole induces seizure activity in pyridoxine-deficient rats. *J. Neurolog. Sci.* 121: 1–9 (1994).
39. J.S. Teitelbaum, R.J. Zatorre, S. Carpenter, D. Gendron, A.C. Evans, A. Gjedde, and N.R. Cashman. Neurologic sequelae of domoic acid intoxication due to ingestion of contaminated mussels. *N. Engl. J. Med.* 322: 1781–1787 (1990).
40. K. Dakshinamurti, S.K. Sharma, and M. Sundaram. Domoic acid induced seizure activity in rats. *Neurosci. Let.* 127: 193–197 (1991).
41. S.M. Strain and R.A.R. Tasker. Hippocampal damage produced by systemic injections of domoic acid in mice. *Neuroscience* 44: 343–352 (1991).
42. K. Dakshinamurti, S.K. Sharma, M. Sundaram, and T. Watanabe. Hippocampal changes in developing postnatal mice following intrauterine exposure to domoic acid. *J. Neurosci.* 13: 4486–4495 (1993).
43. K. Dakshinamurti, S.K. Sharma, and J.D. Geiger. Neuroprotective actions of pyridoxine. *Biochim. Biophys. Acta* 1647: 225–229 (2003).
44. E. Sreekumaran, T. Ramakrishna, T.R. Madhav, D. Anandh, B.M. Prabhu, S. Sulekha, P.N. Bindu, and T.R. Raju. Loss of dendritic connectivity in CA1, CA2 and CA3 neurons in hippocampus in rat under aluminum toxicity: Antidotal effect of pyridoxine, *Brain Res. Bull.* 59: 421–427 (2003).
45. I.K. Huang, K-Y. Yoo, D.H. Kim, B.H. Lee, Y-G. Kwon, and M.O. Won. Time course of changes in pyridoxal phosphate (vitamin B6 active form) and its neuroprotection in experimental ischemic damage. *Experiment. Neurol.* 206: 114–125 (2007).
46. T-T. Yang and S.J. Wang. Pyridoxine inhibits depolarization-evoked glutamate release in nerve terminals from rat cerebral cortex: A possible neuroprotective mechanism? *J. Pharmacol. Exp. Therapeutics* 331: 244–254 (2009).
47. D.H. Yoo, W. Kim, D.W. Kim, K-Y. Yoo, J.Y. Chung, H.Y. Youn, Y.S. Yoon, S.Y. Choi, M-H. Won, and I-K. Hwang. Pyridoxine enhances cell proliferation and neuroblast differentiation by upregulating the GABAergic system in the mouse dentate gyrus. *Neurochem. Res.* 36: 713–721 (2011).
48. D.Y. Yoo, W. Kim, S.M. Nam, J.Y. Chung, J.H. Choi, Y.S. Yoon, M-H. Won, and I.K. Hwang. Combination effects of sodium butyrate and pyridoxine treatment on cell proliferation and neuroblast differentiation in the dentate gyrus of D-galactose–induced aging model mice. *Neurochem. Res.* 37: 223–231 (2012).
49. D.Y. Hoo, W. Kim, S.M. Nam, J.Y. Chung, J.H. Choi, Y.S. Yoon, M-H. Won, and I.K. Hwang. Chronic effects of pyridoxine in the gerbil hippocampal CA1 region after transient forebrain ischemia. *Neurochem. Res.* 37: 1011–1018 (2012).
50. C.S. Paulose, K. Dakshinamurti, S. Packer, and N.L. Stephens. Sympathetic stimulation and hypertension in pyridoxine-deficient adult rat. *Hypertension* 11: 387–391 (1988).
51. K. Dakshinamurti and K.J. Lal. Vitamins and hypertension. *World Rev. Nutrit. Dietet.* 69: 40–73 (1992).
52. K. Dakshinamurti and S. Dakshinamurti. Blood pressure regulation and micronutrients. *Nutrit. Res. Rev.* 14: 3–43 (2001).
53. M. Viswanathan, R. Bose, and K. Dakshinamurti. Increased calcium influx in caudal artery of rats made hypertensive with pyridoxine deficiency. *Am J. Hypertens.* 4: 252–255 (1991).
54. K.J. Lal, K. Dakshinamurti, and J. Thliveris. The effects of vitamin B$_6$ on the systolic blood pressure of rats in various animal models of hypertension. *J. Hypertens.* 14: 355–363 (1996).
55. K. Dakshinamurti, K.J. Lal, and P.K. Ganguly. Hypertension, calcium channels and pyridoxine (vitamin B$_6$). *Mol. Cell. Biochem.* 188: 137–148 (1998).

56. A. Takahara, S. Fujita, K. Moki, Y. Ono, H. Koganei, S. Iwayama, and H. Yamamoto. Neuronal calcium channel blocking action of an antihypertensive drug, Cilinidipine, in IMR-32 human neuroblastoma cells. *Hypertension Research* 26: 743–747 (2003).

57. C.X. Wang, T. Yang, R. Noor, and A. Shuaib. Role of MC-1 alone and in combination with tissue plasminogen activator in focal ischemic brain injury in rats. *J. Neurosurg.* 103: 165–169 (2005).

58. X. Wang, K. Dakshinamurti, S. Musat, and N.S. Dhalla. Pyridoxal 5′-phosphate is an ATP-receptor antagonist in freshly isolated rat cardiomyocytes. *J. Mol. Cell Cardiol.* 31: 1063–1072 (1999).

59. K. Robinson, K. Arheart, and H. Refsum. Low circulating folate and vitamin B_6 concentrations: Risk factors for stroke, peripheral vascular disease, and coronary artery disease. *Circulation* 97: 437–443 (1998).

60. S.J. Chang, H.J. Chuang, and H.H. Chen. Vitamin B_6 down-regulates the expression of human GPIIb gene. *J. Nutr. Sci. Vitaminol.* (Tokyo) 45: 471–479 (1999).

61. S. James, H.H. Vorster, and C.S. Venter. Nutritional status influences fibrinogen concentration: Evidence from the THUSA survey. *Thromb. Res.* 98: 388–394 (2000).

62. S. Friso, P.F. Jacques, P.W. Wilson, I.H. Rosenberg, and J. Selhub. Low circulation vitamin B_6 is associated with elevation of the inflammation marker C-reactive protein independently of plasma homocysteine levels. *Circulation* 103: 2788–2791 (2001).

63. S. Friso, D. Girelli, N. Martinelli, O. Oliveri, V. Lotto, C. Bozzini, F. Pizzolo, G. Faccini, F. Beltrame, and R. Corrocher. Low plasma vitamin B_6 concentrations and modulation of coronary artery disease risk. *Am. J. Clin. Nutr.* 79: 992–998 (2004).

64. S. Friso, D. Girelli, N. Martinelli, O. Olivieri, and R. Corrocher. Reply to J. Dierkes et al. *Am J. Clin. Nutr.* 81: 727–728 (2005).

65. M. Aybak, A. Sermet, M.O. Ayyildiz, and A.Z. Karakilcik. Effect of oral pyridoxine hydrochloride supplementation on arterial blood pressure in patients with essential hypertension. *Arzneim-Forsch./Drug Res.* 45: 1271–1273 (1995).

66. N.S. Dhalla, R. Sethi, and K. Dakshinamurti. U.S. Patent 6,043,259 (2000).

67. W. Zhang, J. Yao, T. Whitney, D. Froese, A.D. Friesen, L. Stang, C. Xu, A. Shuaib, J.M. Diakur, and W. Haque. Pyridoxine as a template for the design of antiplatelet agents. *Bioorg. Med. Chem. Lett.* 14: 4747–4750 (2004).

68. D.B. Tully, V.E. Allgood, and J.A. Cidlowski. Modulation of steroid receptor-mediated gene expression by vitamin B_6. *FASEB J.* 8: 343–349 (1994).

69. T. Oka, N. Komori, M. Kuwahata, Y. Hiroi, T. Shimoda, M. Okada, and Y. Natori. Pyridoxal 5′–phosphate modulates expression of cytosolic aspartate amino-transferase gene by inactivation of glucocorticoid receptor. *J. Nutr. Sci. Vitaminol.* 41: 363–375 (1995).

70. Y. Natori, T. Oka, and M. Kuwahata. Modulation of gene expression by vitamin B_6, In: *Biochemistry and Molecular Biology of Vitamin B_6 and PQQ-Dependent Proteins* (A. Iriarte, H.M. Kagan, and M. Martinez-Carrion, eds.). pp. 301–306, Birkhauser Verlag, Basel (2000).

71. B.A. Davis and B.E. Cowing. Pyridoxal supplementation reduces cell proliferation and DNA synthesis in estrogen-dependent and independent mammary carcinoma cell lines. *Nutr. Cancer* 38: 281–286 (2000).

72. S. Komatsu, N. Yanaka, K. Matsubara, and N. Kato. Antitumor effect of vitamin B_6 and its mechanisms. *Biochim. Biophys. Acta* 1647: 127–130 (2003).

73. M.C. Jansen, H.B Beuno-de-Mesquita, R. Buzina, F. Fidanza, A. Minotti, H. Blackburn, A.M. Nissenen, F.J. Kok, and D. Kromhout. Dietary fiber and plant foods in relation to colorectal cancer mortality: The seven countries study. *Int. J. Cancer*, 81: 174–179 (1999).

74. T.J. Hartman, K. Woodson, R. Stolzenberg-Solomon, J. Virtamo, J. Selhub, M.J. Barrett, and D. Albanes. Association of the B vitamins, pyridoxal 5′-phosphate, B_{12}, and folate with lung cancer risk in older men. *Am. J. Epidemiol.* 153: 688–693 (2001).

75. K. Matsubara, M. Mori, Y. Matsuura, and N. Kato. Pyridoxal 5′-phosphate and pyridoxal inhibit angiogenesis in the serum-free rat aortic ring assay. *Int. J. Mol. Med.* 8: 505–508 (2001).

76. K. Matsubara, H. Matsumoto, Y. Mizushina, J.S. Lee, and N. Kato. Inhibitory effect of pyridoxal 5′-phosphate on endothelial cell proliferation, replicative DNA polymerase and DNA topoisomerase. *Int. J. Mol. Med.* 12: 51–55 (2003).

77. Y. Mizushina, X. Yu, K. Matsubara, C. Murakami, I. Kuriyama, M. Oshiga, M. Takemura, N. Kato, H. Yoshida, and K. Sakaguchi. Pyridoxal 5′-phosphate is a selective inhibitor in vivo of DNA polymerase α and ε. *Biochem. Biophys. Res. Communs.* 312: 1025–1032 (2003).

78. H. Hubscher, G. Mago, and S. Spadari. Eukaryotic DNA polymerase. *Ann. Rev. Biochem.* 71:133–163 (2002).

79. J.J. Vermeersch, S. Christmann-Frank, L.V. Karabashyan, S. Fermandjian, G. Mirambeau, and P. Arsene Der Garabedian. Pyridoxal 5′-phosphate inactivates DNA topoisomerase IB by modifying the lysine general acid. *Nucleic Acid Res.* 32: 5649–5657 (2004).

80. K. Matsubara, S-I. Komatsu, T. Oka, and N. Kato. Vitamin B$_6$-mediated suppression of colon tumorigenesis, cell proliferation and angiogenesis. *J. Nutr. Biochem.* 14: 246–250 (2003).

81. S.C. Larson, N. Orsini, and A. Wolk. Vitamin B6 and risk of colorectal cancer. A meta-analysis of prospective studies. *JAMA* 303: 1077–1083 (2010).

82. M.B. Garg and S.P. Ackland. Pyridoxine to protect from oxyplatin-induced neurotoxicity without compromising antitumor effect. *Cancer Chemother. Pharmacol.* 67: 963–966 (2011).

83. S.K Jain and G. Lim. Pyridoxine and pyridoxamine inhibits superoxide radicals and prevents lipid peroxidation, protein glycosylation and (Na+, K+) −ATPase activity reduction in high glucose-treated human erythrocytes. *Free Radic. Biol. Med.* 30: 232–237 (2001).

84. H-K. Kwak, C.M. Hansen, J.E. Leklem, K. Hardin, and T.D. Shultz. Improved vitamin B$_6$ status is positively related to lymphocyte proliferation in young women consuming a controlled diet. *J. Nutr.* 132: 3308–3313 (2002).

85. R. Roubenoff, R.A. Roubenoff, and J. Selhub. Abnormal vitamin B$_6$ status in rheumatoid cachexia: association with spontaneous tumor necrosis factor alpha production and markers of inflammation. *Arthritis Rheum.* 38: 105–109 (1995).

86. M.K. Baum, E. Mantero-Atienze, G. Shor-Posner, M.A. Fletcher, R. Morgan, C. Eisdorfer, H.E. Sauberlich, P.E. Cornwall, and R.S. Beach. Association of vitamin B$_6$ status with parameters of immune function in early HIV-1 infection. *J. Acqui. Immune Defic. Syndr.* 4: 1122–1132 (1991).

87. L. Guo, N.K. Heinzinger, M. Stevenson, L.M. Schoffer, and J.M. Salfany. Inhibition of gp 120-CD4 interaction and human immunodeficiency virus type 1 infection in vitro by pyridoxal 5-phosphate. *Antimicrob. Agents Chemother.* 38: 2483–2487 (1994).

88. F. Wu, P. Christen, and H. Gehring. A novel approach to inhibit intracellular vitamin B6-dependent enzymes: Proof of principle with human and plasmodium ornithine decarboxylase and human histidine decarboxylase. *FASEB Journal* 25: 2109–2122 (2011).

89. N. Seiler. Thirty years of polyamine-related approaches to cancer therapy, Retrospect and Prospect. Part I. Selective enzyme inhibition. *Curr. Drug Targets* 4: 537–564 (2008).

90. B. Kapes, I. Tews, A. Binter, and P. Macheroux. PLP-dependent enzymes as potential drug targets for protozoan diseases. *Biochim. Biophys. Acta* 1814: 1567–1576 (2011).

91. D. Sato, S. Kobayashi, H. Yasui, N. Shibata, T. Toru, M. Suematsu, and T. Nozaki. Cytotoxic effects of amide derivatives of trifluoromethionine against the enteric proto- zoan parasite *Entamoeba histolytica. Int. J. Antimicrob. Agents* 35: 56–61 (2010).
92. I.B. Muller, F. Wu, B. Bergman, J. Knockel, R.D. Walker, H. Gehring, and C. Wrenger. Poisoning pyridoxal 5-phosphate dependent enzymes: A new strategy to target the malaria parasite *Plasmodium falciparum. PLoS One,* 4: e4406 (2009).
93. L.R. Krauth-Siegel, M.A. Komini, and T. Schlecker. The trypanothione system. *Subcell. Biochem.* 44: 231–251 (2007).

12 Non-Prosthetic Group Functions of Biotin

Krishnamurti Dakshinamurti

CONTENTS

12.1 INTRODUCTION

That biotin should constitute an essential requirement for all organisms would be expected from its obligatory involvement in carbohydrate, lipid, and amino acid metabolism as the prosthetic group of the carboxylases. There are only four biotin-containing enzymes in higher organisms. They are acetyl-coenzyme A (CoA) carboxylase (ACC), propionyl-CoA carboxylase (PCC), pyruvate carboxylase (PC), and β-methyl-crotonyl-CoA carboxylase (MCC). Biotin is covalently bound to a lysine residue of the carboxylase protein in ACC, PCC, MCC, and PC. Each of the biotin-dependent carboxylases catalyzes an adenosine triphosphate (ATP)–dependent CO_2 fixation reaction. Biotin functions as a CO_2 carrier on the surface of the enzyme. There are two variants of ACC: ACC1 is located in the cytosol and is a rate-limiting enzyme of fatty acid synthesis, whereas ACC2 is present on the outer mitochondrial membrane. Malonyl CoA produced by ACC2 has an inhibitory effect on fatty acid transport into mitochondria and thus controls fatty acid oxidation. PC, PCC, and MCC are mitochondrial enzymes. PC catalyzes the incorporation of bicarbonate into oxaloacetate, an intermediate of the tricarboxylic acid cycle. MCC catalyzes a crucial step in the degradation of branched-chain amino acid leucine. PCC catalyzes the incorporation of bicarbonate into propionyl CoA to form methylmalonyl CoA, which enters the tricarboxylic acid cycle after isomerization to succinyl CoA (1–4).

Biotin is synthesized by many microorganisms and plants and is an essential nutrient for higher organisms. The recognition of biotin-responsive multiple carboxylase deficiency syndrome and its treatment with biotin has been a major contribution to the understanding of neonatal disease states (5,6). The effects of experimental biotin deficiency in a variety of species are quite devastating.

Early work on the intracellular fractionation of biotin in a variety of tissues indicated that a significant amount of biotin was associated with the nuclear fraction of the cell (7,8). The biotin content of biotin-deficient rat liver is about one tenth of that of normal rat liver, and a significant 75% of this was present in the nuclear fraction, indicating that nuclear biotin was conserved in the deficient animal (9,10). Nuclear biotin was noncovalently bound to protein (2). The recent identification of the presence of biotin holocarboxylase synthetase (HCS) in the nuclear fraction of various tissues explains the earlier observation (11). However, the fact that nuclei assayed negatively for any of the biotin carboxylases would suggest a function for biotin in the nucleus other than as the prosthetic group of biotin containing carboxylases.

12.2 BIOTIN REQUIREMENT FOR CELLS IN CULTURE AND FOR CELL DIFFERENTIATION

Various early studies had indicated that cells in culture do not require biotin, and it was further suggested that transformed cells might have the ability to synthesize biotin (12). Using biotin-depleted fetal bovine serum and Eagle's minimum essential medium, a requirement for biotin was demonstrated for HeLa cells, human fibroblasts, and Rous sarcoma virus–transformed baby hamster kidney cells (13–15). Mammalian cultured cells, when they do not receive specific signals, which include specific growth factors, come to a halt in a quiescent nongrowing variant of the G1 state, referred to as Go. The synthesis of components of the cell-cycle control system is switched off. Normal cells in G1 arrest due to serine starvation start incorporating [³H]-thymidine into DNA as soon as serine is restored to the medium. Biotin-deficient HeLa cells under similar conditions do not incorporate [³H]-thymidine into DNA even when serine is restored to the medium. However, within 4 h of supplementation of biotin to the biotin-deficient medium, the incorporation of [³H]-thymidine into DNA reaches a maximum (15). By this time, there is a stimulation of protein synthesis. These two phenomena are related, and the growth-promoting effects might be achieved through synthesis of certain proteins.

The L1 subline derived from the mouse fibroblast cell line has the capacity to differentiate in the resting state into an adipose cell type. When the cells reach confluence and start to differentiate, they greatly increase the rate of triglyceride synthesis. This is paralleled by coordinate increases in the activities of the key enzymes of lipogenesis and correlates with the rise in the nuclear runoff transcription rates for the mRNA during differentiation (16). The process of differentiation can be accelerated by increasing the amount of serum in the culture medium or by adding insulin or biotin. The 24–48 h delay in the deposition of triglyceride would suggest the synthesis of some factor required for the synthesis of lipogenic enzymes rather than just for the biotinylation of the apocarboxylase.

Biotin deficiency in mice changes the subpopulations of spleen lymphocytes and decreases the proliferative response of spleenocytes to concanavalin A (17). They have also shown that in experimental biotin deficiency the involution of the thymus is accelerated and thymocyte maturation is arrested in the double negative (DN-4) substage. A specific stage in the T cell maturation process is thus sensitive to biotin deficiency (18).

12.3   DEVELOPMENT OF THE PALATAL PROCESS AND BIOTIN

Congenital malformations have been reported in domestic fowl maintained on a biotin-deficient diet. Even moderate maternal biotin deficiency is teratogenic in mice (19). At midgestation, biotin-deficient embryos weighed less than normal embryos and had external malformations such as micrognathia and micromelia. There was a marked decrease in the size of the palatal process on day 15.5 of gestation, which may be due to altered proliferation of the mesenchyme. The development of the palatal process in culture was investigated (20). After 72 h of organ culture, more than 90% of the explants from normal mouse embryos (biotin replete) were at stage 6 of development. The corresponding figure for explants from biotin-deficient embryos cultured in a biotin-deficient medium was 6.5%. Administration of biotin to biotin-deficient dams 24 h prior to removal of the embryos resulted in 33% of the explants at stage 6 of development when cultured in a biotin-deficient medium and over 50% at stage 6 if cultured in a medium containing 10^{-7} M biotin. There was no detrimental effect of any of the organic acid intermediates or their secondary metabolites on palatal closure of the explants when these compounds were added to the organ culture medium at a concentration of 10^{-4} M. These results indicate the continuous requirement for biotin during proliferation of the mesenchyme, perhaps for the synthesis of growth factors during organogenesis.

12.4   BIOTIN AND THE REPRODUCTIVE SYSTEM

Delayed spermatogenesis and decreased number of spermatozoa due to biotin deficiency have been reported in early literature. In mammals, spermatogenesis is dependent primarily upon testosterone, which is produced in the Leydig cells and acts on the tubular cells of the seminiferous tubules to drive spermatogenesis. Testicular and serum levels of testosterone are decreased in the biotin-deficient rat (21). Biotin deficiency was accompanied by a significant degree of sloughing of the seminiferous tubule epithelium in these rats. Treatment of biotin-deficient rats with gonadotropins or biotin increases the serum levels of testosterone. However, even when testosterone levels are maintained at high levels in biotin-deficient rats by testosterone implants, the increase in serum testosterone does not result in normal spermatogenesis. The administration of biotin alone or biotin in addition to testosterone to biotin-deficient rats leads to normal spermatogenesis, suggesting the involvement of biotin in the formation of local testicular factor(s) that are required in addition to testosterone and follicle-stimulating hormone for the normal interaction among Leydig, Sertoli, and peritubular cells.

The effects of diets with varying biotin contents on the estrus cycle, estradiol and progesterone serum levels, ovarian morphology, the uterine mRNA abundance of estradiol and progesterone nuclear receptors, as well as on the estradiol-degrading enzymes in the liver of BALB/cAnN Hsd female mice were studied (22). Deficiency of biotin was associated with reductions in ovary weight, arrested estrous cycle on the day of diestrus, and notable changes in ovarian morphology. Biotin deficiency abolished primary, antral, and Graafian follicles as well as the corpus luteum. Maturation of the ovarian follicles is associated with hepatic regulation of IGF-1 and systemic concentrations of IGF-1 binding proteins. Biotin deficiency decreases serum IGF-1 concentration, and this might account for the effects observed in ovarian follicles in the biotin-deficient mice.

12.5   NEUROTROPHIC FACTOR AND BIOTIN

Brain-derived neurotrophic factor (BDN) is a member of a family of cell-signaling molecules (neurotrophins) that have an important role in neuronal development and plasticity. BDN has been implicated in development and adult plasticity within the telencephalic brain regions that control learned vocal behavior in songbirds. BDN expression seems to correlate with specific stages of songbird vocal learning. High levels of biotin have been reported in specific telencephalic nuclei (RA and HVC) among juvenile males (23). This might be related to specific up-regulation of biotinylated proteins within the RA and HVC nuclei in juvenile males. The developmental expression of this is correlated with vocal learning. High levels of biotin in the hippocampus, a brain region important for learning and memory, has also been reported (24), emphasizing the important role of biotin-regulated mechanisms in neuronal plasticity.

12.6   BIOTIN-RESPONSIVE BASAL GANGLIA DISEASE

Biotin-responsive basal ganglia disease was first reported in 1998 in a number of families in Saudi Arabia (25). It appeared to be familial in occurrence with possible autosomal recessive inheritance. The neurological presentation was that of subacute encephalopathy at 2–13 years of age. There was loss of developmental milestones with chronic progressive encephalopathy with loss of intellectual abilities and motor function. Symptoms included parkinsonian and pyramidal tract signs. MRI of brain indicated bilateral necrosis in the central parts of the caudate heads and all of the putamen. All patients responded to high doses of biotin (5–10 mg/kg/day) by stabilization or reversal of neurological symptoms. Two European cases of biotin-responsive basal ganglia disease with novel mutations of the SLC19A3 gene resulting in impaired biotin transport have been reported (26).

A variant of the biotin-responsive basal ganglia disease in a neonate was reported (27) presenting with controllable seizures and leading to progressive symptoms of tremor and bradykinesia referable to basal ganglia involvement. MRI changes were in the globi pallidi and not in the caudate and putamen. The infant responded to much lower dose of biotin resulting in only minimal learning difficulties with normal motor function. Carboxylase assays in patients with biotin-responsive basal ganglia disease indicated normal enzyme activities, thus suggesting a specific defect in biotin transport in the central nervous system.

12.7   BIOTIN AND THE SYNTHESIS OF SPECIFIC PROTEINS

Biotin has a role in cellular processes such as survival, differentiation, and development. A role for biotin in the synthesis of specific proteins has been identified (28). Biotin holocarboxylase synthetase (HCS) catalyzes the biotinylation of apocarboxylases. HCS mRNA is significantly reduced in the biotin-deficient rat. A regulatory role for biotin in the control of biotin HCS mRNA levels via signaling cascades involving guanylate cyclase and cGMP-dependent protein kinase has been proposed (29).

Cytokines are secreted by immune cells in response to stimulation by antigens. Cytokines bind to receptors on the surface of target cells, creating an intracellular signaling cascade that controls cellular processes such as growth, proliferation, differentiation, and apoptosis. The expression of genes encoding the cytokine interleukin-2 (IL-2) and IL-2 receptor-gamma correlate with the biotin status of human lymphoid cells (30). The effects of biotin on the metabolism of cytokines might underlie the role of biotin in immune function (17). The activity of various cell signals such as biotinyl-AMP, SP1 and SP3, nuclear factor (NF)-κB, and receptor tyrosine kinase are biotin dependent (31). Biotin deficiency up-regulates TNF-α production and biotin excess down-regulates TNF-α production (32). As TNF-α has an important role in the pathogenesis of inflammatory diseases, the possibility of treating inflammatory diseases with biotin needs to be investigated.

12.8   BIOTIN AND ENZYMES OF GLUCOSE METABOLISM

The initial and rate-limiting step in the metabolic utilization of glucose by the hepatocyte is its phosphorylation by glucokinase. Dietary, nutritional, and hormonal states of the animal influence the activity of glucokinase. Early studies in the 1960s showed that hepatic glucokinase was depressed in biotin-deficient rats fed either high- or low-carbohydrate diets and that insulin or biotin injections restored enzyme activity to normal levels (33, 34). Biotin also played a role in the precocious development of glucokinase in young rats (35). In all of these studies, an increase in enzyme activity was associated with increased protein synthesis. Pharmacological levels of biotin increased the activity of glucokinase in biotin replete animals (28). Hepatic glucokinase of biotin-injected starved rats had increased almost threefold in comparison with the levels in starved rats. The relative amounts of glucokinase mRNA in the liver of biotin-injected starved rats were increased fourfold over the levels seen in normal fed rats and almost twentyfold that seen in starved rats not receiving biotin injection. The induction of glucokinase mRNA by biotin is quite marked and rapid. It correlated with the increase in glucokinase enzyme activity. In "run-on" transcription assays using isolated liver nuclei, biotin administration to the whole animal increased glucokinase gene transcription by the isolated liver nuclei by about sevenfold. This increased transcription of glucokinase was not due to an increase in overall transcriptional efficiency as the transcription of the β-actin gene was not affected (36).

In both fasted and diabetic rats, hepatic phosphoenolpyruvate carboxykinase (PEPCK) activities are markedly increased. Refeeding a high-carbohydrate diet to fasted rats decreased PEPCK mRNA, which is due to the repression of transcription of the PEPCK gene by insulin. Three hours after biotin administration to starved rats, hepatic PEPCK mRNA levels decreased to 15% of the levels seen in non-biotin-injected starved rats. The effect of biotin paralleled the effect of insulin in these animals. In "run on" transcription experiments using isolated hepatic nuclei, biotin suppressed hepatic PEPCK mRNA by 55% at 30 min after biotin administration. The inhibition is dominant over other stimulatory effects (37). There are many similarities between biotin and insulin in their action on enzymes of glucose metabolism. Both induce a key glycolytic enzyme and repress the synthesis of mRNA that encodes for PEPCK, a key gluconeogenic enzyme. In transgenic mice expressing constitutively active Fox 01 in the liver, fasting glucose levels are increased and glucose tolerance is impaired. Genes involved in glucose utilization by glycolysis, pentose phosphate shunt, lipogenesis, and sterol synthetic pathways are suppressed. Thus, Fox 01 proteins promote hepatic glucose production (38). The relationship of biotin to the initiation of the synthesis of Fox 01 proteins has been investigated (39). The alteration of gene expression by biotin administration in the liver of streptozotocin-induced diabetic rats was examined. The mRNA levels of PEPCK and glucose-6-phosphatase were reduced and glucokinase mRNA increased 3 h after administration of biotin. The expression of Fox 01, which is enhanced by insulin, was decreased by biotin, indicating that biotin repressed the gluconeogenic genes through a pathway independent of insulin signaling.

The effect of biotin on pancreatic glucokinase activity and mRNA expression was examined using cultured pancreatic β cells (RIN 1046-38). Biotin had a stimulatory effect after short-term treatment (40). The effects of biotin deficiency on pancreatic islet glucokinase activity and mRNA expression as well as on insulin secretion were examined (41). Biotin stimulated glucokinase activity in rat islets in culture. Treatment with biotin also increased insulin secretion. Islet glucokinase activity and mRNA are reduced by 50% in the biotin-deficient rat. Insulin secretion in response to glucose was also impaired in islets isolated from biotin-deficient rats. Biotin seems to have an effect on the synthesis of glucokinase mRMA and on insulin secretion by pancreatic islets. The effect of insulin on the synthesis and secretion of glucokinase is distinct from the effect of biotin on the synthesis and secretion of glucokinase in the pancreas.

12.9   BIOTIN SENSING

As indicated, biotin has a role in the transcription of multiple genes. The earliest known such function is in the biotin biosynthetic pathway of many microorganisms. The biosynthesis of biotin is encoded by genes bio F, bio A, bio D, and bio B. The biotin regulatory system in *E. coli* has been well characterized. The biosynthesis and transport are regulated by a bifunctional protein Bir A (42, 43). The biotin protein ligase functions both as a biotinylating enzyme for the apocarboxylases and as a transcriptional repressor. This pathway is utilized by a broad range of eubacteria and archebacteria (44). Biotin influences transcription in organisms as diverse as bacteria and humans. In higher eukaryotes biotin status has significant effects on cell cycle and transcription. The biotin protein ligase (HCS) appears to play a central role in this process as well (45, 46). In addition to its role in metabolism, HCS functions in the transcriptional response to biotin. Although HCS is not a site-specific DNA binding protein in the eukaryotes, unlike the bifunctional bacterial enzyme, HCS seems to have a role in the cell nucleus of biotinylation of histones (47). In mammalian cell lines the expression of the genes encoding HCS, the biotin transporter SMVT1 and biotin-acceptor proteins are decreased under conditions of biotin deficiency (48, 49). Addition of biotin or 8-bromo-cGMP to biotin-starved cells led to recovery, although the inhibitions of cGMP-dependent protein kinase nullified the biotin effect. This led to a suggestion that the biotin signaling cascade required guanylate cyclase and cGMP-dependent protein kinase (50, 51). Biotin analogs that do not function as enzymatic cofactors have biotin-like activities on gene transcription (52). Based on this, it was suggested that biotinyl-AMP could be the intracellular signal regulating gene transcription in eukaryotes (51). However, organisms such as yeast lack homologues of guanylate cyclase and c-GMP-dependent protein kinase, questioning the universality of such mechanisms. Biotin protein ligase (HCS) is an important element of the biotin-sensing pathway (53). HCS appears to interact with the methyl-CpG-binding domain protein 2 and also histone methyl transferases, thus creating epigenetic synergies between biotinylation and methylation events (53, 54).

12.10 EFFECTS OF BIOTIN STARVATION ON CARBON METABOLISM GENES

The opposing genetic effects of biotin on the transcription of glucokinase and PEPCK are significant for the regulation of carbon fluxes. To understand how extensively biotin affects the expression of the genes of carbon metabolism, a high-density oligonucleotide micro-array approach was used in a study of three eukaryotes (55, 56). Biotin starvation reduces glucose consumption and energy production in three different enkaryotes, despite the fact that their phylogenetic lines diverged more than 1,000 million years ago. In these three biotin-starved organisms—yeast, nematode, and rat—the genomic expression corresponded to (false) scant glucose conditions, pointing to a strongly selected role of biotin in the control of carbon metabolism. This concept is strengthened by the similarities between biotin and insulin across species (55). As the effects of biotin starvation are suggestive of a condition of insulin resistance, treatment with biotin could have clinical significance (56). Pharmacological doses of biotin appear to ameliorate the diabetic condition in alloxan-induced diabetic rats (57). In genetically diabetic kk mice and in OLETF rats, biotin improved glucose tolerance (58). Similar results were also observed in humans with type I (59) or type II (60) diabetic conditions (61). Supplementation with chromium picolinate resulted in beneficial effects on glucose, insulin, and cholesterol levels in patients with type II diabetes (62). A combination of biotin and chromium picolinate improved glycemic control in poorly controlled diabetics receiving antidiabetic therapy (63, 64).

Various reports indicate that pharmacological doses of biotin reduce hyperlipidemia induced in healthy volunteers (65) and in patients with hyperlipidemia (66, 67). Spontaneous symptoms of biotin deficiency were seen in rats genetically prone to the development of hyperlipidemia (68). Pharmacological doses of biotin decrease serum triglyceride concentrations and lipogenic gene expression in liver and adipose tissue of male BALB/c AnN Hsd mice (69). A combination of chromium picolinate and biotin reduced the atherogenic index of plasma in patients with type 2 diabetes mellitus (70). Administration of pharmacological doses of biotin also decreased the systolic blood pressure in the stroke-prone spontaneously hypertensive rat via NO-independent direct activation of soluble guanylate cyclase (71).

12.11   BIOTIN-BINDING PROTEINS

The enzymes biotinidase and holocarboxylase synthetase have specific roles in biotin metabolism. Holocarboxylase synthetase has a role in the cell nucleus in addition to its role in the biotinylation of the apocarboxylases. Biotinidase participates in the "biotin cycle" of the cell designed to conserve biotin (4). In the egg-laying hen there are three biotin-binding proteins. In addition to avidin, which is present in the albumen of the egg, there are two biotin-binding proteins, BBP1 and BBP2, in egg yolk. These proteins have a role in the development of the embryo. Proteins of the avidin family have a high affinity for biotin. The avidin-biotin system, in association with the monoclonal antibody for biotin (72), has been used as an essential biotechnological tool in many fields of research over the past few decades (73). Apart from strepavidin, isolated from streptomyces species, avidin-like proteins have been isolated from other prokaryote and fungal sources (74). Avidins are tetrameric proteins with four biotin-binding sites. The high-affinity characteristics are closely related to the quarternary structure, as dissociation of avidins into the dimeric or monomeric state dramatically reduces the affinity for biotin. A dimeric member of the avidin family, rhizavidin, with high affinity for biotin, was isolated from *Rhizobium etli* (75). Another dimeric member, schwanavidin, from the marine proteobacterium *Shewanella denitrificans*, has an innate dimetric structure with high affinity for biotin (74). It is interesting that these two high-affinity avidins have been isolated from bacteria inhabiting very diverse environments. *R. etli*, which produces rhizavidin, is found in the soil rhizosphere habitat, whereas the marine bacterim *S. dentrificans* producing shwanavidin is found in the Gotland basin region of the Baltic Sea at a depth of 120–130 m (73). The unique features of rhizavidin and schwanavidin might help in designing a high-affinity monovalent biotin-binding system.

One of the most devastating rice pathogens in the world is the fungus *Magnaporthe oryzae*, which is dependent on biotin for its growth (76). Tamavidin 1, an avidin-like biotin-binding protein from the mushrooms *Pleurotus cornucopiae* restricts the growth of *M. oryzae*. The possibility of reducing the availability of biotin to *M. oryzae* in rice by expression of the gene that encodes for tamavidin 1 was investigated (77). The positive results indicate a role for the strategy to engineer disease resistance to higher plants through the pathogen's auxotrophy. Biotin-binding proteins expressed in transgenic plants are insecticidal to a wide variety of insects. The extreme stability and resistance to proteolysis makes them very valuable in transgenic crop protection. Although their role in the protection of food crops such as rice has been established with some caveats, their use in protection of non-food crops such as fiber, forestry, and biofuel crops is boundless (78).

REFERENCES

1. Mistry SP and Dakshinamurti K. Biochemistry of biotin. *Vitam. Horm.* 22: 1–55 (1964).
2. Dakshinamurti K and Bhagavan HN. eds. Biotin. *Ann. N.Y. Acad. Sci.* 447: 1–441 (1985).
3. Dakshinamurti K and Chauhan J. Regulation of biotin enzymes. *Annu. Rev. Nutr.* 8: 211–233 (1988).
4. Dakshinamurti K and Chauhan J. Biotin. *Vitam. Horm.* 45: 337–384 (1989).
5. Roth KS, Yang W, Allan L, Saunders M, Gravel RA, and Dakshinamurti K. Prenatal administration of biotin in biotin-responsive multiple carboxylase deficiency. *Pediatr. Res.* 16: 126–129 (1982).
6. Sweetman L and Nyhan WL. Inheritable biotin-treatable disorders and associated phenomena. *Annu. Rev. Nutr.* 6: 317–343 (1986).
7. Dakshinamurti K and Mistry SP. Tissue and intracellular distribution of biotin-[14]COOH in rats and chick. *J. Biol. Chem.* 238: 294–296 (1963).
8. Dakshinamurti K and Mistry SP. Amino acid incorporation and biotin deficiency. *J. Biol. Chem.* 238: 297–301 (1963).
9. Boeckx RLD and Dakshinamurti K. Biotin-medicated protein synthesis. *Biochem. J.* 140: 549–556 (1974).
10. Boeckx RLO and Dakshinamurti K. Effect of biotin on ribonucleic acid synthesis. *Biochem. Biophys. Acta* 383: 282–289 (1975).
11. Gravel R and Narang M. Molecular genetics of biotin metabolism: Old vitamins, new science. *J. Nutr. Biochem.* 16: 428–431 (2005).
12. Keranen AJA. The biotin synthesis of the HeLa cells in vitro. *Cancer Res.* 32: 119–124 (1972).
13. Chalifour LE and Dakshinamurti K. The requirement of human fibroblasts in culture. *Biochem. Biophys. Res. Communs.* 104: 1047–1053 (1982).
14. Bhullar RP and Dakshinamurti K. The effects of biotin on cellular functions in HeLa cells. *J. Cell Physiol.* 122: 425–430 (1985).
15. Dakshinamurti K, Chalifour LE, and Bhullar RP. Requirement for biotin and the function of biotin in cell culture. *Ann. N.Y. Acad Sci.* 447: 38–55 (1985).
16. Bernlohr DA, Bolanowski MS, Kelly Jr. IJ, and Lane MD. Evidence for an increase in transcription of specific mRNA during differentiation of 3T3-L1 preadiposites. *J. Biol. Chem.* 260: 5563–5567 (1985).
17. Baez-Saldana A, Diaz G, Espinoza B, and Ortega E. Biotin deficiency induces changes in subpopulations of spleen lymphocytes in mice. *Am. J. Clin. Nutr.* 67: 431–437 (1998).
18. Baez-Saldana A and Ortega E. Biotin deficiency blocks thymocyte maturation, accelerates thymus involution and decreases nose-rump length in mice. *J. Nutr.* 134: 1970–1977 (2004).
19. Watanabe T. Micronutrients and congenital anomalies. *Congenit. Anom.* 30: 79–92 (1990).
20. Watanabe T, Dakshinamurti K, and Persaud TVN. Effect of biotin on palatal development of mouse embryos in organ culture. *J. Nutr.* 125: 2114–2121 (1995).
21. Paulose CS, Thliveris T, Viswanathan M, and Dakshinamurti K. Testicular function in biotin-deficient adult rats. *Horm. Metab. Res.* 21: 661–665 (1989).
22. Baez-Saldana A, Camacho-Arroyo J, Espinosa-Aguirre JJ, Neri-Gomez T, Rojas-Ochoa A, Guerra-Araiza C, Larrieta E, Vital P, Diaz G, Chavira R, and Fernandez-Mejia C. Biotin deficiency and biotin excess: Effects on the female reproductive system. *Steroids* 74: 863–869 (2009).
23. Johnson R, Norstrom E, and Soderstrom K. Increased expression of endogenous biotin, but not BDNF, in telencephalic song regions during zebra finch vocal learning. *Dev. Brain Res.* 120: 113–123 (2000).

24. Wang H and Prevsner J. Detection of endogenous biotin in various tissues: Novel functions in the hippocampus and implications for its use in avidin-biotin technology. *Cell Tissue Res.* 296: 511–516 (1999).

25. Ozand PT, Gascon GG, and Al Essa M. Biotin-responsive basal ganglia disease: A novel entity. *Brain* 121: 1267–1279 (1998).

26. Debs R, Depienne C, Rastetter A, Bellanger A, Degos B, Galanaud D, Keren B, Lyon-Caen O, Brice A, and Sedel F. Biotin-responsive basal ganglia disease in ethnic Europeans with novel SLC19A3 mutations. *Arch. Neurol.* 67: 126–130 (2010).

27. El-Hajj TI, Karam PE, and Mikati MA. Biotin-responsive basal ganglia disease: Case report and review of literature. *Neuropediatrics* 39: 268–271 (2008).

28. Dakshinamurti K. Biotin: A regulator of gene expression. *J. Nutr. Biochem.* 16: 419–423 (2005).

29. Rodriguez-Melandez R, Cane S, Mendez ST, and Velazques A. Biotin regulates the genetic expression of holocarboxylase synthetase and mitochondrial carboxylases in rats. *J. Nutr.* 131: 1909–1913 (2001).

30. Wiedmann S, Eudy JD, and Zempleni J. Biotin supplementation increases expression of genes encoding interferon-gamma, interleukin-1β and 3-methylcotonyl-CoA carboxylase and decreases expression of the gene encoding interleukin-4 in human peripheral blood mononuclear cells. *J. Nutr.,* 133: 716–719 (2003).

31. Zempleni J. Uptake, localization and noncarboxylase roles of biotin. *Ann. Rev. Nutr.* 25: 175–196 (2005).

32. Kuroishi T, Endo Y, Muramoto K, and Sugawara S. Biotin deficiency up-regulates TNF-α production in marine macrophages. *J. Leukocyte Biol.* 83: 912–920 (2008).

33. Dakshinamurti K and Cheah-Tan C. Liver glucokinase of the biotin deficient rat. *Can. J. Biochem.* 46: 75–80 (1968).

34. Dakshinamurti K and Cheah-Tan C. Biotin-mediated synthesis of hepatic glucokinase in the rat. *Arch. Biochem. Biophys.* 127: 17–21 (1968).

35. Dakshinamurti K and Hong HC. Regulation of key glycolytic enzymes. *Enzymol. Biol. Clin.* 11: 422–428 (1969).

36. Chauhan J and Dakshinamurti K. Transcriptional regulation of the glucokinase gene by biotin in starved rats. *J. Biol. Chem.* 266: 10035–10038 (1991).

37. Dakshinamurti K and Li W. Transcriptional regulation of liver phosphenolpyruvate carboxykinase by biotin in diabetic rats. *Mol. Cell. Biochem.* 132: 127–132 (1994).

38. Zhang W, Patil S, Chauhan B, Cuo S, Powell DR, Klotsat A, Matika R, Xiao X, Franks R, Heidenreich KA, Sajan MP, Farese RV, Stolz DB, Tso P, Koo S-H, Montonini M, and Unterman TC. Fox o1 regulates multiple metabolic pathways in the liver: Effects on gluconeogenic, glycolytic and lipogenic gene expression. *J. Biol. Chem.* 281: 10105–10117 (2006).

39. Sugita Y, Shirakawa H, Sugimoto R, Furakawa Y, and Komai M. Effect of biotin treatment on hepatic gene expression in streptozotocin-induced diabetic rats. *Bio. Sci. Biotechnol. Biochem.* 72: 1290–1298 (2008).

40. Borboni P, Magnaterra R, Rabini RA, Staffolini R, Porzio O, Sesti G, Fusco A, Mazzanti L, Lauro R, and Marlier LNJL. Effect of biotin on glucokinase activity, m RNA expression and insulin release in cultured beta-cells. *Acta. Diabetol.* 33: 154–158 (1996).

41. Romero-Navarro G, Cabrera-Valladaros G, German MS, Matchinsky FM, Velazquez A, Wang J, and Fernandez-Mejia C. Biotin regulation of pancreatic glucokinase and insulin in primary cultured rat islets and in biotin-deficient rats. *Endocrinology* 140: 4595–4600 (1999).

42. Cronan JE Jr. The *E coli* bio operon: Transcriptional repression by an essential protein modification enzyme. *Cell* 58: 427–429 (1989).

43. Bower S, Perkin J, and Yocum RR. Cloning and characterization of *Bacillus subtilis* bir A gene encoding a repressor of the biotin operon. *J. Bacteriol.* 177: 2572–2575 (1995).

44. Rodionov DA, Mironov AA, and Gelfand MS. Conservation of the biotin regulon and the bir A regulatory signal in Eubacteria and Archea. *Genome Res.* 12: 1507–1516 (2002).

45. Beckett D. Biotin sensing: Universal influence of biotin status on transcription. *Annu. Rev. Genet.* 41: 443–464 (2007).

46. Beckett D. Biotin sensing at the molecular level. *J. Nutr.* 139: 167–170 (2009).

47. Narang MA, Dumas R, Ayer LM, and Gravel RA. Reduced histone biotinylation in multiple carboxylase deficiency patients: A nuclear role for holocarboxylase synthetase. *Hum. Mol. Genet.* 13: 15–23 (2004).

48. Pacheco-Alvarez D, Solorzano-Vargas RS, Gravel RA, Cervantes-Roldan R, Velazquez A, and Leon-Del Rio A. Paradoxical regulation of biotin utilization in brain and liver and implications for inherited multiple carboxylase deficiency. *J. Biol. Chem.* 279: 52312–52318 (2004).

49. Pacheco-Alvarez D, Solorzano-Vargas RS, Gonzales-Noriega A, Michalak C, Zempleni J, and Leon-Del Rio. Biotin availability regulates expression of the sodium-dependent multivitamin transporter and the rate of biotin uptake in Hep G2 cells. *Mol. Genet. Metab.* 85: 301–307 (2005).

50. Vesely DL, Wormser HC, and Abramson HN. Biotin analogs activate guanylate cyclase. *Mol. Cell. Biochem.* 60: 109–114 (1984).

51. Singh I and Dakshinamurti K. Stimulation of guanylate cyclase and RNA polymerase II activities in HeLa cells and fibroblasts by biotin. *Mol. Cell. Biochem.* 79: 47–55 (1988).

52. Rodriguez-Melandez R, Lewis B, MeMohan RJ, and Zempleni J. Diamino biotin and desthio biotin have biotin-like activities in Jurkat cells. *J. Nutr.* 133: 1259–1264 (2003).

53. Pirner HM and Stolz J. Biotin sensing in *Saccharomyces cerevisiae* is mediated by a conserved DNA element and requires the activity of biotin-protein ligase. *J. Biol. Chem.* 281: 12381–12389 (2006).

54. Zempleni J, Teixeira DC, Kuroishi T, Cordonier EL, and Baier S. Biotin requirements for DNA damage prevention. *Mutation Res.* 733: 58–60 (2012).

55. Ortega-Cuellar D, Hernandez-Mendoza A, Morene-Arriola E, Carvajal-Aguilera K, Perez-Vazquez V, Gonzalez-Alvarez R, and Velazquez-Arellano A. Biotin starvation with adequate glucose provision causes paradoxical changes in fuel metabolism gene expression similar in rat (*Rattus norvegicus*), nematode (*Caenorhabditis elegans*) and yeast (*Saccharomyces cerevisiae*). *J. Nutrigenet. Nutrigenomics* 3: 18–30 (2010).

56. Velazquez-Arellano A, Ortega-Cuellar D, Hernandez-Mendoza A, and Moreno-Arriola E. A heuristic model for paradoxical effects of biotin starvation on carbon metabolism genes in the presence of abundant glucose. *Mol. Genet. Metab.* 102: 69–77 (2011).

57. Dakshinamurti K, Terrago-Litvak L, and Hong HC. Biotin and glucose metabolism. *Can. J. Biochem.* 48: 493–500 (1970).

58. Reddi A, DeAngelis B, Frank O, Lasker N, and Baker H. Biotin supplementation improves glucose and insulin tolerance in genetically diabetic KK mice. *Life Sci.* 42: 1323–1330 (1988).

59. Coggeshall JC, Heggers JP, Robson MC, and Baker H. Biotin status and plasma glucose levels in diabetics. *Ann. NY Acad. Sci.* 447: 389–392 (1985).

60. Maebashi M, Makino R, Furukawa Y, Ohinata K, Kimura S, and Takeo S. Therapeutic evalution of the effects of biotin on hyperglycemia in patients with non-insulin diabetic mellitus. *J. Clin. Biochem. Nutr.* 14: 211–218 (1993).

61. Fernandez-Meija C. Pharmacological effects of biotin. *J. Nutr. Biochem.* 16: 424–427 (2005).

62. Cefalu WT and Hu FB. Role of chromium in human health and in diabetes. *Diabetes Care* 27: 2741–2751 (2004).

63. Singer GM and Geshas J. The effect of chromium picolinate and biotin supplementation on glycemic control in poorly controlled patients with type II diabetes mellitus: A placebo-controlled, double-blinded, randomized trial. *Diabetes Technol. Ther.* 8: 636–643 (2006).

64. Albarracin CA, Fuqua BC, Evans JL, and Goldfine ID. Chromium picolinate and biotin combination improves glucose metabolism in treated, uncontrolled overweight to obese patients with type II diabetes. *Diabetes Metab. Res. Rev.* 24: 41–51 (2008).
65. Marshall MW, Kliman PG, Washington VA, Mackin JF, and Weinland BT. Effects of biotin on lipids and on other constituents of plasma of healthy men and women. *Artery* 7: 330–351 (1980).
66. Dokusova OK and Krivoruchenko IV. The effect of biotin on blood cholesterol levels of atherosclerotic patients in idiopathic hyperlipidemia. *Kardiologia* 12: 113 (1972).
67. Revilla-Monsalve C, Zendejas-Ruiz I, Islas-Andrade S, Baez-Saldana A, Palomino-Garibay MA, Hernandez-Quiroz PM, and Fernandez-Mejia C. Biotin supplementation reduces plasma triacyl glycerol and VLDL in type 2 diabetic patients and in nondiabetic subjects with hypertriglyceridemia. *Biomed. Pharmacother.* 60: 182–185 (2006).
68. Marshall MW, Smith BP, and Lehman RP. Dietary response of two genetically different lines of inbred rats: Lipids in serum and liver. *Proc. Soc. Exp. Biol. Med.* 131: 1271–1277 (1969).
69. Larrieta E, Velasco F, Vital P, Lopez-Aceres T, Lazo-de-da-Vega-Monroy M, Rojas A, and Fernandez-Mejia C. Pharmacological concentrations of biotin reduce serum triglycerides and the expression of lipogenic genes. *Euro. J. Pharmacol.* 644: 263–268 (2010).
70. Geohas J, Daly A, Juturu V, Finch M and Komorowski JR. Chromium picolinate and biotin combination reduces atherogenic index of plasma in patients with type 2 diabetes mellitus: A placebo controlled, double-blinded, randomized clinical trial. *Am. J. Med. Sci.* 333: 145–153 (2007).
71. Watanabe-Kamiyama M, Kamiyama S, Hotiuchi K, Ohinata K, Shirakawa H, Furakawa Y, and Komai M. Antihypertensive effect of biotin in stroke-prone spontaneously hypertensive rats. *Brit. J. Nutr.* 99: 756–763 (2008).
72. Dakshinamurti and Rector ES. Monoclonal antibody to biotin. *Methods Enzymol.* 184: 111–119 (1990).
73. Bayer EA and Wilchek M. Applications of avidin-biotin technology to affinity based separations. *J. Chromatogr.* 510: 3–11 (1990).
74. Meir A, Bayer EA, and Linvah O. Structural adaptation of a thermostable biotin-binding protein in a psychrophilic environment. *J. Biol. Chem.* 287: 17951–17965 (2012).
75. Helppolainen SH, Nurminen KP, Maatta JA, Halling KK, Slotte JP, Huhtala T, Liimatanen T, Yla-Herttula S, Airenne KJ, Narvanen A, Janis J, Vainiotalo P, Valjakka J, Kuloma MS, and Nordlund HR. Rhizavidin from Rhizobium etli: The first natural dimer in the avidin protein family. *Biochem J.* 405: 397–405 (2005).
76. Skamnioti P and Gurr SJ. Against the grain: Safeguarding rice from rice blast disease. *Trends Biotechnol.* 27: 141–150 (2009).
77. Takakura Y, Oka N, Suzuki J, Tsukamoto H, and Ishida Y. Intracellular production of Tamavidin 1, a biotin-binding protein from Tamogitake mushroom, confers resistance to blast fungus *Magnaporthe oryzae* in transgenic rice. *Mol. Biotechnol.* 51: 9–17 (2012).
78. Christeller JT, Markwick NP, Burgess EPJ, and Malone LA. The use of biotin-binding proteins for insect control. *J. Econ. Entomol.* 103: 497–508 (2010).

13 Mechanisms of Gene Transcriptional Regulation through Biotin and Biotin-Binding Proteins in Mammals

Janos Zempleni, Dandan Liu, Daniel Camara Teixeira, and Mahendra P. Singh

CONTENTS

13.1 INTRODUCTION

Mammals cannot synthesize biotin and depend on a regular dietary supply of this water-soluble vitamin (Zempleni et al., 2009). The Adequate Intake for biotin in adults is 30 µg/d (National Research Council, 1998). The classical role of biotin in mammalian intermediary metabolism is to serve as a covalently bound coenzyme in five carboxylases (Zempleni et al., 2009). Both the cytoplasmic acetyl-CoA carboxylase 1 (ACC1) and the mitochondrial acetyl-CoA carboxylase 2 (ACC2) catalyze the binding of bicarbonate to acetyl-CoA to generate malonyl-CoA, but the two isoforms have distinct functions in intermediary metabolism (Kim et al., 1997). ACC1 produces malonyl-CoA for the synthesis of fatty acid synthesis in the cytoplasm; ACC2

219

is an important regulator of fatty acid oxidation in mitochondria. The malonyl-CoA produced by ACC2 inhibits mitochondrial uptake of fatty acids for β-oxidation.

Pyruvate carboxylase (PC), propionyl-CoA carboxylase (PCC), and 3-methyl-crotonyl-CoA carboxylase (MCC) localize in mitochondria (Zempleni et al., 2009). PC is a key enzyme in gluconeogenesis. PCC catalyzes an essential step in the metabolism of propionyl-CoA, which is produced in the metabolism of some amino acids, the cholesterol side chain, and odd-chain fatty acids. MCC catalyzes an essential step in leucine metabolism. Both PCC and MCC are composed of nonidentical subunits, that is, biotinylated α subunits and non-biotinylated β subunits, which are encoded by distinct genes.

Holocarboxylase synthetase (HLCS) is the sole ligase in the human proteome that can catalyze the covalent binding of biotin to carboxylases in vivo (Suzuki et al., 1994). The enzyme biotinidase also has biotinyl protein transferase activity in vitro (Hymes et al., 1995), but this activity apparently lacks relevance in vivo (Camporeale et al., 2006). Evidence for a role of biotin in gene regulation, in addition to its role as a coenzyme, was provided more than forty years ago (Dakshinamurti and Cheah-Tan, 1968b, Dakshinamurti and Cheah-Tan, 1968a). However, the mechanisms of gene regulation by biotin remained unknown until about ten years ago, when we began to understand the pathways of gene regulation through biotin and HLCS. These gene regulatory mechanisms are reviewed in this chapter.

13.2 BIOTIN-SIGNALING PATHWAYS

Biotin affects cell signaling by transcription factors such as cGMP, NF-κB, Sp1 and Sp3, receptor tyrosine kinases; by calcium, nitric oxide, the biotin intermediate biotinyl-5′-adenosine monophosphate (biotinyl-5′-AMP), microRNAs (miRs); and by HLCS-dependent signaling. Biotin also affects gene expression at the posttranscriptional level (Collins et al., 1988), which is not the focus of this chapter.

13.2.1 cGMP

Spence and Koudelka (1984) were the first to demonstrate that the addition of biotin to the culture medium of primary rat hepatocytes increased their content of cGMP 3-fold within 1 h. The elevated levels of cGMP (or the synthetic, non-hydrolyzable analog 8-bromo-cGMP) in biotin-supplemented cells were associated with a four-fold increase in the activity of glucokinase 6 h following the addition of biotin to the medium, consistent with previous reports of biotin-dependent expression of glucokinase (Dakshinamurti and Cheah-Tan, 1968b). The above effects were seen at nanomolar levels of biotin in culture media (Spence and Koudelka, 1984), that is, concentrations that can easily be achieved by using low to moderate doses of biotin supplements (Mock et al., 1995, Zempleni et al., 2001). The biotin dependence of cGMP signaling was subsequently expanded to demonstrate a crucial role of biotinyl-5′-AMP in this pathway (see following) (Solorzano-Vargas et al., 2002, Pacheco-Alvarez et al., 2005).

13.2.2   NF-κB

NF-κB is activated in response to cell stress and has been linked with cancer, inflammatory processes, cell survival, and autoimmune diseases (Baeuerle and Henkel, 1994, Baldwin, 1996). Evidence suggests that when cell stress is induced by biotin depletion, the nuclear translocation and activity of NF-κB increase in human lymphoid (Jurkat) cells (Rodriguez-Melendez et al., 2004). This effect is caused by phosphorylation and subsequent degradation of inhibitor of NF-κB (IκB) alpha, leading to an increase in the nuclear import of the p50 and p65 proteins from the NF-κB family in biotin-depleted cells compared with biotin-supplemented cells. Consistent with this observation, biotin depletion is associated with the activation of cell survival mechanisms, as evidenced by high expression of the anti-apoptotic gene Bfl-1/A1, low activity of the apoptotic enzyme caspase-3 in response to treatment with tumor necrosis factor alpha, and a low rate of cell death in response to serum starvation compared to biotin-supplemented cells. These observations might have clinical relevance given that biotin-starved lymphoma cell cultures have a greater resistance to antineoplastic drugs compared with biotin-supplemented cells (Griffin and Zempleni, 2005).

13.2.3   Sp1 AND Sp3

Members of the Sp/Krüppel-like factor family of transcription factors (e.g., the ubiquitous Sp1 and Sp3) play important roles in the expression of numerous mammalian genes (Black et al., 2001). The activity of Sp1 and Sp3 depends on biotin in Jurkat cells (Griffin et al., 2003). Specifically, the binding of Sp1 and Sp3 to their DNA-binding sites (GC box and CACCC box) is 76%–149% greater in nuclear extracts from biotin-supplemented cells compared with biotin-deficient cells, as determined by electrophoretic mobility shift assays. The increased binding to response elements in supplemented cells translates into a 50% increase in the activity of a reporter gene vector driven by Sp1 binding sites. Effects of biotin on Sp1/Sp3 signaling are mediated by an increased expression of the two transcription factors rather than their dephosphorylation in biotin-supplemented cells compared with biotin-depleted cells.

13.2.4   Receptor Tyrosine Kinases

The use of high-throughput immunoblots revealed that biotin alters the abundance of ~5% of the proteins quantified in human hepatocarcinoma (HepG2) cells (Rodriguez-Melendez et al., 2005). Biotin-dependent proteins cluster in pathways related to cell signaling, particularly receptor tyrosine kinase-mediated signaling. The abundance of receptor tyrosine kinases is greater in biotin-depleted cells than in cells cultured in biotin-normal medium. The high levels of receptor tyrosine kinases are associated with increased DNA-binding activities of transcription factors Fos and Jun in biotin-depleted cells compared with normal controls, as judged by electromobility shift assay. Consistent with this observation, Fos/Jun-dependent reporter gene plasmids have a 45% greater activity in biotin-deficient HepG2 cells compared with normal controls.

It has been proposed that Fos and Jun mediate increased expression of the biotin transporters SMVT and MCT1 in response to biotin deficiency (Manthey et al., 2002, Crisp et al., 2004). The regulatory region of the human SMVT gene contains 4 AP1 sites and 1 AP1-like site (Dey et al., 2002), whereas the regulatory region of the human MCT1 gene contains 2 AP1 sites (Hadjiagapiou et al., 2005). In addition to its putative effects on biotin transporter expression, nuclear translocation of Fos and Jun might enhance stress resistance of biotin-deficient cells (Inoue et al., 2001, Ramet et al., 2002, Sluss et al., 1996) in analogy to effects of biotin on NF-κB (Rodriguez-Melendez et al., 2004). It is possible that the increased nuclear abundance of both Fos/Jun and NF-κB observed in biotin-deficient cells might contribute to the proinflammatory and antiapoptotic effects of biotin deficiency.

13.2.5 CALCIUM

Biotin supplementation studies in healthy adults and human cell cultures suggest that biotin supplementation decreases the transport of Ca^{2+} from the cytoplasm into the endoplasmic reticulum (ER). When healthy adults were supplemented with a typical over-the-counter biotin supplement (8.8 µmol/d biotin for 21 d), the expression of the sarco/endoplasmic reticulum Ca^{2+}-ATPase 3 (SERCA3), isoform a, decreased by >80% in post-supplementation lymphocytes compared with pre-supplementation lymphocytes (Wiedmann et al., 2004). SERCAs 1, 2, and 3 transport calcium from the cytoplasm into the ER, thereby maintaining a resting concentration of free Ca^{2+} in the ER that is three to four orders of magnitude greater than in the cytoplasm (Burk et al., 1989, Lytton et al., 1992). SERCA3 is abundantly expressed in lymphoid cells (Martin et al., 2002), suggesting a pivotal role for this transporter in the calcium homeostasis in immune cells.

Studies in human Jurkat cells produced results consistent with the observations made in human lymphocytes in the previously discussed biotin supplementation studies. Both the transcriptional activity of a SERCA3 reporter gene and the abundance of SERCA3 mRNA were 50% lower in biotin-supplemented Jurkat cells than in biotin-depleted controls (Griffin et al., 2006). The low expression of SERCA3 caused a moderate accumulation of Ca^{2+} in the cytoplasm of biotin-supplemented cells. The perturbation of cellular calcium homeostasis was associated with ER stress, as evidence by decreased protein secretion, and up-regulation of ER stress response systems ubiquitin activating enzyme 1, growth arrest and DNA damage 153 gene, X-box binding protein 1, and phosphorylated eukaryotic translation initiation factor 2α in biotin-supplemented cell cultures.

It has been proposed that the effects of biotin on the expression of SERCA3 are mediated by Sp1 and Sp3. The 5′-flanking region of the SERCA3 gene contains twenty-five consensus binding sites for the biotin-dependent transcription factors Sp1 and Sp3 (Dode et al., 1998). Sp1 and Sp3 may act as transcriptional activators or repressors, depending on the context (Birnbaum et al., 1995, Black et al., 2001). Based on these data, it seems likely that the decreased expression of SERCA3 in biotin-supplemented cells is mediated by increased abundance of Sp1 and Sp3.

13.2.6 NITRIC OXIDE

Nitric oxide (NO) signaling depends on biotin in cell culture (Rodriguez-Melendez and Zempleni, 2009). It has been proposed that biotin-dependent NO signaling is an early event in the biotin-dependent activation of cyclic guanosine monophosphate, cGMP (see following). When Jurkat cells were cultured in biotin-defined media, the concentration of NO decreased by 18% in cells cultured in medium containing 0.025 nmol/L biotin compared with physiological controls; the concentration of NO increased by 27% in cells cultured in medium containing 10 nmol/L biotin compared with physiological controls. Differences among treatment groups were abolished if biotin-defined cells were treated with the NO synthase (NOS) inhibitor N^G-monomethyl-L-arginine or the NO donor SNI-1. The effect of biotin on NO signaling is caused by increased expression of both endothelial and neuronal NOS in Jurkat cells, whereas the expression of inducible NOS appears not to depend on biotin. Studies with the NOS inhibitor N^G-monomethyl-L-arginine and the protein kinase G substrate BPDEtide (Arg-Lys-Ile-Ser-Ala-Ser-Glu-Phe-Asp-Arg-Pro-Leu-Arg) established a causal link between biotin-dependent signaling through NO and cGMP signaling.

13.2.7 BIOTINYL-5'-AMP

Biotinyl-5'-AMP is an intermediate in the HLCS-dependent biotinylation of proteins (Camporeale and Zempleni, 2006). Evidence suggests that the biotin-dependent expression of HLCS, ACC1, and PCC (alpha chain) depends on the generation of cGMP by soluble guanylate cyclase in HepG2 cells (Solorzano-Vargas et al., 2002). As described before, NO appears to play a role in the biotin-dependent regulation of cGMP.

13.2.8 MIR SIGNALING

Evidence is emerging that miRs also play roles in the regulation of genes by biotin. Specifically, the abundance of miR-539 depends on biotin in primary human fibroblasts and various human cell lines, thereby affecting the expression of HLCS (Bao et al., 2010). Also, the abundance of miR-153 depends on biotin in human kidney carcinoma (HEK-293) cells, but the effect might depend on synergies with DNA methylation events (Bao et al., 2012). This line of research is in its infancy, and it is uncertain whether the biotin-dependent effects of miR-539 and miR-153 take place at the level of mRNA degradation or translational inhibition.

13.2.9 HLCS-DEPENDENT SIGNALING AT THE CHROMATIN LEVEL

HLCS translocates to the cell nucleus (Narang et al., 2004) where it binds to chromatin (Camporeale et al., 2006, Singh et al., 2011). Human HLCS is a single copy gene, which maps to chromosome 21q22.1 (Suzuki et al., 1994) and codes for a full-length protein of 726 amino acid with a predicted molecular weight of 81 kDa (Yang et al., 2001). Three HLCS transcripts plus additional splicing variants originate in exons 1, 2, and 3 of the gene (Yang et al., 2001); methionines-1, -7, and -58 in exons 6 and 7 have been identified as possible translation start sites (Hiratsuka et al., 1998). There is consensus that HLCS with a translation start site in methionine-58 can enter the cell nucleus and that HLCS with a translation start site in methionine-7 cannot enter the nucleus (Bailey et al., 2010, Bao et al., 2011b). The claim that HLCS with a translation start site in methionine-1 cannot enter the nucleus (Bailey et al., 2010) has been challenged (Bao et al., 2011b).

Nuclear HLCS catalyzes the binding of biotin to histones H1, H3, H4, and, to a much lesser extent, H2A (Stanley et al., 2001, Camporeale et al., 2004, Kobza et al., 2005, Chew et al., 2006, Kobza et al., 2008, Bao et al., 2011a, Kuroishi et al., 2011). Biotinylated histones are enriched in repressed loci in the human genome (Camporeale et al., 2007, Chew et al., 2008, Pestinger et al., 2011, Rios-Avila et al., 2012), where biotinylation appears to contribute to chromatin condensation (Filenko et al., 2011). However, less than 0.001% of human histones H3 and H4 are biotinylated, raising concerns that the abundance of biotinylated histones might be too low to explain the substantial biological effects of histone biotinylation that were observed in vivo (Stanley et al., 2001, Bailey et al., 2008, Kuroishi et al., 2011). Based on these concerns, a revised model for biotin-dependent epigenetic effects has been proposed. In this new model, HLCS interacts physically with other chromatin proteins to form a multiprotein gene repression complex (Kuroishi et al., 2011). Gene repression depends on the presence of HLCS, whereas biotinylation of histones is a mere mark for HLCS docking sites in human chromatin. Evidence is starting to emerge in support of this model. For example, HLCS appears to interact physically with the methylated cytosine binding protein MeCP2, the nuclear corepressor N-CoR, and histone deacetylases (Xue and Zempleni, 2011, Liu and Zempleni, 2012). Importantly, both the expression and the nuclear translocation of HLCS depends on biotin, thereby creating a direct link between biotin supply and gene repression by epigenetic mechanisms (Gralla et al., 2008, Kaur Mall et al., 2010).

REFERENCES

AEUERLE, P., & HENKEL, T. (1994) Function and activation of NF-kB in the immune system. *Annu. Rev. Immunol.,* 12, 141–179.

BAILEY, L. M., IVANOV, R. A., WALLACE, J. C., & POLYAK, S. W. (2008) Artifactual detection of biotin on histones by streptavidin. *Anal. Biochem.,* 373, 71–77.

BAILEY, L. M., WALLACE, J. C., & POLYAK, S. W. (2010) Holocarboxylase synthetase: Correlation of protein localisation with biological function. *Arch. Biochem. Biophys.,* 496, 45–52.

BALDWIN, A. S. (1996) The NF-kB and IkB proteins: New discoveries and insights. *Annu. Rev. Immunol.,* 14, 649–681.

BAO, B., PESTINGER, V., HASSAN, Y. I., BORGSTAHL, G. E. O., KOLAR, C., & ZEMPLENI, J. (2011a) Holocarboxylase synthetase is a chromatin protein and interacts directly with histone H3 to mediate biotinylation of K9 and K18. *J. Nutr. Biochem.,* 22, 470–475.

BAO, B., RODRIGUEZ-MELENDEZ, R., WIJERATNE, S. S., & ZEMPLENI, J. (2010) Biotin regulates the expression of holocarboxylase synthetase in the miR-539 pathway in HEK-293 cells. *J. Nutr.,* 140, 1546–1551.

BAO, B., RODRIGUEZ-MELENDEZ, R., & ZEMPLENI, J. (2012) Cytosine methylation in *miR-153* gene promoters increases the expression of holocarboxylase synthetase, thereby increasing the abundance of histone H4 biotinylation marks in HEK-293 human kidney cells. *J. Nutr. Biochem.,* 23, 635–639.

BAO, B., WIJERATNE, S. S., RODRIGUEZ-MELENDEZ, R., & ZEMPLENI, J. (2011b) Human holocarboxylase synthetase with a start site at methionine-58 is the predominant nuclear variant of this protein and has catalytic activity. *Biochem. Biophys. Res. Commun.,* 412, 115–120.

BIRNBAUM, M. J., WIJNEN, A. J. V., ODGREN, P. R., LAST, T. J., SUSKE, G., STEIN, G. S., & STEIN, J. L. (1995) Sp1 trans-activation of cell cycle regulated promoters is selectively repressed by Sp3. *Biochemistry,* 34, 16503–16508.

BLACK, A., BLACK, J. D., & AZIZKHAN-CLIFFORD, J. (2001) Sp1 and Krüppel-like factor family of transcription factors in cell growth regulation and cancer. *J. Cell. Physiol.,* 188, 143–160.

BURK, S. E., LYTTON, J., MACLENNAN, D. H., & SHULL, G. E. (1989) cDNA cloning, functional expression, and mRNA tissue distribution of a third organellar Ca2+ pump. *J. Biol. Chem.,* 264, 18561–18568.

CAMPOREALE, G., GIORDANO, E., RENDINA, R., ZEMPLENI, J., & EISSENBERG, J. C. (2006) *Drosophila* holocarboxylase synthetase is a chromosomal protein required for normal histone biotinylation, gene transcription patterns, lifespan and heat tolerance. *J. Nutr.,* 136, 2735–2742.

CAMPOREALE, G., OOMMEN, A. M., GRIFFIN, J. B., SARATH, G., & ZEMPLENI, J. (2007) K12-biotinylated histone H4 marks heterochromatin in human lymphoblastoma cells. *J. Nutr. Biochem.,* 18, 760–768.

CAMPOREALE, G., SHUBERT, E. E., SARATH, G., CERNY, R., & ZEMPLENI, J. (2004) K8 and K12 are biotinylated in human histone H4. *Eur. J. Biochem.,* 271, 2257–2263.

CAMPOREALE, G., & ZEMPLENI, J. (2006) Biotin. In BOWMAN, B. A., & RUSSELL, R. M. (Eds.) *Present Knowledge in Nutrition.* 9th ed. Washington, D.C., International Life Sciences Institute.

CHEW, Y. C., CAMPOREALE, G., KOTHAPALLI, N., SARATH, G., & ZEMPLENI, J. (2006) Lysine residues in N- and C-terminal regions of human histone H2A are targets for biotinylation by biotinidase. *J. Nutr. Biochem.,* 17, 225–233.

CHEW, Y. C., WEST, J. T., KRATZER, S. J., ILVARSONN, A. M., EISSENBERG, J. C., DAVE, B. J., KLINKEBIEL, D., CHRISTMAN, J. K., & ZEMPLENI, J. (2008) Biotinylation of histones represses transposable elements in human and mouse cells and cell lines, and in *Drosophila melanogaster. J. Nutr.,* 138, 2316–2322.

COLLINS, J. C., PAIETTA, E., GREEN, R., MORELL, A. G., & STOCKERT, R. J. (1988) Biotin-dependent expression of the asialoglycoprotein receptor in HepG2. *J. Biol. Chem.,* 263, 11280–11283.

CRISP, S. E. R. H., CAMPOREALE, G., WHITE, B. R., TOOMBS, C. F., GRIFFIN, J. B., SAID, H. M., & ZEMPLENI, J. (2004) Biotin supply affects rates of cell proliferation, biotinylation of carboxylases and histones, and expression of the gene encoding the sodium-dependent multivitamin transporter in JAr choriocarcinoma cells. *Eur. J. Nutr.,* 43, 23–31.

DAKSHINAMURTI, K., & CHEAH-TAN, C. (1968a) Biotin-mediated synthesis of hepatic glucokinase in the rat. *Arch Biochem Biophys,* 127, 17–21.

DAKSHINAMURTI, K., & CHEAH-TAN, C. (1968b) Liver glucokinase of the biotin deficient rat. *Can. J. Biochem.,* 46, 75–80.

DEY, S., SUBRAMANIAN, V. S., CHATTERJEE, N. S., RUBIN, S. A., & SAID, H. M. (2002) Characterization of the 5′ regulatory region of the human sodium-dependent multivitamin transporter, hSMVT. *Biochim. Biophys. Acta,* 1574, 187–192.

DODE, L., GREEF, C. D., MOUNTIAN, I., ATTARD, M., TOWN, M. M., CASTEELS, R., & WUYTACK, F. (1998) Structure of the human sarco-endoplasmic reticulum Ca^{2+}-ATPase 3 gene. *J. Biol. Chem.*, 273, 13982–13994.

FILENKO, N. A., KOLAR, C., WEST, J. T., HASSAN, Y. I., BORGSTAHL, G. E. O., ZEMPLENI, J., & LYUBCHENKO, Y. L. (2011) The role of histone H4 biotinylation in the structure and dynamics of nucleosomes. *PLoS ONE,* 6, e16299.

GRALLA, M., CAMPOREALE, G., & ZEMPLENI, J. (2008) Holocarboxylase synthetase regulates expression of biotin transporters by chromatin remodeling events at the SMVT locus. *J. Nutr. Biochem.*, 19, 400–408.

GRIFFIN, J. B., RODRIGUEZ-MELENDEZ, R., DODE, L., WUYTACK, F., & ZEMPLENI, J. (2006) Biotin supplementation decreases the expression of the *SERCA3* gene (*ATP2A3*) in Jurkat cells, thus, triggering unfolded protein response. *J. Nutr. Biochem.,* 17, 272–281.

GRIFFIN, J. B., RODRIGUEZ-MELENDEZ, R., & ZEMPLENI, J. (2003) The nuclear abundance of transcription factors Sp1 and Sp3 depends on biotin in Jurkat cells. *J. Nutr.,* 133, 3409–3415.

GRIFFIN, J. B., & ZEMPLENI, J. (2005) Biotin deficiency stimulates survival pathways in human lymphoma cells exposed to antineoplastic drugs. *J. Nutr. Biochem.,* 16, 96–103.

HADJIAGAPIOU, C., BORTHAKUR, A., DAHDAL, R. Y., GILL, R. K., MALAKOOTI, J., RAMASWAMY, K., & DUDEJA, P. K. (2005) Role of USF1 and USF2 as potential repressor proteins for human intestinal monocarboxylate transporter 1 (MCT1) promoter. *Am. J. Physiol. Gastrointest. Liver Physiol.,* 288, G1118–26.

HIRATSUKA, M., SAKAMOTO, O., LI, X., SUZUKI, Y., AOKI, Y., & NARISAWA, K. (1998) Identification of holocarboxylase synthetase (HCS) proteins in human placenta. *Biochim. Biophys. Acta,* 1385, 165–171.

HYMES, J., FLEISCHHAUER, K., & WOLF, B. (1995) Biotinylation of histones by human serum biotinidase: Assessment of biotinyl-transferase activity in sera from normal individuals and children with biotinidase deficiency. *Biochem. Mol. Med.,* 56, 76–83.

INOUE, H., TATENO, M., FUJIMURA-KAMADA, K., TAKAESU, G., ADACHI-YAMADA, T., NINOMIYA-TSUJI, J., IRIE, K., NISHIDA, Y., & MATSUMOTO, K. (2001) A *Drosophila* MAPKKK, D-MEKK1, mediates stress responses through activation of p38 MAPK. *EMBO J.,* 20, 5421–5430.

KAUR MALL, G., CHEW, Y. C., & ZEMPLENI, J. (2010) Biotin requirements are lower in human Jurkat lymphoid cells but homeostatic mechanisms are similar to those of HepG2 liver cells. *J. Nutr.,* 140, 1086–1092.

KIM, K.-H., MCCORMICK, D. B., BIER, D. M., & GOODRIDGE, A. G. (1997) Regulation of mammalian acetyl-coenzyme A carboxylase. *Ann. Rev. Nutr.,* 17, 77–99.

KOBZA, K., CAMPOREALE, G., RUECKERT, B., KUEH, A., GRIFFIN, J. B., SARATH, G., & ZEMPLENI, J. (2005) K4, K9, and K18 in human histone H3 are targets for biotinylation by biotinidase. *FEBS J.,* 272, 4249–4259.

KOBZA, K., SARATH, G., & ZEMPLENI, J. (2008) Prokaryotic BirA ligase biotinylates K4, K9, K18 and K23 in histone H3. *BMB Reports,* 41, 310–315.

KUROISHI, T., RIOS-AVILA, L., PESTINGER, V., WIJERATNE, S. S. K., & ZEMPLENI, J. (2011) Biotinylation is a natural, albeit rare, modification of human histones. *Mol. Genet. Metab.,* 104, 537–545.

LIU, D., & ZEMPLENI, J. (2012) Holocarboxylase synthetase (HLCS) interacts physically with nuclear corepressor (N-CoR) and histone deacetylases (HDACs) to mediate gene repression. Oral presentation in minisymposium "Nutrition and Epigenetics" (Chairs: Zempleni J., & Ross, S.A.), Experimental Biology Meeting; San Diego, CA, April 22.

LYTTON, J., WESTLIN, M., BURK, S. E., SHULL, G. E., & MACLENNAN, D. H. (1992) Functional comparisons between isoforms of the sarcoplasmic or endoplasmic reticulum family of calcium pumps. *J. Biol. Chem.,* 267, 14483–14489.

MANTHEY, K. C., GRIFFIN, J. B., & ZEMPLENI, J. (2002) Biotin supply affects expression of biotin transporters, biotinylation of carboxylases, and metabolism of interleukin-2 in Jurkat cells. *J. Nutr.,* 132, 887–892.

MARTIN, V., BREDOUX, R., CORVAZIER, E., GORP, R. V., KOVACS, T., GELEBART, P., & ENOUF, J. (2002) Three novel sarco/endoplasmic reticulum Ca^{2+}-ATPase (SERCA) 3 isoforms. *J. Biol. Chem.,* 277, 24442–24452.

MOCK, D. M., LANKFORD, G. L., & MOCK, N. I. (1995) Biotin accounts for only half of the total avidin-binding substances in human serum. *J. Nutr.,* 125, 941–946.

NARANG, M. A., DUMAS, R., AYER, L. M., & GRAVEL, R. A. (2004) Reduced histone biotinylation in multiple carboxylase deficiency patients: A nuclear role for holocarboxylase synthetase. *Hum. Mol. Genet.,* 13, 15–23.

NATIONAL RESEARCH COUNCIL. (1998) *Dietary Reference Intakes for Thiamin, Riboflavin, Niacin, Vitamin B6, Folate, Vitamin B12, Pantothenic Acid, Biotin, and Choline.* Washington, DC, National Academy Press.

PACHECO-ALVAREZ, D., SOLORZANO-VARGAS, R. S., GONZALEZ-NORIEGA, A., MICHALAK, C., ZEMPLENI, J., & LEON-DEL-RIO, A. (2005) Biotin availability regulates expression of the sodium-dependent multivitamin transporter and the rate of biotin uptake in HepG2 cells. *Mol. Genet. Metab.,* 85, 301–307.

PESTINGER, V., WIJERATNE, S. S. K., RODRIGUEZ-MELENDEZ, R., & ZEMPLENI, J. (2011) Novel histone biotinylation marks are enriched in repeat regions and participate in repression of transcriptionally competent genes. *J. Nutr. Biochem.,* 22, 328–333.

RAMET, M., LANOT, R., ZACHARY, D., & MANFRUELLI, P. (2002) JNK signaling pathway is required for efficient wound healing in *Drosophila. Dev. Biol.,* 241, 145–156.

RIOS-AVILA, L., PESTINGER, V., WIJERATNE, S. S. K., & ZEMPLENI, J. (2012) K16-biotinylated histone H4 is overrepresented in repeat regions and participates in the repression of transcriptionally competent genes in human Jurkat lymphoid cells. *J. Nutr. Biochem.,* 23, 1559–1564.

RODRIGUEZ-MELENDEZ, R., GRIFFIN, J. B., SARATH, G., & ZEMPLENI, J. (2005) High-throughput immunoblotting identifies biotin-dependent signaling proteins in HepG2 hepatocarcinoma cells. *J. Nutr.,* 135, 1659–1666.

RODRIGUEZ-MELENDEZ, R., SCHWAB, L. D., & ZEMPLENI, J. (2004) Jurkat cells respond to biotin deficiency with increased nuclear translocation of NF-kB, mediating cell survival. *Int. J. Vitam. Nutr. Res.,* 74, 209–216.

RODRIGUEZ-MELENDEZ, R., & ZEMPLENI, J. (2009) Nitric oxide signaling depends on biotin in Jurkat human lymphoma cells. *J. Nutr.,* 139, 429–433.

SINGH, D., PANNIER, A. K., & ZEMPLENI, J. (2011) Identification of holocarboxylase synthetase chromatin binding sites using the DamID technology. *Anal. Biochem.,* 413, 55–59.

SLUSS, H. K., HAN, Z., BARRETT, T., DAVIS, R. J., & IP, Y. T. (1996) A JNK signal transduction pathway that mediates morphogenesis and an immune response in *Drosophila. Genes Dev.,* 10, 2745–2758.

SOLORZANO-VARGAS, R. S., PACHECO-ALVAREZ, D., & LEON-DEL-RIO, A. (2002) Holocarboxylase synthetase is an obligate participant in biotin-mediated regulation of its own expression and of biotin-dependent carboxylases mRNA levels in human cells. *Proc. Natl. Acad. Sci. USA,* 99, 5325–5330.

SPENCE, J. T., & KOUDELKA, A. P. (1984) Effects of biotin upon the intracellular level of cGMP and the activity of glucokinase in cultured rat hepatocytes. *J. Biol. Chem.,* 259, 6393–6396.

STANLEY, J. S., GRIFFIN, J. B., & ZEMPLENI, J. (2001) Biotinylation of histones in human cells: Effects of cell proliferation. *Eur. J. Biochem.,* 268, 5424–5429.

SUZUKI, Y., AOKI, Y., ISHIDA, Y., CHIBA, Y., IWAMATSU, A., KISHINO, T., NIIKAWA, N., MATSUBARA, Y., & NARISAWA, K. (1994) Isolation and characterization of mutations in the human holocarboxylase synthetase cDNA. *Nat. Genet.,* 8, 122–128.

WIEDMANN, S., RODRIGUEZ-MELENDEZ, R., ORTEGA-CUELLAR, D., & ZEMPLENI, J. (2004) Clusters of biotin-responsive genes in human peripheral blood mononuclear cells. *J. Nutr. Biochem.,* 15, 433–439.

XUE, J., & ZEMPLENI, J. (2011) Epigenetic synergies between methylation of cytosines and biotinylation of histones in gene repression. *Experimental Biology 2011.* Washington, DC [abstract].

YANG, X., AOKI, Y., LI, X., SAKAMOTO, O., HIRATSUKA, M., KURE, S., TAHERI, S., CHRISTENSEN, E., INUI, K., KUBOTA, M., OHIRA, M., OHKI, M., KUDOH, J., KAWASAKI, K., SHIBUYA, K., SHINTANI, A., ASAKAWA, S., MINOSHIMA, S., SHIMIZU, N., NARISAWA, K., MATSUBARA, Y., & SUZUKI, Y. (2001) Structure of human holocarboxylase synthetase gene and mutation spectrum of holocarboxylase synthetase deficiency. *Hum. Genet.,* 109, 526–534.

ZEMPLENI, J., HELM, R. M., & MOCK, D. M. (2001) in vivo biotin supplementation at a pharmacologic dose decreases proliferation rates of human peripheral blood mononuclear cells and cytokine release. *J. Nutr.,* 131, 1479–1484.

ZEMPLENI, J., WIJERATNE, S. S., & HASSAN, Y. I. (2009) Biotin. *Biofactors,* 35, 36–46.

14 Folate Receptor-Mediated Therapeutics

Folate Receptor-Mediated Particle Systems for Drug and Gene Delivery in Cancer Therapy

Yoshie Maitani

CONTENTS

14.1 INTRODUCTION

Cancer is a serious medical and social problem throughout the world. However, there remain many problems with cancer chemotherapy agents; foremost of these is their high toxicity. After systemic injection, anticancer drugs distribute throughout the body because low molecular weight drugs can readily extravasate from blood vessels. On the other hand, particulated drugs cannot. Blood vessels near a tumor are leaky, and, therefore, particles up to a diameter of ~200 nm can extravasate. Long-circulating particles can accumulate only at tumors as a result of the enhanced permeability and retention (EPR) effect. Because polyethylene glycolated (PEGylated) particles have prolonged circulation half-lives, an encapsulated drug in PEGylated particles also increases with a corresponding increase in the duration of drug exposure. Therefore, this strategy is able to evade rapid clearance by the reticuloendothelial system (RES), mainly represented by Kupffer cells in the liver and spleen macrophages (Gabison et al. 1994). Therapies utilizing this approach have now been licensed, including liposomal doxorubicin and antifungal agents.

A further strategy is known as "active targeting," in which particles with ligands are designed to interact with tumor cells. This is therapeutically attractive because it involves biomolecularly specific recognition, but no actively targeted agent has received regulatory approval to date except for monoclonal antibodies. To deliver drugs to specific cancer cell types, receptor-mediated endocytosis is a promising approach.

Folate as a targeting ligand has been examined as a tumor-targeted drug carrier. Folic acid (FA), which is a high-affinity ligand for folate receptors (FR), retains its FR-binding and endocytosis properties even if it is covalently linked to a wide variety of molecules and particles. However, until now, high therapeutic efficacy was reported only for ascites leukemia models by intraperitoneal injection (Shmeeda et al. 2006). Recently, intravenous injection of folate-linked particles has been reported to exhibit an antitumor effect. Here we focus on folate-linked particles for drug and gene delivery in cancer therapy. Particles include liposomes, emulsions, polymer micelles, nanoparticles, and so forth (Figure 14.1). In this chapter, nanoparticles represent a classification of particles in both the nano-size range and with a spherical structure.

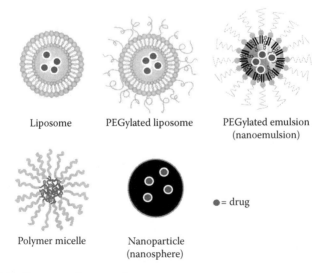

Liposome   PEGylated liposome   PEGylated emulsion
                                                         (nanoemulsion)

Polymer micelle   Nanoparticle   ● = drug
                             (nanosphere)

FIGURE 14.1   Structure of nanocarriers.

14.2 FOLATE-LINKED PARTICLES

Folate-linked particles are formed with folate-conjugated polymers alone, or preformed particles are conjugated to folate polymers. In particular, lipid-based nanoparticles, including liposomes, have been modified in many cases using folate-polyethylene glycol (PEG)-lipid (Lee and Low 1994; Gabizon et al. 1999; Yamada et al. 2008) (Figure 14.2). As an alternative, the potential use of a novel conjugate, folate-poly(L-lysine) by coating anionic liposomes by ionic interaction was explored as a tumor targeting device for injection (Watanabe et al. 2012). For chemotherapy, the particles that constituted the carriers were generally ~200 nm in size and had negative zeta-potential.

14.2.1 FOLATE-LINKED NANOEMULSIONS

In folate-linked nanoemulsions, ligands are usually attached to the distal end of a PEG chain to extend outside of the PEG surface layer (Figure 14.3B). The importance of an longer PEG chain length in the ligand was for recognition by FR, as demonstrated by Lee and Low in vitro (1994).

For the optimal folate-linker PEG length of nanoemulsions, cellular uptake of folate-linked nanoemulsions loading with aclacinomycin A was increased by a longer PEG-linker in an FR-overexpressing human nasopharyngeal cell line, KB cells, and such an increment corresponded well with enhanced cytotoxicity (Shiokawa et al. 2005). With regard to the density of folate modification, cellular uptake of folate-linked nanoemulsions was increased by a lower density of folate in vitro, but such an increment did not correspond with enhanced cytotoxicity in vivo. A longer folate-linker PEG length was most effective via FR-mediated endocytosis among folate-linked nanoemulsions in vitro and in vivo, and exhibited slightly higher tumor suppression compared to nontargeted PEGylated nanoemulsions in tumor-bearing mice (Shiokawa et al. 2005).

14.2.2 FOLATE-LINKED POLYMER MICELLES

In order to overcome multidrug resistance in solid tumors, the surface of pH-sensitive polymer micelles loaded with doxorubicin (DXR) was decorated with folate

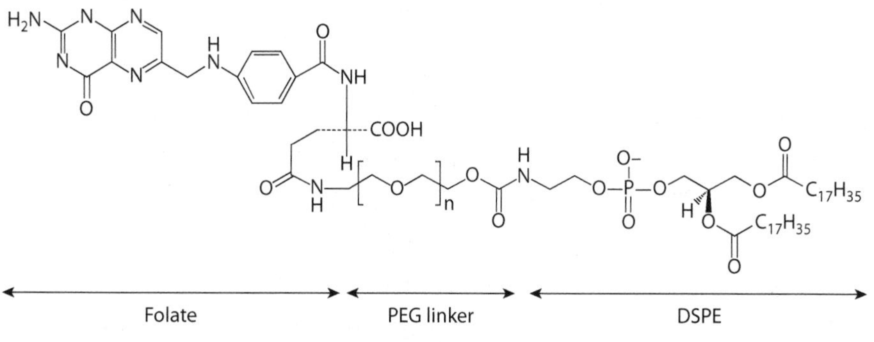

FIGURE 14.2 Structure of folate-linked PEG lipid. PEG lipid: polyethylene glycol derivative of distearoylphosphatidylethanolamine (PEG-DSPE).

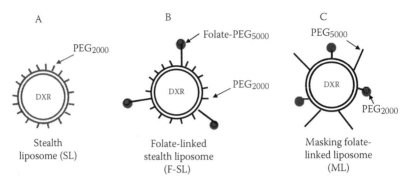

FIGURE 14.3 Design of PEG-linker length and folate density of liposomes. SL: stealth liposome, PEGylated (2000) liposome (A), F-SL: 0.25 mol% folate-PEG (5000) linked PEGylated (2000) liposome (B), ML: masking liposome, 0.25 mol% folate-PEG (2000) linked PEGylated (5000) liposome (C).

(PHSM/f) (Lee et al. 2005). In mice bearing DXR-resistant MCF-7 xenografts, PHSM/f exhibited significantly higher inhibition of tumor growth.

When polymer micelles were modified with folate-PEG lipid, polymer micelles with longer folate-linker PEG length showed high cellular uptake in KB cells, similar to nanoemulsion (Hayama et al. 2008). In these cases, nanoemulsions and polymer micelles are essentially composed of PEG lipid and PEG block copolymers, respectively. The effect of folate-PEG on FR-mediated endocytosis was not clear because of the covering with PEG moieties. Because conventional liposomes can be formed without PEG lipid (Figure 14.1), optimal folate modification could be examined.

14.2.3 FOLATE-LINKED LIPOSOMES

Three kinds of folate-linked liposomes to increase FR targeting were designed and prepared: conventional liposomes without PEG coating; liposomes with PEG coating, called stealth liposomes; and masking liposomes (Figure 14.3) (Yamada et al. 2008). The folate ligand is "masked" by adjacent PEG chains while in the circulation. It was expected that these PEG chains can ultimately be lost to allow presentation of the targeting ligand at the site of the tumor. In the cellular uptake of folate-linked liposomal DXR, the optimal folate-PEG linker length was longer and the density was higher in vitro (Figure 14.3B). In vitro cytotoxicity of folate-linked liposomal DXR was higher in the order: folate-linked liposomes > folate-linked stealth liposomes > stealth liposomes > masking folate-linked liposomes. Masking liposomes showed lower toxicity because the PEG_{2000} layers inhibited interaction with cells, and, therefore, it was thought to show a cytotoxicity reduction. After intravenous injection of folate-linked liposomal DXR into mice, stealth liposomes showed significantly longer blood circulation time among the liposomes examined. Even masking liposomes could not achieve such a long circulation period in the blood. Compared with the in vitro result, when the antitumor effect of folate-linked liposomal DXR was evaluated in M109 (FR [+]) solid tumors following intravenous injection, folate-linked stealth liposomes and masking liposomes showed significantly higher antitumor effect than folate-linked liposomes. A long circulation period is needed for folate-linked particles to obtain higher antitumor effects in vivo. The accumulation of FR-targeted particles depends on the dual effects of passive and active targeting.

Folic acid and a cell-penetrating peptide, TAT-peptide, were conjugated with modified chitosan-cholesterol polymeric liposomes (FA-TATp-PLs) to generate both a targeting effect and transmembrane penetration ability. Paclitaxel-loaded FA-TATp-PLs exhibited a superior antitumor effect in vitro and in vivo after intravenous injection as compared with that with Taxol&Reg (Zhao et al. 2010).

14.2.4 FOLATE-LINKED LIPOSOMES COMBINED WITH ANTI-ANGIOGENESIS REAGENTS

In the distribution of liposomal DXR, masking liposomes accumulated tumor in vessels more effectively than stealth liposomes, although they showed similar antitumor activity. To increase the perfusion of particles into tumor cells with FR, an anti-angiogenesis reagent was combined with folate-linked liposomal drugs in order to change the tumor microenvironment. It was reported that transforming growth factor (TGF)-β inhibitor enhanced the accumulation of nanoparticles in tumors by decreasing the number of pericytes (Kano et al. 2007). To apply this information to folate-linked stealth liposomes (F-SL; Figure 14.3B), a new TGF-β inhibitor, A-83-01, was used (Taniguchi et al. 2010). Coadministration of A83-01 enhanced the tumor accumulation of F-SL more than stealth liposomes (SL) injected intravenously and induced significantly higher antitumor activity than that without A-83-01 in vivo (Figure 14.4).

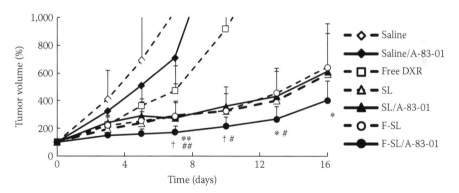

FIGURE 14.4 Coadministration of transforming growth factor (TGF)-β type I receptor inhibitor A-83-01 enhanced the antitumor activity of a single injection of folate-linked liposomal doxorubicin (DXR) in CDF1 mice bearing M109 tumors. Free DXR, liposomal DXR (8 mg/kg), or saline was administered intravenously in a single bolus with or without intraperitoneal injection of A-83-01 (0.5 mg/kg) to mice (n = 6–8). SL: PEGylated liposomal DXR, F-SL: 0.25 mol% of folate-PEG (5000) linked 2.25 mol% PEGylated (2000) liposomal DXR. #: $P < 0.05$, ##: $P < 0.01$ vs SL, †: $P < 0.05$ vs SL/A-83-01, *: $P < 0.05$, **: $P < 0.01$ vs. F-SL. Each value represents the mean ± S.D. (From Taniguchi, Y., Kawano, K., Minowa, T. et al. Enhanced antitumor efficacy of folatelinked liposomal doxorubicin with TGF-β type I receptor inhibitor. *Cancer Sci.*, 101, 2207–13, 2010.)

14.2.5 FOLATE-LINKED LIPOSOMES WITH ENDOSOMAL RELEASE

The release of folate-linked liposomal drugs after endocytosis is also of importance for activity as well as cellular uptake. For endosomal escape, in many cases,
folate-linked pH-sensitive liposomes using pH-sensitive lipids, such as cholesteryl
hemisuccinate (CHEMS) and oleyl alcohol, have been investigated. Following endocytosis, rapid endosomal acidification (~2–3 min) occurs due to a vacuolar proton
ATPase-mediated proton influx. As a result, the pH levels of early endosomes, sorting endosomes, and multivesicular bodies drop rapidly to pH <6.0 (Yang et al. 2007).
Efficient intracellular drug and plasmid DNA delivery using FR-targeted pH-sensitive
liposomes composed of phosphatidylcholine, pH-sensitive lipids, and Tween-80 was
observed in vitro (Shi et al. 2002). Although pH-triggered mechanisms designed to
be activated during liposome endocytosis have been shown to interfere with liposome stability in vitro, the pH-triggered mechanism does not compromise liposome
stability in vivo.

Other strategies have been investigated recently using in vitro studies. Folate-
linked liposomes with heterogeneous membranes are designed to release encapsulated DXR in response to a decrease in local pH by forming domains with "leaky"
interfaces (Mamasheva et al. 2011). Folate-linked liposomes with high fluidity membranes could readily release drugs in endosomes (Kawano et al. 2009). Liposomal
membranes composed of hydrogenated soybean phosphatidylcholine (HSPC) showed
lower fluidity than those composed of dimyristoyl phosphatidylcholine (DMPC).
Folate-linked HSPC liposomes showed higher KB cellular uptake but lower antitumor activity than folate-linked DMPC liposomes when entrapped with mitoxantrone
in vitro. These findings suggest higher bioexposure of cells to the therapeutic agent
possibly via faster and more extensive release from the liposomes, which is the limiting stage for cytotoxicity.

14.2.6 FOLATE-LINKED NANOPARTICLES

FR-targeted nanoparticles consist of a heparin-folate paclitaxel (HFT) backbone
with additional paclitaxel loaded in their hydrophobic core. Intravenous injection
of the nanoparticles markedly enhanced the antitumor efficacy of paclitaxel in a
drug-resistant KB-8-5 xerograph model, indicating that targeted nanoparticles could
be a specific and efficient drug delivery system to overcome P-gp-mediated drug
resistance in human cancers (Wang et al. 2011). Organic nanotubes were modified
with folate-PEG lipids by simply mixing with folate-PEG lipid, and they exhibited
FR targeting in vitro (Wakasugi et al. 2011).

14.3 DIAGNOSTIC AGENTS

FR-mediated MRI contrast agents, such as folate-linked Gd-chelates (Corot et al.
2008), folate-conjugated superparamagnetic nanoparticles (Sun et al. 2006; Meier et
al. 2010), and folate-linked liposomes (Kamaly et al. 2009) have been investigated.
When folate-linked nanoparticles were maintained in the circulation by the addition of PEG lipid, they accumulated significantly in tumor tissues without altering

other biodistribution and exhibited enhanced tumor MR signal intensity compared with nonmodified nanoparticles (Nakamura et al. submitted). Recently, in attempts to develop theranostic nanoparticles, both therapeutic and imaging moieties have been incorporated within the same construct.

14.4   CANCER GENE THERAPY

Cancer gene therapy in clinical trials includes strategies that involve augmentation of chemotherapeutic and immunotherapeutic approaches. These strategies include inactivation of oncogene expression, gene replacement for tumor suppressor genes, drug sensitization with genes for prodrug delivery, and cytokine transfer. Tumor-targeted delivery of particulated anticancer drugs cannot be proved to suppress tumor growth because drugs can penetrate cancer cells even if the drugs were released before reaching the tumor. However, tumor targeting can be proved for gene delivery by gene expression because the gene itself cannot penetrate the cell membrane.

Activated liver-derived macrophages (Kupffer cells) (Paulos et al. 2004) and the apical membrane of proximal tubules in kidneys of mice express FR, and are responsible for capturing folate-linked nanoparticles. Therefore, FR-targeted delivery of therapeutic genes will damage normal cells in the liver and kidney. In this regard, using a tumor-specific promoter to restrict expression transcriptionally to the target cancer cells has promise.

14.4.1   Delivery of Plasmid DNA and NFκB Decoy In Vitro

In the human prostate, a high-affinity folate binding protein was characterized (Holm et al. 1993), and folic acid binds to the membrane fraction that cross-reacts with the anti-prostate-specific membrane antigen (PSMA) antibody. Folate-linked nanoparticles, NP-F composed of DC-Chol as a cationic lipid, Tween 80, and folate-PEG lipid showed greater transfection efficiency in human prostate cancer LNCaP cells and human cervix carcinoma HeLa cells than in KB cells (Hattori and Maitani 2005). Folate-linked nanoparticles may bind to PMSA and then be taken up via an endocytic mechanism by LNCaP cells (Hattori and Maitani 2004).

Folate conjugated with either cholesterol or DSPE was incorporated into lipoplexes formulated with DOTAP/cholesterol (wt/wt: 31/69) that possess cholesterol nanodomains. The presence of the folate ligand within the cholesterol domain promotes more productive transfection in cultured cells, and intracellular trafficking of the lipoplexes after entry into cells plays a crucial role in gene delivery (Xu and Anchordoquy 2010).

The inhibition of nuclear factor kappa B (NFκB) translocation into the nucleus in activated macrophages with FR is effective for the suppression of the inflammatory reaction in rheumatoid arthritis (Morishita et al. 2004). NP-F could effectively deliver an NFκB decoy into the cytoplasm of activated macrophages via FR and was detected in the cytoplasm of activated murine macrophage-like RAW264.7 cells. NP-F also exhibited an inhibitory effect on the translocation of NFκB into the nucleus (Hattori et al. 2006).

14.4.2   DELIVERY OF PLASMID DNA AND ANTISENSE OLIGONUCLEOTIDE FOR THERAPY

Folate-linked liposomes showed efficient FR-dependent cellular uptake and transfection in vitro but have not delivered plasmid DNA successfully in vivo as part of gene therapy experiments (Hofland et al. 2002; Reddy et al. 2002). The major limitation of in vivo gene therapy using liposomes is the low transfection efficiency. A size for gene transfer vectors of less than 300 nm is required for extravasation and intratumoral diffusion, which could be limiting for targeted delivery.

The maximum in vivo transfection activity of a reporter gene (luciferase) occurred with intraperitoneally administered folate-linked liposomes in a disseminated intraperitoneal L1210A tumor model (Reddy et al. 2002). When NP-F nanoplexes of the luciferase plasmid were injected directly into KB xenografts, NP-F showed about 100-fold more luciferase activity than a commercial transfection reagent, Tfx20 (Hattori and Maitani 2005). Dysfunction of p53 and the presence of mutant p53 are associated with the progression of tumorigenesis and an unfavorable prognosis, respectively, for many human cancers. The systemic delivery of p53 into tumors resulted in the efficient expression of functional p53, which sensitized the tumors to chemotherapy and radiotherapy (Xu et al. 1999).

Suicide gene therapy is a strategy whereby a gene is introduced into cancer cells that makes them sensitive to a drug that is normally nontoxic. One of the most frequently used suicide genes is the herpes simplex virus thymidine kinase (HSV-tk) gene, which phosphorylates a prodrug, ganciclovir (GCV), into a toxic form, GCV triphosphate. Injection of NP-F nanoplexes with the HSV-tk plasmid into KB tumor xenografts significantly inhibited tumor growth, and HSV-tk plus Cx43 plasmids introduced into LNCaP tumor xenografts increased the mean survival time of mice (Hattori and Maitani 2005). FR-targeted LPDI lipoplexes of the HSV-tk gene significantly increased survival time when administered systemically to mice bearing murine breast adenocarcinoma 410.4 tumor xenografts (Bruckheimer et al. 2004). FA-associated lipoplexes also showed greater transfection efficacy than FA-conjugated lipoplexes in vitro and successfully suppressed tumor growth when used in suicide gene therapy in vivo (Duarte et al. 2012).

The combination of intravenously administered folate-liposome–anti-HER-2 antisense ODN (AS HER-2 ODN) complex and docetaxel resulted in a marked inhibition of xenograft growth in an aggressive breast cancer model that does not overexpress HER-2, even after treatment ended (Rait et al. 2002).

14.4.3   DELIVERY OF siRNA FOR THERAPY

Double-stranded RNA (dsRNA) suppresses the expression of a target gene by triggering specific degradation of the complementary mRNA sequence. To promote their gene inhibition effect, folate-linked cationic nanoparticles have been employed to form complexes with negatively charged synthetic small interfering RNA (siRNA). NP-F/Her-2 siRNA nanoplexes efficiently mediated intracellular delivery of siRNA to KB cells, resulting in a significant down-regulation of HER-2 expression and cell growth inhibition (70% cell viability compared with that of control siRNA) (Yoshizawa et al. 2008) (Figure 14.5).

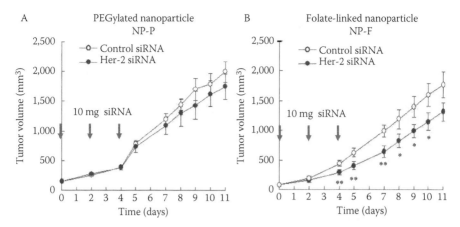

FIGURE 14.5 In vivo gene therapy of KB tumor xenografts with Her-2 siRNA in mice. Mice were divided into two groups: group I, control-M siRNA (10 µg); group II, Her-2 siRNA (10 µg). PEGylated nanoparticle (A) or folate-linked nanoparticle (NP-F) complexes (B) with siRNAs formed at a charge ratio (+/−) of 2/1 in 5 mM NaCl were injected directly into the tumor three times (days 0, 2, and 4). The results indicate the mean ± SE ($n = 4$–6). *$P < 0.05$ and **$P < 0.01$. (Yoshizawa et al. 2008).

MYCN amplification is strongly associated with malignancy and correlates with poor prognosis in patients with neuroblastoma. MYCN mRNA expression was knocked down 53.1% in tumor tissues with peritoneal injection of liposome-encapsulated MYCN siRNA compared with control siRNA. Furthermore, targeted delivery of MYCN siRNA delivered by folate-linked liposomes into LA-N-5 cells was efficacious and capable of suppressing MYCN expression in vitro and in vivo after peritoneal injection (Feng et al. 2010).

14.5 CONCLUSIONS

As a result of the delicate balance among the amount of FR-targeting ligand to promote effective targeting, the PEG shield to avoid immune detection and the endosomal release in cells that have taken up the particles, methods that enable the prediction and characterization of particle surface composition and the content of particle surface density have become essential for the translation of these particles into clinical practice. In addition, temporarily covered folate-linked particles may have great potential to exert their function and may contribute FR-mediated cancer therapy.

REFERENCES

Bruckheimer, E., Harvie, P., Orthel, J., et al. 2004. In vivo efficacy of folate-targeted lipid-protamine-DNA (LPD-PEG-Folate) complexes in an immunocompetent syngeneic model for breast adenocarcinoma. *Cancer Gene Ther.* 11: 128–34.

Corot, C., Robert, P., Lancelot, E., et al. 2008.Tumor imaging using P866, a high-relaxivity gadolinium chelate designed for folate receptor targeting. *Magn Reson Med.* 60: 1337–46.

Duarte, S., Faneca, H., de Lima, M. C. 2012. Folate-associated lipoplexes mediate efficient gene delivery and potent antitumoral activity in vitro and in vivo. *Int J Pharm.* 28 (423): 365–77.

Feng, C., Wang, T., Tang, R., et al. 2010. Silencing of the MYCN gene by siRNA delivered by folate receptor-targeted liposomes in LA-N-5 cells. *Pediatr Surg Int.* 26: 1185–91.

Gabizon, A., Catane, R., Uziely, B., et al. 1994. Prolonged circulation time and enhanced accumulation in malignant exudates of doxorubicin encapsulated in polyethylene-glycol coated liposomes. *Cancer Res.* 54: 987–92.

Gabizon, A., Horowitz, A. T., Goren, D. et al. 1999. Targeting folate receptor with folate linked to extremities of poly(ethyleneglycol)-grafted liposomes: In vitro studies. *Bioconjug Chem.* 10: 289–98.

Hattori, Y., Maitani, Y. 2004. Enhanced in vitro DNA transfection efficiency by novel folate-linked nanoparticles in human prostate cancer and oral cancer. *J Control Release* 97: 173–83.

Hattori, Y., Maitani, Y. 2005. Folate-linked nanoparticle-mediated suicide gene therapy in human prostate cancer and nasopharyngeal cancer with herpes simplex virus thymidine kinase. *Cancer Gene Ther.* 12: 796–809.

Hattori, Y., Sakaguchi, M., Maitani, Y. 2006. Folate-linked lipid-based nanoparticles deliver a NFkappaB decoy into activated murine macrophage-like RAW264.7 cells. *Biol Pharm Bull.* 29: 1516–20.

Hayama, A., Yamamoto, T., Yokoyama, M. et al. 2008. Polymeric micelles modified by folate-PEG-lipid for targeted drug delivery to cancer cells in vitro. *J Nanosci Nanotechnol.* 8: 3085–90.

Hofland, H. E,, Masson, C., Iginla, S. et al. 2002. Folate-targeted gene transfer in vivo. *Mol Ther.* 5: 739–44.

Holm, J., Hansen, S. I., Høier-Madsen, M. 1993. High-affinity folate binding in human prostate. *Biosci Rep.* 13: 99–105.

Kamaly, N., Kalber, T., Thanou, M., et al. 2009. Folate receptor targeted bimodal liposomes for tumor magnetic resonance imaging. *Bioconjug Chem.* 20: 648–55.

Kano, M. R., Bae, Y., Iwata, C. et al. 2007. Improvement of cancer-targeting therapy, using nanocarriers for intractable solid tumors by inhibition of TGF-beta signaling. *Proc Natl Acad Sci USA* 104: 3460–65.

Kawano, K., Onose, E., Hattori, Y. et al. 2009. Higher liposomal membrane fluidity enhances the in vitro antitumor activity of folate-targeted liposomal mitoxantrone. *Mol Pharm.* 6: 98–104.

Lee, R. J., Low, P. S. 1994. Delivery of liposomes into cultured KB cells via folate receptor-mediated endocytosis. *J Biol Chem.* 269: 3198–204.

Lee, E. S., Na, K., Bae, Y. H. 2005. Doxorubicin loaded pH-sensitive polymeric micelles for reversal of resistant MCF-7 tumor. *J Control Release* 103: 405–18.

Mamasheva, E., O'Donnell, C., Bandekar, A. et al. 2011. Heterogeneous liposome membranes with pH-triggered permeability enhance the in vitro antitumor activity of folate-receptor targeted liposomal doxorubicin. *Mol Pharmaceutics* 8: 2224–32.

Meier, R., Henning, T. D., Boddington, S., et al. 2010. Breast cancers: MR imaging of folate-receptor expression with the folate-specific nanoparticle P1133. *Radiology* 255: 527–35.

Morishita, R., Tomita, N., Kaneda, Y. et al. 2004. Molecular therapy to inhibit NFkappaB activation by transcription factor decoy oligonucleotides. *Curr Opin Pharmacol.* 4: 139–46.

Nakamura, T., Kawano, K., Shiraishi, K.. et al. Folate-targeted gadolinium-lipid based nanoparticles as magnetic resonance contrast agent for tumor imaging. *Biol Pharm Bull.*, submitted.

Paulos, C. M., Turk, M. J., Breur, G. J. et al. 2004. Folate receptor-mediated targeting of therapeutic and imaging agents to activated macrophages in rheumatoid arthritis. *Adv Drug Deliv Rev.* 56: 1205–17.

Rait, A. S., Pirollo, K. F., Xiang, L. et al. 2002. Tumor-targeting, systemically delivered antisense HER-2 chemosensitizes human breast cancer xenografts irrespective of HER-2 levels. *Mol Med.* 8: 475–86.

Reddy, J. A., Abburi, C., Hofland, H., et al. 2002. Folate-targeted, cationic liposome-mediated gene transfer into disseminated peritoneal tumors. *Gene Ther.* 9: 1542–50.

Shi, G., Guo, W., Stephenson, S. M. et al. 2002. Efficient intracellular drug and gene delivery using folate receptor-targeted pH-sensitive liposomes composed of cationic/anionic lipid combinations. *J Control Release.* 80: 309–19.

Shiokawa, T., Hattori, Y., Kawano, K. et al., 2005. Effect of polyethylene glycol linker chain length of folate-linked microemulsions loading aclacinomycin A on targeting ability and antitumor effect in vitro and in vivo. *Clin Cancer Res.* 11: 2018–25.

Shmeeda, H., Mak, L., Tzemach, D., et al. 2006. Intracellular uptake and intracavitary targeting of folate-conjugated liposomes in a mouse lymphoma model with up-regulated folate receptors. *Mol Cancer Ther.* 5: 818–24.

Sun, C., Sze, R., Zhang, M. 2006. Folic acid-PEG conjugated superparamagnetic nanoparticles for targeted cellular uptake and detection by MRI. *J Biomed Mater Res A.* 78: 550–57.

Taniguchi, Y., Kawano, K., Minowa, T. et al. 2010. Enhanced antitumor efficacy of folate-linked liposomal doxorubicin with TGF-β type I receptor inhibitor. *Cancer Sci.* 101: 2207–13.

Wakasugi, A., Asakawa, M., Kogiso, M. et al. 2011. Organic nanotubes for drug loading and cellular delivery. *Int J Pharm.* 413: 271–8.

Wang, Z. J., Boddington, S., Wendland, M. et al. 2008. MR imaging of ovarian tumors using folate-receptor-targeted contrast agents. *Pediatr Radiol.* 38: 529–537.

Wang, X., Li, J., Wang, Y. et al. 2011. A folate receptor-targeting nanoparticle minimizes drug resistance in a human cancer model. *ACS Nano* 5: 6184–94.

Watanabe, K., Kaneko, M., Maitani, Y. 2012. Functional coating of liposomes using a folate-polymer conjugate to target folate receptors. *Int J Nanomedicine* 7:3679–88.

Xu, L., Pirollo, K. F., Rait, A. et al. 1999. Systemic p53 gene therapy in combination with radiation results in human tumor regression. *Tumor Targeting* 4: 92–104.

Xu, L., Anchordoquy, T. J. 2010. Effect of cholesterol nanodomains on the targeting of lipid-based gene delivery in cultured cells. *Mol Pharm.* 7: 1311–7.

Yamada, A., Taniguchi, Y., Kawano, K. et al. 2008. Design of folate-linked liposomal doxorubicin to its antitumor effect in mice. *Clin Cancer Res.* 14: 8161–8.

Yang, J., Chen, H., Vlahov, I. R., Cheng, J., Low, P. S. 2007. Characterization of the pH of folate receptor-containing endosomes and the rate of hydrolysis of internalized acid-labile folate-drug conjugates. *J Pharmcol Exp Ther.* 321: 462–468.

Yoshizawa, T., Hattori, Y., Hakoshima, M. et al. 2008. Folate-linked lipid-based nanoparticles for synthetic siRNA delivery in KB tumor xenografts. *Eur J Pharm Biopharm.* 70: 718–25.

Zhao, P., Wang, H., Yu, M. et al. 2010. Paclitaxel-loaded, folic-acid-targeted and tat-peptide-conjugated polymeric liposomes: In vitro and in vivo evaluation. *Pharm Res.* 27: 1914–26.

15 Vitamin B$_{12}$ Derivatives and Preferential Targeting of Tumors

Evelyne Furger and Eliane Fischer

CONTENTS

15.1 INTRODUCTION

Cancer cells have a higher proliferation rate than normal cells. In consequence, their demand for various nutrients and vitamins is significantly increased. Efficient uptake of these vital factors is often met by up-regulation of their specific receptors. Several vitamins have therefore been explored as carrier molecules for drug delivery to cancer cells. Most prominently, derivatives of the vitamin folic acid have been widely used as radiotracers for diagnosis or as vehicles to deliver cytotoxicity (Müller 2012, Vlahov and Leamon 2012). Together with folic acid, B$_{12}$ is essential as a cofactor for various steps in central metabolism, and cancer cells have an increased requirement of both vitamins. However, compared to folic acid, the use of B$_{12}$ derivatives for tumor targeting is more challenging for several reasons: (1) Chemically, B$_{12}$ is the most complex vitamin. (2) In the human body, B$_{12}$ has a very complicated transport and uptake system, consisting of three transport proteins and several membrane-bound receptors. This constrains the design of tumor-targeting drugs, because they need to retain binding affinity to the transport proteins. (3) Even rapidly dividing cells require only very small amounts of B_{12}, which limits the intracellular concentration of B_{12} derivatives that can be delivered to tumor cells. However, B_{12} interacts with extremely high affinity ($K_D \sim 10^{-15}$ M) with each transport protein. This allows the body to efficiently capture and distribute minute amounts of the vitamin.

Finally, B_{12} is taken up by the liver and kidneys, which have a presumed storage function for the vitamin. Thus, B_{12} derivatives are prone to accumulation in these organs, leading to cytotoxic side effects or background uptake of an imaging tracer.

Despite these challenges for B_{12} derivatives as tumor-targeting agents, there are several approaches in which they have shown promise both in vitro and in vivo. In this chapter, we will outline the complexity of B_{12} transport systems and the consequences for uptake of B_{12} derivatives in cultured tumor cells, animal models of cancer, and cancer patients.

15.2   VITAMIN B_{12} AND ITS TRANSPORT PROTEINS

15.2.1   Cobalamins

Cobalamins are organometallic compounds with a central positively charged cobalt ion. All cobalamins show the same core structure consisting of a chiral corrin ring coordinating a cobalt ion through four nitrogen atoms and possessing seven amide side chains (a–g) (Figure 15.1). A 5,6-dimethylbenzimidazole ribonucleotide moiety is linked to side chain f and coordinates cobalt as the lower axial (α-side) ligand. The upper axial (β-side) ligand is exchangeable and can be a cyano (CN), hydroxyl (OH), methyl (Me), or 5′-deoxyadenosyl (Ado) group.

Cyanocobalamin (CNCbl) is the most important synthetic form of the cobalamins despite its lack of direct physiological function. It is used in many pharmaceuticals and food additives because of its stability and lower cost (Kräutler 1998, Nielsen et al. 2012, Roth et al. 1996). In the human body, cobalamin is modified to two biologically active forms to act in the cytosol (methylcobalamin, MeCbl) or in the mitochondria (5′-deoxy-5′-adenosylcobalamin, AdoCbl) (Kräutler 2012). These two forms of B_{12} are cofactors for two human enzymes: the cytoplasmic methionine synthase (Utley et al. 1985) and mitochondrial methylmalonyl-CoA mutase (Calafat et al. 1995). MeCbl mediates transfer of the methyl group from the inactive form of folate (methyltetrahydrofolate) to homocysteine, thereby catalyzing methylation of homocysteine to methionine (Carmel et al. 2003). Indirectly, this important reaction contributes to nucleotide synthesis. AdoCbl is involved in catabolism of different organic compounds, such as branched chain amino acids and odd-chain fatty acids, by acting as a cofactor in the conversion of methylmalonyl-CoA to succinyl-CoA (Banerjee and Ragsdale 2003). Due to this central role for human catabolism, a B_{12} deficiency has severe consequences, including anemia and impaired function of the nervous system (Wolters et al. 2004).

15.2.2 B_{12} Transport Proteins and Their Receptors

Because B_{12} is exclusively produced by bacteria, humans are dependent on nutritional intake from dietary sources including, for example, meat, fish, and eggs. The

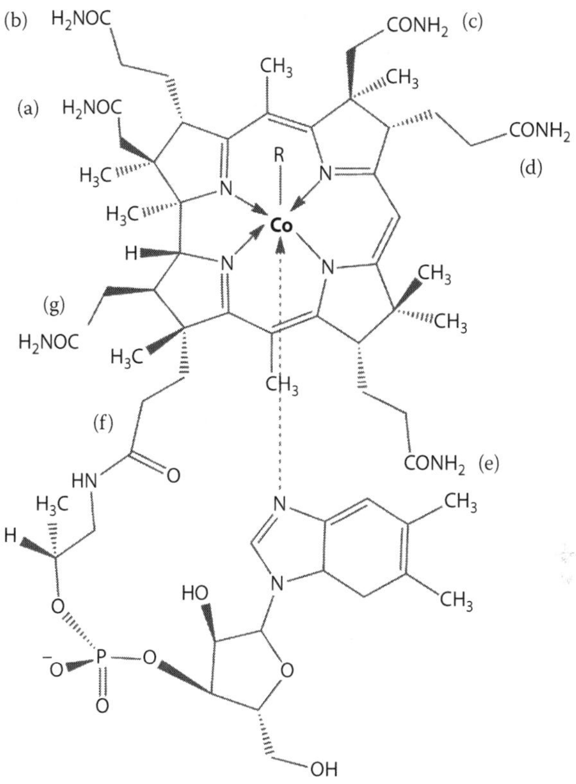

FIGURE 15.1 B$_{12}$ structure. The core of B$_{12}$ consists of a corrin ring that coordinates a central cobalt ion and possesses seven amide side chains (a–g). The upper axial ligand (R) is variable and can be exchanged in B$_{12}$ derivatives. Enzymatically active forms carry either a methyl or a 5'-deoxyadenosyl group as upper axial ligand.

limited available amount is compensated by a complex and efficient transport and cell-uptake system involving three transport proteins as well as a number of receptors (Allen 1975, Fedosov 2012, Moestrup 2006, Nexo 1998, Nielsen et al. 2012). The three transport proteins are intrinsic factor (IF), transcobalamin (TC, also known as transcobalamin II), and haptocorrin (HC, previously referred to as transcobalamin I, transcobalamin III, or R-binder). They share a similar overall architecture whereby each protein can carry a single B$_{12}$ molecule that is tightly bound and only released after degradation of the transport protein.

The molecular weights of the protein backbones are very similar, about 45 kDa. In contrast to TC, IF and HC are glycosylated and therefore show higher apparent molecular weights of 55 kDa (IF) and 65 kDa (HC) (Allen 1975, Nexo 1998). The glycans of these two proteins protect them from proteolysis in the intestine and prevent HC from glomerular filtration by increasing its molecular radius. However, they are not directly involved in B$_{12}$- or specific receptor binding (Fedosov 2012). IF ensures the uptake of B$_{12}$ in the intestinal enterocytes via the specific receptor cubilin. Internalization is facilitated by amnionless, a transmembrane protein. Together, cubilin and amnionless form the receptor complex cubam (Fyfe et al. 2004), which only recognizes the IF-B_{12} complex but not IF or free B_{12} alone (Birn et al. 1997, Kozyraki et al. 1998).

TC is the evolutionarily oldest of the three transport proteins and is present in all investigated species (Greibe et al. 2012). The essential role of TC is to transport absorbed B_{12} in the bloodstream to cells of the body. Cell uptake of TC-B_{12} is achieved via the transmembrane protein CD320 (TCblR) (Quadros et al. 2009) by endocytosis. This receptor is present on the cell surface of virtually all tissues (Amagasaki et al. 1990, Cho et al. 2008, Park et al. 2009) and is regulated according to the proliferative and differentiation status of the cell (Amagasaki et al. 1990). A different receptor mediates the uptake of TC-B_{12} in the kidney. There, megalin prevents urinary loss of B_{12} in the renal glomerular filtrate by reabsorption of TC-B_{12} (Birn et al. 2002, Moestrup et al. 1996). HC is present in blood plasma and also in secretions, including saliva, milk, and tears (Morkbak et al. 2007). It is responsible for binding B_{12} in the upper part of the gastrointestinal tract. HC binds a considerable amount of B_{12} and inactive forms of B_{12} (so called analogues) in plasma. Depending on the source of synthesis (probably liver and granulocytes), different glycoforms of HC have been described (Yang et al. 1982). HC in the blood that originates from granulocytes lacks terminal sialic acid and is known to be rapidly cleared by the asialoglycoprotein receptor in the liver (Burger et al. 1975). For HC with high amounts of sialic acids from assumed liver origin, no receptor has been identified so far, and this glycoform has been shown to clear very slowly from the blood (Allen 1975).

15.2.3 B_{12} TRANSPORT AND UPTAKE IN THE HUMAN BODY

A complex transport and uptake system involving the previously described proteins ensures efficient assimilation of B_{12} from dietary sources (Fedosov 2012, Nielsen et al. 2012) (Figure 15.2). After entering the human body, released B_{12} from food is captured by salivary or gastric HC and escorted down the upper part of the gastrointestinal tract. B_{12} usually encounters HC in the stomach, because it needs first to be released from binding proteins by proteolytic activity of pepsin (DelCorral and Carmel 1990). Although the concentration of IF in the stomach is much higher than that of HC, B_{12} preferentially binds to the latter protein at acidic pH. In the duodenum, B_{12} is released after degradation of HC by pancreatic enzymes. It transfers next to IF, which carries it to the terminal ileum. There, IF-B_{12} is absorbed by endocytosis via the cubam receptor complex. For export of B_{12} from epithelial cells, the multidrug resistance protein 1 (MRP1) has been identified as one possible element (Beedholm-Ebsen et al. 2010). Upon release of B_{12} into the bloodstream, B_{12} is transported by TC and readily taken up by cells of liver and other tissues by the transmembrane protein CD320 (Quadros et al. 2009). Inside the target cells, TC-B_{12} is degraded in lysosomes. The free B_{12} molecule is released from the lysosomes by a specific transporter and then processed to the two cofactors AdoB$_{12}$ and MeB$_{12}$ (Banerjee et al. 2009).

15.3 VITAMIN B_{12} DERIVATIVES AND PROTEIN BINDING

Given the complexity of B_{12} transport and uptake, a B_{12} derivative needs to bind to at least one transport protein in order to be retained by the human body. Recent

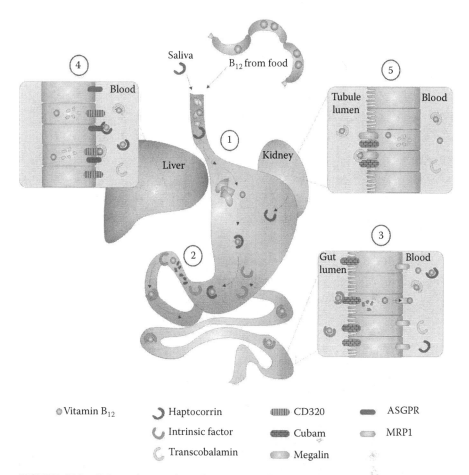

Vitamin B$_{12}$ ⟩ Haptocorrin   CD320   ASGPR

Intrinsic factor   Cubam   MRP1

Transcobalamin   Megalin

FIGURE 15.2 Schematic overview of transport and uptake of B$_{12}$ in humans. B$_{12}$ released from food is first bound to HC present in saliva and gastric fluids (1). In the duodenum, HC is degraded by pancreatic enzymes and B$_{12}$ transferred to IF (2). The B$_{12}$-IF complex is then endocytosed by Cubam, a receptor complex (3). In the intestinal cells, IF is degraded and B$_{12}$ is released to blood plasma from the basolateral side of the cell by MRP1, a multidrug resistance protein. In the blood plasma, B$_{12}$ is bound either to HC or TC, whereas only TC is responsible for the B$_{12}$ uptake into cells (4). This uptake is mediated by the TC receptor CD320. In the liver, glycoforms of HC are taken up by the asialoglycoprotein receptor (ASGPR). In the kidney, megalin instead of CD320 is the receptor responsible for uptake of B$_{12}$ (5).

progress in solving the crystal structures and studying the binding characteristics of the three transport proteins now facilitates the rational design of novel B$_{12}$ derivatives. Further, these studies contribute to explain the differences in selectivity of the three transport proteins for binding of "true" B$_{12}$, natural B$_{12}$ analogues, or synthetic B$_{12}$-derivatives.

The first crystal structure of a mammalian B$_{12}$ transport protein was solved in 2006 with the structure of recombinant human and bovine TC in complex with aquacobalamin (H_2OCbl) (Wuerges et al. 2006) followed by the crystal structure of IF (Mathews et al. 2007). Both proteins are composed of two domains, α and β, connected by a flexible linker. The N-terminal "α-domain" contains twelve α-helices, whereas the C-terminal "β-domain" is built of a five- and a three-stranded β sheet and one α-helix. B_{12} is captured in the interface of the two domains via hydrogen bonds and hydrophobic interactions (Wuerges et al. 2006). In contrast to the enzymatic B_{12} binding proteins, the transport proteins bind B_{12} in its "base-on" state, where the 5,6-dimethylbenzimidazole moiety is coordinated to the Co ion.

Crystallization of TC with the ligand H_2OCbl showed a coordination bond of the Co ion to a histidine side chain (H173) at the β side. The solvent accessibility of H_2OCbl is only about 7% after binding to TC. It is primarily the ribose group of the nucleotide moiety that maintains its contact to the bulk solvent and therefore becomes a potential site for derivatization of B_{12} (Wuerges et al. 2006). The His-Co^{3+} interaction is conserved among TC from other species but is not present in IF, where H_2O remains coordinated and the β side of B_{12} becomes accessible for solvent or external ligands, resulting in a solvent accessibility of about 17% (Randaccio et al. 2010). Notably, stronger β-axial ligands, including CN- or OH-, are not displaced by His upon binding to TC.

Although human TC and IF share only 27% sequence identity, a very similar overall structure was found for IF in complex with CNCbl solved at a resolution of 2.6 Å (Mathews et al. 2007). Apparently, the linker region was cleaved during the crystallization process, but both domains retained their ability to bind to CNCbl, and the resulting complex was still able to bind to the cubam receptor complex (Fedosov et al. 2004, Fedosov et al. 2005, Mathews et al. 2007). Similar to TC-CNCbl, IF-CNCbl shows a solvent accessible surface of about 19% with the ribose 5'-hydroxyl group and the β side of B_{12} being exposed to solvent (Mathews et al. 2007).

Because the crystal structure of HC is not yet available, multiple sequence alignments and comparative modeling of HC based on the structure of human and bovine TC or IF are so far the only tools for structural predictions of the protein. HC and TC also share only about 25% sequence identity, but it is assumed that the overall architecture and in particular the α domain of HC is very similar to that of TC (Wuerges et al. 2007). With the recently reported successful expression of recombinant human HC (Furger et al. 2012), the crystal structure of the third B_{12} transport protein is expected to be solved in the near future.

Binding of B_{12} to the transport proteins in mammals occurs both very fast and exceptionally tight, whereupon all three transport proteins show a dissociation constant of about 10^{-15} M toward B_{12} (Fedosov et al. 2002). However, the three proteins differ in their specificity for B_{12} analogues to be ranked in the order HC $\ll$ TC < IF. HC is the least specific transport protein and is able to bind major structural different B_{12} analogues that even lack the nucleotide moiety, such as cobinamide. Kinetic studies using a fluorescent B_{12} conjugate allowed more detailed characterization of the binding mechanisms to B_{12} transport proteins and the resulting specificity for natural B_{12} analogues (Fedosov et al. 2007).

There are four described sites for B_{12} conjugation generally resulting in the recognition of all three transport proteins: (1) the e-propionamide of the peripheral corrin ring, (2/3) the 2'- and 5'-hydroxyl groups of the ribose unit at the α side of B_{12}, and

(4) the cobalt ion (Clardy et al. 2011). Tumor-targeting B$_{12}$ derivatives are usually designed to retain binding to TC. The drugs are injected intravenously, where they bind to TC in blood plasma and are then taken up by tumor cells expressing high levels of CD320. Pathare et al. (1996) systematically synthesized B$_{12}$-biotin conjugates by modification of each reactive site of B$_{12}$ and assessed binding to TC. Based upon their observations, the preferable sites of conjugation to ensure binding to TC are the ribose 5' hydroxyl group and derivatization on the Co itself by exchanging the β-axial ligand. On the other hand, conjugation to the various propionamide side chains resulted in diminished binding to TC. These observations are in good agreement with the conclusions from crystal structures, which demonstrated solvent accessibility of the ribose 5' hydroxyl group and β-axial ligands.

By introducing appropriate linkers, conjugation at the *b*-propionamide site of the corrin ring may also result in TC-binding B$_{12}$ derivatives. This approach requires controlled hydrolysis of the propionamide chains and isolation of the *b*-monocarboxylic acid prior to conjugation. For ^{99m}Tc-PAMA derivatives, a long spacer of >5 CH$_2$ ensured binding to TC, while derivatives with shorter spacers were still able to bind HC and IF but not TC (Allis et al. 2010, Waibel et al. 2008).

15.4   CELLULAR UPTAKE IN VITRO AND CYTOTOXICITY

The main receptor responsible for cellular uptake of B$_{12}$ has recently been identified as CD320. It binds TC-B$_{12}$ efficiently but has a much lower affinity for the apo-form of TC. Upon binding of TC-B$_{12}$, the complex is internalized and B$_{12}$ or a B$_{12}$ derivative is released from TC within the lysosomes. The TC receptor is thought to recycle to the cell surface for a new cycle of TC internalization (Figure 15.3).

Expression of CD320 has been shown to be cell-cycle dependent and is probably highest immediately before cell division takes place (Jiang et al. 2010). However, even in rapidly dividing cells, the maximum density of CD320 on the cellular surface is relatively low. For example, between 3,000 and 6,000 receptors have been determined on leukemia cells K562 or HL-60 (Amagasaki et al. 1990). This number is extremely low when compared to other receptors used for drug delivery, for example, the transferrin receptor (200,000 receptors/cell) or folate receptor (10^5–10^6 receptors/cell; Parker et al. 2005). McLean et al. demonstrated that leukemia cell growth is modulated by B$_{12}$ analogues in vitro (McLean et al. 1997a). In another study (McLean et al. 1997b), the authors could show that this mechanism is indeed TC dependent and could be blocked by antibodies against TC. These studies underline the importance of efficient B$_{12}$ uptake for cell proliferation.

Various B$_{12}$ derivatives have been developed with the aim to deliver cytotoxic moieties into tumor cells by selective uptake through the receptor CD320. These conjugates need to be designed in a way that they (1) still bind to TC with high affinity and (2) only become cytotoxic once they are internalized. This can, for example, be achieved by introduction of an acid-labile linker between B$_{12}$ and the attached cytotoxic moiety. Further, the toxicity of the drug must be high enough for it to kill cells even at the low concentrations that can be reached by targeted delivery of B$_{12}$ conjugates. One example of a cytotoxic drug that has been linked to B$_{12}$ is colchicine, an inhibitor of tubulin polymerization that cannot be systemically administered to

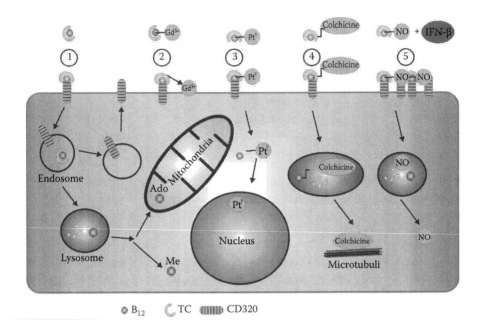

FIGURE 15.3 Cell uptake and cytotoxicity of experimental B_{12} derivatives. TC-B_{12} binds to CD320 and internalizes by endocytosis (1). After dissociation of TC-B_{12} from the receptor, CD320 recycles to the cellular membrane, while TC is degraded in the lysosome and B_{12} is exported to the cytoplasm. It acts as a cofactor in the mitochondria (AdoCbl) or the cytosol (MeCbl). Cytotoxic metals like Gd^{3+} (2) or Pt^{II} (3) have been conjugated to B_{12}. Gd^{3+} ions are thought to be released near the cytoplasmic membranes, while Pt^{II} is putatively released from B_{12} after adenosylation and introduces DNA damage in the nucleus. Colchicine is released in the lysosomes by cleavage of an acid-labile linker (4). Similarly, NO is thought to dissociate from NOB_{12} in the lysosomes (5). Expression of CD320 was up-regulated by coadministration of IFN-β.

patients due to severe side effects (Bagnato et al. 2004). Targeted delivery of a B_{12}-colchicine conjugate to rapidly proliferating tumor cells could potentially improve the therapeutic index of this drug. Colchicine is relatively potent, with an LC_{50} in the sub-micromolar range for many cell lines. Further, the target structure is tubulin, which resides in the cytosol and can be accessed by the drug without the need of nuclear import. Colchicine was coupled to B_{12} at the β-axial site via an acid-labile hydrazone linkage. This leads to release of the drug from B_{12} in the acidic environment of the lysosomes. The authors could demonstrate cytotoxicity of the conjugate at slightly higher concentrations than free colchicine in three different cell lines, and cytotoxicity was strongly dependent on the acid-labile linker.

Similarly, the cytotoxic metal Gd^{3+} has been used for B_{12}-mediated delivery into leukemia cells (Siega et al. 2009). The authors conjugated two different chelators for Gd^{3+} ions (DTPA and TTHA) to CNCbl by esterification of the ribose 5′-OH group. High cell uptake of Gd^{3+} and a cytotoxic effect were shown with the less stable DTPA conjugate, but not with TTHA-CNCbl. The authors postulated the release of Gd^{3+} ions already near the cell surface as soon as the TC-bound conjugate is bound to the CD320.

Platinum-based chemotherapeutics are part of the standard repertoire in current cancer therapy. Recently, Ruiz-Sanchez et al. (2011) described novel B$_{12}$-CN-Pt^{II} conjugates by binding of cisplatin-like Pt^{II} complexes to the cyanide group of CNCbl. These conjugates were still converted to AdoCbl by the corresponding enzyme (Ruiz-Sanchez et al. 2008). Upon adenosylation, the cytotoxic Pt^{II} complex is released inside the cell. The authors could demonstrate that this approach resulted in cytotoxicity and, more specifically, introduction of DNA damage by the Pt^{II} complex, as could be shown by the detection of γ-H2AX foci in the treated cells. The uptake mechanism was shown to be TC dependent. Thus, the concentration of delivered Pt^{II} complex in this study was high enough to induce DNA damage, even though cytotoxicity of the B$_{12}$ conjugate was reduced when compared to free cisplatin. In other examples, potentially cytotoxic agents have been released from B$_{12}$ derivatives by cleavage of the weak Co-C bond through photolysis or sonolysis (Howard et al. 1997).

For all these approaches, the amounts of toxic agents that can be delivered to a cell seem to be limited when depending on CD320-mediated uptake. Combination therapy that results in synergetic mechanisms of drug treatment is therefore an interesting option to enhance the efficacy of B$_{12}$ derivatives. For example, Bauer et al. (2002) used an NO-Cbl conjugate in combination with IFN-β, which potentially upregulates CD320 expression. Indeed, they observed higher uptake of ^{57}Co-Cbl and enhanced toxicity with NO-Cbl. In a second study, NO-Cbl was shown to enhance the cytotoxic effect of standard chemotherapeutics (Bauer et al. 2007).

While all of these studies aimed at targeting CD320 with TC-bound B$_{12}$ derivatives, other reports describe targeting of cubilin with an IF-bound B$_{12}$-Re^{I} conjugate in a placental choriocarcinoma (Viola-Villegas et al. 2009) and a lung carcinoma cell line (Vortherms et al. 2011). Based on this observation, IF-binding B$_{12}$ derivatives could have the potential to specifically target cubilin-expressing tumors in vivo.

15.5 IN VIVO TRANSPORT AND TUMOR ACCUMULATION OF B$_{12}$ DERIVATIVES

Several of the previously described approaches succeeded in uptake of B$_{12}$ derivatives to a level where cytotoxic effects could be demonstrated in a variety of cancer cell lines. Despite these promising results, the in vivo efficacy of most of these compounds still needs to be demonstrated. In this section, we discuss how B$_{12}$ derivatives distribute in human patients and in preclinical animal models of cancer. For this purpose, B$_{12}$ derivatives are labeled with radionuclides that can be imaged with a γ-camera, including, for example, ^{99m}Tc and ^{111}In. ^{57}Co is an interesting alternative to radiolabel B$_{12}$ or B$_{12}$ derivatives without affecting structure and protein binding. However, the long half-life of the radionuclide limits the dose that can be administered to patients, and low specific activity of ^{57}Co-B$_{12}$ decreases imaging sensitivity.

15.5.1 Tumor Accumulation of B$_{12}$ Derivatives in Human Patients

While assimilation of dietary B$_{12}$ depends on binding to IF and subsequent uptake by cubam in the gastrointestinal tract, B$_{12}$ derivatives for tumor targeting are usually administered intravenously. Therefore, they do not necessarily need to bind to IF, which is strongly selective for "true" B_{12}. In the blood, B_{12} derivatives bind to circulating TC and HC. If binding to these transport proteins is abolished or if the B_{12}-binding proteins in the blood are saturated, free B_{12} derivatives are rapidly excreted by the kidneys due to their low molecular weight. TC-B_{12} is efficiently taken up by the peripheral tissue or by tumor cells through the receptor CD320. However, TC-B_{12} is also recognized by megalin, which is highly expressed in the kidney proximal tubule and in several epithelia (Christensen and Birn 2002). Upon binding to megalin, TC-B_{12} is internalized and B_{12} accumulates in the lysosomes of the kidney cells. In consequence, any TC-binding B_{12} derivative will accumulate in the kidneys and eventually release a cytotoxic drug when an acid-labile linker is used, resulting in nephrotoxicity. In contrast, HC-bound B_{12} derivatives circulate in the blood or are eventually removed by the asialoglycoprotein receptor in the liver.

In a first clinical trial, cancer patients have been imaged by single-photon emission computed tomography (SPECT) imaging after injection of an ^{111}In-DTPA derivatized B_{12} (^{111}In-DAC; Collins et al. 1999, Collins et al. 2000). While it was indeed possible to image tumors in many of the patients, background uptake in some healthy organs was high, especially in the liver and spleen and in lacrimal, nasal, and salivary glands. ^{111}In-DAC has binding affinity toward both TC and HC. After intravenous injection, both transport proteins may act as carriers for the B_{12} derivative and contribute to its final tissue distribution.

15.5.2 ANIMAL MODELS OF CBL DELIVERY

Rodent cancer models are widely used in drug discovery and are usually the first models to test new compounds in vivo. However, there are several differences between rodents and humans with respect to B_{12} transport that need to be considered. Importantly, mice and rats both lack HC. Instead, there is a single plasma B_{12} transport protein in rodents that has a form that resembles both HC and TC (Hygum et al. 2011). For example, mouse TC is able to bind incomplete cobalamin (cobinamide) and various other B_{12} analogues, just like human HC. This needs to be considered when using rodent models to study the targeting properties of new B_{12} derivatives, and the binding characteristics to mouse TC need to be studied. Furthermore, mice are very often held on a B_{12}-deficient diet prior to biodistribution experiments with radiolabeled B_{12} derivatives in order to reduce accumulation of the tracer in the kidney. The kidney serves as a storage organ in rodents, and uptake is higher in the presence of high B_{12} levels.

15.5.3 B_{12} DERIVATIVES FOR PRECLINICAL BIODISTRIBUTION AND IMAGING

Pharmacokinetics and biodistribution of several radiolabeled B_{12} derivatives have been described in mice. TC-binding B_{12} derivatives labeled with [99mTc] (Kunze et al. 2004, van Staveren et al. 2004, Yang et al. 2005), [125I] (Wilbur et al. 1996) or [111In] (Collins and Hogenkamp 1997) have been developed. Similar to [57Co]-B_{12}, TC-binding B_{12} derivatives predominantly accumulate in the kidneys. This finding is clearly different from the situation in human patients described for [111In]-DAC.

Interestingly, ^{99m}Tc-labeled B$_{12}$ derivatives that do not bind to TC have recently been shown to accumulate in tumors without the high background radioactivity in kidneys and other organs in mice. However, the mechanism by which these derivatives accumulate in tumors is still not clear. One explanation could be that these derivatives bind to HC, which is produced by the human tumor xenograft. Melanoma, lung adenocarcinoma, fibrolamellar hepatocellular carcinoma (Lildballe et al. 2011), and breast carcinoma are tumors in which high levels of HC are expressed. It is still unclear if the described HC-selective B$_{12}$ derivatives accumulate within the tumor cells or reside in the extracellular space of the tumor. The application of cytotoxic B$_{12}$ derivatives is only useful if accumulation within the tumor cells can be demonstrated. On the other hand, accumulation of B$_{12}$ derivatives in the extracellular space still allows tumor diagnosis and may also be considered for radiotherapeutic approaches. There, the conjugate does not necessarily need to be internalized, but neighboring cells can be irradiated by the therapeutic radioisotope.

A second category of B$_{12}$ derivatives that can potentially be used for in vivo tumor imaging are fluorescently labeled conjugates. Fluorophores, including fluorescein, naphthofluorescein, or Oregon Green have been conjugated at the upper axial ligand position of B$_{12}$ (Smeltzer et al. 2001). A second generation of fluorescent B$_{12}$ derivatives has been described in which a rigid linker introduced at the ribose 5′OH contributed to an increased quantum yield (Smeltzer et al. 2001). Rhodamine-labeled polymers have been conjugated to B$_{12}$, folate, and biotin and targeted to tumor cells both in vitro and in mice (Russell-Jones et al. 2011). The authors demonstrated increased uptake of the vitamin-conjugated polymers. Fluorescent B$_{12}$ derivatives may have potential in intraoperative imaging, similar to folate conjugates that have already proven success in clinical trials (van Dam et al. 2011).

15.5.4 PRECLINICAL TUMOR THERAPY WITH B$_{12}$ DERIVATIVES

In contrast to imaging, therapy studies with B$_{12}$ derivatives have so far been limited. However, the previously described NO-Cbl has been used in a variety of therapy or combination therapy studies in mice (Bauer et al. 2002, Chawla-Sarkar et al. 2003) and dogs with spontaneous tumors (Bauer et al. 2010). The authors could demonstrate therapeutic efficacy of NO-Cbl. Generally, therapy with cytotoxic B$_{12}$ derivatives is probably hampered by the accumulation in healthy tissue, which may cause severe side effects.

15.6 CONCLUSIONS

There are a variety of interesting studies that successfully used B$_{12}$ derivatives for selective targeting of tumor cells. In vitro, tumor cells have been specifically killed by cytotoxic B$_{12}$ derivatives, involving a TC- and CD320-dependent uptake mechanism. In preclinical in vivo studies, radiolabeled or fluorescent B$_{12}$ derivatives have been shown to accumulate in tumors. Consequently, tumors could be imaged in patients with an ^{111}In-labeled B$_{12}$ derivative. However, the high uptake of B$_{12}$ derivatives in healthy tissue poses a challenge for the successful use of cytotoxic B$_{12}$ derivatives for cancer therapy.

Combination therapy with chemotherapeutics is a promising strategy to improve the therapeutic potential of B_{12} derivatives. Recent insights into the complexity of B_{12} transport, including the crystal structures of the transport proteins and the identification of their receptors, may lead to rational design of a second generation of B_{12} derivatives with higher selectivity for tumor cells or improved pharmacokinetics.

REFERENCES

Allen, R. H. 1975. Human vitamin B12 transport proteins. *Prog Hematol,* 9, 57–84.

Allis, D. G., Fairchild, T. J., & Doyle, R. P. 2010. The binding of vitamin B12 to transcobalamin(II): Structural considerations for bioconjugate design—a molecular dynamics study. *Mol Biosyst,* 6, 1611–18.

Amagasaki, T., Green, R., & Jacobsen, D. W. 1990. Expression of transcobalamin II receptors by human leukemia K562 and HL-60 cells. *Blood,* 76, 1380–6.

Bagnato, J. D., Eilers, A. L., Horton, R. A., & Grissom, C. B. 2004. Synthesis and characterization of a cobalamin-colchicine conjugate as a novel tumor-targeted cytotoxin. *J Org Chem,* 69, 8987–96.

Banerjee, R., Gherasim, C., & Padovani, D. 2009. The tinker, tailor, soldier in intracellular B12 trafficking. *Curr Opin Chem Biol,* 13, 484–91.

Banerjee, R., & Ragsdale, S. W. 2003. The many faces of vitamin B12: Catalysis by cobalamin-dependent enzymes. *Annu Rev Biochem,* 72, 209–47.

Bauer, J. A., Frye, G., Bahr, A., Gieg, J., & Brofman, P. 2010. Anti-tumor effects of nitrosylcobalamin against spontaneous tumors in dogs. *Invest New Drugs,* 28, 694–702.

Bauer, J. A., Lupica, J. A., Schmidt, H., Morrison, B. H., Haney, R. M., Masci, R. K., Lee, R. M., Didonato, J. A., & Lindner, D. J. 2007. Nitrosylcobalamin potentiates the anti-neoplastic effects of chemotherapeutic agents via suppression of survival signaling. *PLoS One,* 2, e1313.

Bauer, J. A., Morrison, B. H., Grane, R. W., Jacobs, B. S., Dabney, S., Gamero, A. M., Carnevale, K. A., Smith, D. J., Drazba, J., Seetharam, B., & Lindner, D. J. 2002. Effects of interferon beta on transcobalamin II-receptor expression and antitumor activity of nitrosylcobalamin. *J Natl Cancer Inst,* 94, 1010–9.

Beedholm-Ebsen, R., Van De Wetering, K., Hardlei, T., Nexo, E., Borst, P., & Moestrup, S. K. 2010. Identification of multidrug resistance protein 1 (MRP1/ABCC1) as a molecular gate for cellular export of cobalamin. *Blood,* 115, 1632–9.

Birn, H., Verroust, P. J., Nexo, E., Hager, H., Jacobsen, C., Christensen, E. I., & Moestrup, S. K. 1997. Characterization of an epithelial approximately 460-kDa protein that facilitates endocytosis of intrinsic factor-vitamin B12 and binds receptor-associated protein. *J Biol Chem,* 272, 26497–504.

Birn, H., Willnow, T. E., Nielsen, R., Norden, A. G., Bonsch, C., Moestrup, S. K., Nexo, E., & Christensen, E. I. 2002. Megalin is essential for renal proximal tubule reabsorption and accumulation of transcobalamin-B(12). *Am J Physiol Renal Physiol,* 282, F408–16.

Burger, R. L., Schneider, R. J., Mehlman, C. S., & Allen, R. H. 1975. Human plasma R-type vitamin B12-binding proteins. II. The role of transcobalamin I, transcobalamin III, and the normal granulocyte vitamin B12-binding protein in the plasma transport of vitamin B12. *J Biol Chem,* 250, 7707–13.

Calafat, A. M., Taoka, S., Puckett, J. M., Jr., Semerad, C., Yan, H., Luo, L., Chen, H., Banerjee, R., & Marzilli, L. G. 1995. Structural and electronic similarity but functional difference in methylmalonyl-CoA mutase between coenzyme B12 and the analog 2′,5′-dideoxyadenosylcobalamin. *Biochemistry,* 34, 14125–30.

Carmel, R., Green, R., Rosenblatt, D. S., & Watkins, D. 2003. Update on cobalamin, folate, and homocysteine. *Hematology Am Soc Hematol Educ Program,* 62–81.

Chawla-Sarkar, M., Bauer, J. A., Lupica, J. A., Morrison, B. H., Tang, Z., Oates, R. K., Almasan, A., Didonato, J. A., Borden, E. C., & Lindner, D. J. 2003. Suppression of NF-kappa B survival signaling by nitrosylcobalamin sensitizes neoplasms to the antitumor effects of Apo2L/TRAIL. *J Biol Chem,* 278, 39461–9.

Cho, W., Choi, J., Park, C. H., Yoon, S. O., Jeoung, D. I., Kim, Y. M., & Choe, J. 2008. Expression of CD320 in human B cells in addition to follicular dendritic cells. *BMB Rep,* 41, 863–7.

Christensen, E. I., & Birn, H. 2002. Megalin and cubilin: Multifunctional endocytic receptors. *Nat Rev Mol Cell Biol,* 3, 256–66.

Clardy, S. M., Allis, D. G., Fairchild, T. J., & Doyle, R. P. 2011. Vitamin B12 in drug delivery: Breaking through the barriers to a B12 bioconjugate pharmaceutical. *Expert Opin Drug Deliv,* 8, 127–40.

Collins, D. A., & Hogenkamp, H. P. 1997. Transcobalamin II receptor imaging via radiolabeled diethylene-triaminepentaacetate cobalamin analogs. *J Nucl Med,* 38, 717–23.

Collins, D. A., Hogenkamp, H. P., & Gebhard, M. W. 1999. Tumor imaging via indium 111-labeled DTPA-adenosylcobalamin. *Mayo Clin Proc,* 74, 687–91.

Collins, D. A., Hogenkamp, H. P., O'Connor, M. K., Naylor, S., Benson, L. M., Hardyman, T. J., & Thorson, L. M. 2000. Biodistribution of radiolabeled adenosylcobalamin in patients diagnosed with various malignancies. *Mayo Clin Proc,* 75, 568–80.

Del Corral, A., & Carmel, R. 1990. Transfer of cobalamin from the cobalamin-binding protein of egg yolk to R binder of human saliva and gastric juice. *Gastroenterology,* 98, 1460–6.

Fedosov, S. N. 2012. Physiological and Molecular Aspects of Cobalamin Transport. *In:* Stanger, O. (ed.) *Water Soluble Vitamins.* Springer Science+Business Media.

Fedosov, S. N., Berglund, L., Fedosova, N. U., Nexo, E., & Petersen, T. E. 2002. Comparative analysis of cobalamin binding kinetics and ligand protection for intrinsic factor, transcobalamin, and haptocorrin. *J Biol Chem,* 277, 9989–96.

Fedosov, S. N., Fedosova, N. U., Berglund, L., Moestrup, S. K., Nexo, E., & Petersen, T. E. 2004. Assembly of the intrinsic factor domains and oligomerization of the protein in the presence of cobalamin. *Biochemistry,* 43, 15095–102.

Fedosov, S. N., Fedosova, N. U., Berglund, L., Moestrup, S. K., Nexo, E., & Petersen, T. E. 2005. Composite organization of the cobalamin binding and cubilin recognition sites of intrinsic factor. *Biochemistry,* 44, 3604–14.

Fedosov, S. N., Fedosova, N. U., Krautler, B., Nexo, E., & Petersen, T. E. 2007. Mechanisms of discrimination between cobalamins and their natural analogues during their binding to the specific B12-transporting proteins. *Biochemistry,* 46, 6446–58.

Furger, E., Fedosov, S. N., Launholt Lildballe, D., Waibel, R., Schibli, R., Nexo, E., & Fischer, E. 2012. Comparison of recombinant human haptocorrin expressed in human embryonic kidney cells and native haptocorrin. *PLoS One,* 7, e37421.

Fyfe, J. C., Madsen, M., Hojrup, P., Christensen, E. I., Tanner, S. M., De La Chapelle, A., He, Q., & Moestrup, S. K. 2004. The functional cobalamin (vitamin B12)-intrinsic factor receptor is a novel complex of cubilin and amnionless. *Blood,* 103, 1573–9.

Greibe, E., Fedosov, S., & Nexo, E. 2012. The cobalamin-binding protein in zebrafish is an intermediate between the three cobalamin-binding proteins in human. *PLoS One,* 7, e35660.

Howard, W. A., Jr., Bayomi, A., Natarajan, E., Aziza, M. A., El-Ahmady, O., Grissom, C. B., & West, F. G. 1997. Sonolysis promotes indirect Co-C bond cleavage of alkylcob(III) alamin bioconjugates. *Bioconjug Chem,* 8, 498–502.

Hygum, K., Lildballe, D. L., Greibe, E. H., Morkbak, A. L., Poulsen, S. S., Sorensen, B. S., Petersen, T. E., & Nexo, E. 2011. Mouse transcobalamin has features resembling both human transcobalamin and haptocorrin. *PLoS One,* 6, e20638.

Jiang, W., Sequeira, J. M., Nakayama, Y., Lai, S. C., & Quadros, E. V. 2010. Characterization of the promoter region of TCblR/CD320 gene, the receptor for cellular uptake of transcobalamin-bound cobalamin. *Gene,* 466, 49–55.

Kozyraki, R., Kristiansen, M., Silahtaroglu, A., Hansen, C., Jacobsen, C., Tommerup, N., Verroust, P. J., & Moestrup, S. K. 1998. The human intrinsic factor-vitamin B12 receptor, cubilin: Molecular characterization and chromosomal mapping of the gene to 10p within the autosomal recessive megaloblastic anemia (MGA1) region. *Blood,* 91, 3593–600.

Kräutler, B. 1998. B12-Coenzymes, the Central Theme. *In:* Kräutler, B., Arigoni, D., & Golding, B. T. (eds.) *Vitamin B12 and B12 Proteins.* Weinheim: Wiley-VCH.

Kräutler, B. 2012. Biochemistry of B12-Cofactors in Human Metabolism. *In:* Stanger, O. (ed.) *Water Soluble Vitamins.* Springer Science+Business Media.

Kunze, S., Zobi, F., Kurz, P., Spingler, B., & Alberto, R. 2004. Vitamin B12 as a ligand for technetium and rhenium complexes. *Angew Chem Int Ed Engl,* 43, 5025–9.

Lildballe, D. L., Nguyen, K. Q., Poulsen, S. S., Nielsen, H. O., & Nexo, E. 2011. Haptocorrin as marker of disease progression in fibrolamellar hepatocellular carcinoma. *Eur J Surg Oncol,* 37, 72–9.

Mathews, F. S., Gordon, M. M., Chen, Z., Rajashankar, K. R., Ealick, S. E., Alpers, D. H., & Sukumar, N. 2007. Crystal structure of human intrinsic factor: Cobalamin complex at 2.6-A resolution. *Proc Natl Acad Sci U S A,* 104, 17311–6.

Mclean, G. R., Pathare, P. M., Wilbur, D. S., Morgan, A. C., Woodhouse, C. S., Schrader, J. W., & Ziltener, H. J. 1997a. Cobalamin analogues modulate the growth of leukemia cells in vitro. *Cancer Res,* 57, 4015–22.

Mclean, G. R., Quadros, E. V., Rothenberg, S. P., Morgan, A. C., Schrader, J. W., & Ziltener, H. J. 1997b. Antibodies to transcobalamin II block in vitro proliferation of leukemic cells. *Blood,* 89, 235–42.

Moestrup, S. K. 2006. New insights into carrier binding and epithelial uptake of the erythropoietic nutrients cobalamin and folate. *Curr Opin Hematol,* 13, 119–23.

Moestrup, S. K., Birn, H., Fischer, P. B., Petersen, C. M., Verroust, P. J., Sim, R. B., Christensen, E. I., & Nexo, E. 1996. Megalin-mediated endocytosis of transcobalamin-vitamin-B12 complexes suggests a role of the receptor in vitamin-B12 homeostasis. *Proc Natl Acad Sci U S A,* 93, 8612–7.

Morkbak, A. L., Poulsen, S. S., & Nexo, E. 2007. Haptocorrin in humans. *Clin Chem Lab Med,* 45, 1751–9.

Müller, C. 2012. Folate based radiopharmaceuticals for imaging and therapy of cancer and inflammation. *Curr Pharm Des,* 18, 1058–83.

Nexo, E. 1998. Cobalamin Binding Proteins. *In:* Kräutler, B., Arigoni, D., & Golding, B. T. (eds.) *Vitamin B12 and B12 Proteins.* Weinheim: Wiley-VCH.

Nielsen, M. J., Rasmussen, M. R., Andersen, C. B., Nexo, E., & Moestrup, S. K. 2012. Vitamin B(12) transport from food to the body's cells–a sophisticated, multistep pathway. *Nat Rev Gastroenterol Hepatol,* 9(6), 345–54.

Park, H. J., Kim, J. Y., Jung, K. I., & Kim, T. J. 2009. Characterization of a novel gene in the extended MHC region of mouse, NG29/Cd320, a homolog of the human CD320. *Immune Netw,* 9, 138–46.

Parker, N., Turk, M. J., Westrick, E., Lewis, J. D., Low, P. S., & Leamon, C. P. 2005. Folate receptor expression in carcinomas and normal tissues determined by a quantitative radioligand binding assay. *Anal Biochem,* 338, 284–93.

Pathare, P. M., Wilbur, D. S., Heusser, S., Quadros, E. V., McLoughlin, P., & Morgan, A. C. 1996. Synthesis of cobalamin-biotin conjugates that vary in the position of cobalamin coupling. Evaluation of cobalamin derivative binding to transcobalamin II. *Bioconjug Chem,* 7, 217–32.

Quadros, E. V., Nakayama, Y., & Sequeira, J. M. 2009. The protein and the gene encoding the receptor for the cellular uptake of transcobalamin-bound cobalamin. *Blood,* 113, 186–92.

Randaccio, L., Geremia, S., Demitri, N., & Wuerges, J. 2010. Vitamin B12: Unique metalorganic compounds and the most complex vitamins. *Molecules,* 15, 3228–59.

Roth, J. R., Lawrence, J. G., & Bobik, T. A. 1996. Cobalamin (coenzyme B12): Synthesis and biological significance. *Annu Rev Microbiol,* 50, 137–81.

Ruiz-Sanchez, P., Konig, C., Ferrari, S., & Alberto, R. 2011. Vitamin B as a carrier for targeted platinum delivery: In vitro cytotoxicity and mechanistic studies. *J Biol Inorg Chem,* 16, 33–44.

Ruiz-Sanchez, P., Mundwiler, S., Spingler, B., Buan, N. R., Escalante-Semerena, J. C., & Alberto, R. 2008. Syntheses and characterization of vitamin B12-Pt(II) conjugates and their adenosylation in an enzymatic assay. *J Biol Inorg Chem,* 13, 335–47.

Russell-Jones, G., McTavish, K., & McEwan, J. 2011. Preliminary studies on the selective accumulation of vitamin-targeted polymers within tumors. *J Drug Target,* 19, 133–9.

Siega, P., Wuerges, J., Arena, F., Gianolio, E., Fedosov, S. N., Dreos, R., Geremia, S., Aime, S., & Randaccio, L. 2009. Release of toxic Gd3+ ions to tumour cells by vitamin B12 bioconjugates. *Chemistry,* 15, 7980–9.

Smeltzer, C. C., Cannon, M. J., Pinson, P. R., Munger, J. D., Jr., West, F. G., & Grissom, C. B. 2001. Synthesis and characterization of fluorescent cobalamin (CobalaFluor) derivatives for imaging. *Organ Lett,* 3, 799–801.

Utley, C. S., Marcell, P. D., Allen, R. H., Antony, A. C., & Kolhouse, J. F. 1985. Isolation and characterization of methionine synthetase from human placenta. *J Biol Chem,* 260, 13656–65.

Van Dam, G. M., Themelis, G., Crane, L. M., Harlaar, N. J., Pleijhuis, R. G., Kelder, W., Sarantopoulos, A., De Jong, J. S., Arts, H. J., Van Der Zee, A. G., Bart, J., Low, P. S., & Ntziachristos, V. 2011. Intraoperative tumor-specific fluorescence imaging in ovarian cancer by folate receptor-alpha targeting: First in-human results. *Nat Med,* 17, 1315–9.

Van Staveren, D. R., Mundwiler, S., Hoffmanns, U., Pak, J. K., Spingler, B., Metzler-Nolte, N., & Alberto, R. 2004. Conjugation of a novel histidine derivative to biomolecules and labelling with [99mTc(OH2)3(CO)3]+. *Org Biomol Chem,* 2, 2593–603.

Viola-Villegas, N., Rabideau, A. E., Bartholoma, M., Zubieta, J., & Doyle, R. P. 2009. Targeting the cubilin receptor through the vitamin B(12) uptake pathway: Cytotoxicity and mechanistic insight through fluorescent Re(I) delivery. *J Med Chem,* 52, 5253–61.

Vlahov, I. R., & Leamon, C. P. 2012. Engineering folate-drug conjugates to target cancer: From chemistry to clinic. *Bioconjug Chem,* 23, 1357–69.

Vortherms, A. R., Kahkoska, A. R., Rabideau, A. E., Zubieta, J., Andersen, L. L., Madsen, M., & Doyle, R. P. 2011. A water soluble vitamin B12-ReI fluorescent conjugate for cell uptake screens: Use in the confirmation of cubilin in the lung cancer line A549. *Chem Commun (Camb),* 47, 9792–4.

Waibel, R., Treichler, H., Schaefer, N. G., Van Staveren, D. R., Mundwiler, S., Kunze, S., Kuenzi, M., Alberto, R., Nuesch, J., Knuth, A., Moch, H., Schibli, R., & Schubiger, P. A. 2008. New derivatives of vitamin B12 show preferential targeting of tumors. *Cancer Res,* 68, 2904–11.

Wilbur, D. S., Hamlin, D. K., Pathare, P. M., Heusser, S., Vessella, R. L., Buhler, K. R., Stray, J. E., Daniel, J., Quadros, E. V., McLoughlin, P., & Morgan, A. C. 1996. Synthesis and nca-radioiodination of arylstannyl-cobalamin conjugates. Evaluation of aryliodo-cobalamin conjugate binding to transcobalamin II and biodistribution in mice. *Bioconjug Chem,* 7, 461–74.

Wolters, M., Strohle, A., & Hahn, A. 2004. Cobalamin: A critical vitamin in the elderly. *Prev Med,* 39, 1256–66.

Wuerges, J., Garau, G., Geremia, S., Fedosov, S. N., Petersen, T. E., & Randaccio, L. 2006. Structural basis for mammalian vitamin B12 transport by transcobalamin. *Proc Natl Acad Sci U S A,* 103, 4386–91.

Wuerges, J., Geremia, S., Fedosov, S. N., & Randaccio, L. 2007. Vitamin B12 transport proteins: Crystallographic analysis of beta-axial ligand substitutions in cobalamin bound to transcobalamin. *IUBMB Life,* 59, 722–9.

Yang, J. Q., Li, Y., Lu, J., & Wang, X. B. 2005. Preparation and biodistribution in mice of 99mTc-DTPA-b-cyanocobalamin. *J Radioanal Nucl Chem,* 265, 467–72.

Yang, S. Y., Coleman, P. S., & Dupont, B. 1982. The biochemical and genetic basis for the microheterogeneity of human R-type vitamin B12 binding proteins. *Blood,* 59, 747–55.

16 Ascorbic Acid
Binding Proteins and Pathophysiology

Fryad Rahman and Michel Fontés

CONTENTS

16.1 INTRODUCTION

In humans, L-ascorbic acid (AA) is essential for life (which is why it is called a vitamin) by virtue of its antioxidant properties, which protect cells against oxidative stress (Padayatty et al., 2003). AA deficiency is termed scurvy (Wilson, 1975). James Lind, in 1747, was the first to conclude that eating fruits could prevent scurvy. During the eighteenth and nineteenth centuries, foods able to prevent scurvy were termed antiscorbutic, although the chemical principle was not identified. Albert

Szent-Gyorgyi's group first isolated the antiscorbutic principle and named it ascorbic acid (he received the Nobel Prize in Physiology or Medicine in 1937).

Ascorbic acid is a sugar acid with the same furanose ring as ribose. Most animals and plants synthesize AA from glucose through a four-enzyme pathway (Chatterjee, 1973; Smirnoff, 2001; Linster and Schaftingen, 2007). However, primates (as well as guinea pigs) have lost the ability to synthesize ascorbic acid because they lack one enzyme, L-gulonolactone oxidase, in the biosynthetic process (Burns, 1956, 1957). Therefore, normal primate diets should include foods containing ascorbic acid (e.g., fruits, sauerkraut).

Ascorbic acid is liberated in the digestive tract by digestion and transported into the blood by SVCT1 (SLC23A1) (Tsukaguchi et al., 1999; Daruwala et al., 1999; Eck et al., 2004), a glycoprotein located on the apical and basal faces of the epithelial cells (Boyer et al., 2005). SVCT1 is also located in the kidney (Castro et al., 2008; Lee et al., 2006), where it participates in reabsorption into the circulation. SVCT2 (SLC23A2) is a membrane protein located at the apical poles of polarized cells (Bianchi et al., 1985). It transports AA into cells via a Na-dependent mechanism, which allows the intracellular AA concentration to exceed the extracellular concentration. AA binding to SVCT2 is very specific, as most closely related analogues are not properly transported by it (Rumsey et al., 1999). The intracellular distribution of AA and exact biological role of SVCT2 have not been fully elucidated. These points will be further discussed later.

The metabolic and biochemical properties of AA have been extensively documented. Most of them are associated with the chemical nature of the molecule, indicating its antioxidant properties. Surprisingly, until recently, AA was not considered a potential signaling molecule as, for example, vitamins A and D are.

We will thus discuss, in the following sections, the different functions of AA, including its role on gene expression, the transporters of AA, and their roles. Finally, we will discuss the potential implications of AA in different disorders.

16.2 FUNCTIONS OF ASCORBIC ACID

The first clear function of AA was to prevent scurvy. Evidence that ascorbic acid was the molecule necessary to prevent the appearance of this fatal disease has already been described, without any indication of a role this molecule could play.

16.2.1 BIOCHEMICAL FUNCTIONS

Ascorbic acid's chemical structure makes it an electron donor and therefore a reducing agent. AA has thus been involved in two different biochemical functions: redox/antioxidant properties and enzymatic cofactor. AA has been demonstrated to be an electron donor for different enzymes. Among these enzymes, three are involved in collagen hydroxylation (Bates et al., 1972; Levene et al., 1972). Two are involved in carnitine synthesis (Nelson et al., 1981; Dunn et al., 1984). The remaining are respectively involved in norepinephrine synthesis (Kuo, 1979) and tyrosine synthesis (La Duand Zannoni, 1964). Deficiency in AA has thus been associated with extracellular matrix defects that are probably involved in vascular problems observed in scurvy.

This enzymatic property has certainly had an impact on cell metabolism that could influence gene expression. However, these mechanisms are probably not directly involved in the control of gene expression and cell signaling. The first explanation came from recent results of our group. As we described before, AA treatment lowers the expression of the gene *PMP22*, overexpressed in CMT1A patients (Kaya et al., 2007). This is likely the basis of phenotypic correction observed in a CMT animal model. Expression of *PMP22* is under the control of cAMP, via fixation of CREB on a site located on the promoter (1.5 kb from the initiation of transcription) (Saberan-Djoneidi et al., 2000; Hai et al., 2001; Orfali et al., 2005). Incubating Schwann cell lines with increasing concentrations of AA results in a dosage-dependent inhibition of *PMP22* expression. In addition, an evaluation of the intracellular cAMP pool demonstrated that this pool is decreased when cells are incubated with increasing concentration of AA (Kaya et al., 2007). This inhibition is specific to AA and is not shared by other antioxidants (Kaya et al., 2008a). Recently, using classical enzymatic experiments, it has been demonstrated that AA is a competitive inhibitor of adenylate cyclase (Kaya et al., 2008b), probably because AA and ATP possess both a furanic ribose ring (with different side chains). Therefore, AA could be a molecule that represses the expression of a gene under the control of the cAMP-dependent pathway.

16.2.2 Ascorbic Acid and Development

Two lines of research indicate that AA could be involved in cell differentiation. The first regards Schwann cell differentiation and myelination. It has been demonstrated that AA should be added to Schwann cells/axons coculture to induce the formation of myelin (Carey and Todd, 1987; Eldridge et al., 1987; Plant et al., 2002). However, no clear mechanism except antioxidant properties has been demonstrated. The second line of research has been more recently initiated and regards embryonic stem cell (ES) differentiation. Using high throughput screening of a chemical library, Takahashi et al. (2003) found that the only molecule present in a chemical library able to promote differentiation of murine ES into cardiomyocytes was AA. They also demonstrated that this differentiation is specific to AA and is not shared with other antioxidants, suggesting that antioxidant mechanisms are probably not involved in this process.

Finally, Sotiriou et al. (2002) describe the phenotype of a mouse mutant in which the gene coding for SVCT2 has been invalidated. Mutant animals, homozygotes, as well as heterozygotes present a normal AA concentration in blood, but the molecule is unable to enter cells. Embryos died in a perinatal period with anomalies of lung and vessels.

These different observations clearly ask the question of AA as a signaling molecule in development and differentiation of mammalians. These will be fascinating new topics that will probably develop in the near future.

16.3 ASCORBIC ACID TRANSPORT, TRANSPORTING PROTEINS, AND RECEPTORS

Although numerous receptors of vitamin A, D, and E have been described for a long time, there is no publication related to proteins that could act as a receptor of AA.

However, at least two transporters of AA have been reported. Thus we will describe regulation of the flux of AA entering cells and proteins involved in this process. Finally, we will describe the possibility that SVCT2 could act as a receptor of AA.

16.3.1 PASSIVE TRANSPORT

Reduced AA exists predominantly as the ascorbate anion in most body fluids. Molecules that are comparably water soluble diffuse rapidly through nonspecific pathways in cell membranes, especially through the lipid bilayer. In contrast, ascorbate, because of its size and charge, does not readily permeate the lipid bilayer. Simple diffusion of DHAA into cells is negligible probably due to the structure and water nucleation around the molecule. Weak organic acids can enter cells by simple diffusion of their undissociated forms. Once in the cytoplasm, these acids dissociate into organic ions and protons. However, it has been demonstrated that this process did not occur for AA (Wilson and Dixon, 1989).

Further studies showed that these cells increase their intracellular concentration of AA through plasma membrane transport systems that translocate either the ascorbate anion (Wilson and Dixon, 1989) or DHAA (Qutob et al., 1998). Therefore, simple diffusion across plasma membrane at physiological pH comprises only a slow component of AA accumulation and may be negligibly small in those cells that contain transport systems with high affinity for ascorbate or DHAA. Specific proteins mediate the entry and exit of AA in cells by facilitated diffusion or active transport. Facilitated diffusion achieves net movement only in the direction of a chemical or electrochemical gradient of the transported solute, whereas active transport can move solute against this gradient by using energy derived from cellular metabolism.

16.3.2 UPTAKE OF DEHYDROASCORBIC ACID INTO CELLS

When ascorbate acts as an antioxidant or enzyme cofactor, it becomes oxidized to DHAA. Ascorbate and DHAA possess roughly equivalent bioavailability. Bioavailability is determined by the rates of absorption, distribution, and metabolism within the body, and by excretion. Ascorbate and DHAA are absorbed along the entire length of the human intestine (Malo and Wilson, 2000). For both the DHAA and ascorbate transport systems, initial rates of uptake saturate with increasing external substrate concentration, reflecting high-affinity interactions that can be described by Michaelis-Menten kinetics.

The mechanism of DHAA uptake by luminal membranes of human jejunum has pharmacological characteristics that clearly differ from those of ascorbate uptake. Sodium-independent carriers take up DHAA by facilitated diffusion, and these are distinct from the sodium-dependent transporters of ascorbate. Glucose inhibits ascorbate uptake but not DHAA uptake, which raises the possibility that glucose derived from food may increase the bioavailability of DHAA relative to ascorbate (Malo and Wilson, 2000). Human enterocytes contain reductases that convert DHAA to ascorbate (Buffinton and Doe, 1995). This conversion keeps the intracellular level of DHAA low, and the resulting concentration gradient favors uptake of oxidized AA across the enterocyte plasma membrane.

DHAA competes with glucose for uptake through the mammalian facilitative glucose transporters GLUT1, GLUT3, and GLUT4 (Korcok et al., 2003; Rumsey et al., 1997, 2000; Vera et al., 1998). GLUT proteins do not transport ascorbic acid or ascorbate. The results obtained with mammalian transporters that have been expressed experimentally in *Xenopus* oocytes are generally consistent with the hypothesis that GLUT1 and GLUT3 transport DHAA with affinities and maximal capacities similar to their transport of glucose. The hypothesis is also supported by a case report that uptake of DHAA by erythrocytes of a patient with GLUT1 deficiency was impaired to the same degree as that of the glucose analog, 3-*O*-methylglucose (Klepper et al., 1999).

The separate transporters for DHAA and ascorbate can be regulated independently of each other. For instance, colony-stimulating factors enhance DHAA uptake in human neutrophils that lack sodium-ascorbate cotransporters (Vera et al., 1998). Bone-derived osteoblasts provide another intriguing example: Transforming growth factor-β increases the maximal rate of ascorbate uptake through sodium-ascorbate cotransporters, but does not stimulate DHAA uptake (Wilson and Dixon, 1995), whereas insulin increases the maximal rate of DHAA uptake through glucose transporters without changing sodium-ascorbate cotransport activity (Qutob et al., 1998). Thus, transforming growth factor-β and insulin both increase intracellular ascorbate concentration in osteoblasts but act through different AA transport systems.

16.3.3 ACTIVE TRANSPORT OF ASCORBIC ACID

Ascorbate is absorbed from the lumen of the human intestine by sodium-ascorbate cotransport in enterocytes, as has been shown by measuring transport activities in luminal (brush-border) membrane vesicles (Malo and Wilson, 2000). This phenomenon couples ascorbate uptake to the concentration gradient of sodium ion across the plasma membrane that is maintained by sodium/potassium-ATPase. It is likely because of the limited capacity of enterocytes for sodium-ascorbate cotransport that large oral doses of ascorbate are absorbed less completely than are small doses. Absorption sites for ascorbate are found along the entire length of the small intestine (Malo and Wilson, 2000).

Besides intestinal absorption, another important determinant of bioavailability is secondary active transport of ascorbate in the kidney. Most AA circulates in the blood in the form of the ascorbate anion. The ascorbate in the blood plasma is freely filtered at the renal glomerulus, but much of it is reabsorbed in the proximal tubule. Ascorbate uptake across the luminal membranes of renal proximal tubule cells occurs through sodium-ascorbate cotransport. The amount of ascorbate lost in the urine rises when the plasma ascorbate concentration exceeds the renal threshold. Above this threshold the tubular reabsorptive capacity is overwhelmed. The renal threshold for AA is reported to be slightly higher in men than in women (plasma ascorbate concentrations of 86 and 71 µM, respectively), but the underlying mechanism and physiological importance of this difference are unknown (Oreopoulos et al., 1993).

Sodium-ascorbate cotransporters are remarkably specific for L-ascorbate (Dixon and Wilson, 1992; Franceschi et al., 1995; Liang et al., 2002; Malo and Wilson, 2000;

Wilson and Dixon, 1989). Among the molecules that have been tested and found not to be substrates for these cotransporters are ascorbate-2-O-phosphate, DHAA, glucose, 2-deoxyglucose, xanthine, hypoxanthine, L-gulono-lactone, formate, lactate, pyruvate, gluconate, oxalate, malonate, succinate, and an assortment of nucleosides and nucleotides. The cotransporters stereoselectivity has been demonstrated as a greater affinity for L-ascorbate over the epimer D-isoascorbate, which leads to higher intracellular ascorbate concentrations of L-ascorbate than D-isoascorbate at steady state (Franceschi et al., 1995). The cotransporters are absolutely dependent on sodium and translocation of at least two sodium cations with each ascorbate anion is required.

16.3.4   Cotransporters SVCT1 and SVCT2

16.3.4.1   Gene Structure and Expression

Two human isoforms have been cloned (Daruwala et al., 1999; Rajan et al., 1999). SVCT1 and SVCT2 are encoded by the SLC23A1 and SLC23A2 genes. The open reading frames (ORFs) of the two genes are of comparable size (1,791 and 1,953 for SLC23A1 and SLC23A2, respectively) and the exon-intron borders are in similar positions in the two genes (Eck et al., 2004). Homology among transcripts from different species (mouse, rat, pig, and human) is about 86%–95% and 89%–95% for SVCT1 and SVCT2 mRNAs, respectively (Daruwala et al., 1999; Clark et al., 2002; Obrenovich et al., 2006). These data have suggested a duplication of a common ancestral gene, which has been calculated to occur prior to the divergence of bony fish and tetrapods, an event dating 450 million years ago (Kumar and Hedges, 1998; Eck et al., 2004). The hypothesis of gene duplication is sustained by the observation that neighboring genes of both SLC23A1 and SLC23A2 are conserved in human and mouse (Eck et al., 2004). We may note that SLC23A2 is one of the most conserved genes during evolution (about 70% identity between human and *Drosophila*).

Both genes have been screened for genetic variants and, interestingly, only SLC23A1 has been found to contain non-synonymous single nucleotide polymorphisms. This suggests that the coding region of the larger gene could be under a more conservative selective pressure (Eck et al., 2004). Implications will be discussed in the final section.

16.3.4.2   Protein Structure

A putative structure for SVCT1 and SVCT2 has been predicted by hydropathy analysis. Kyte-Doolittle plots for the two transporters are superimposable (Faaland et al., 1998; Liang et al., 2001) and indicate that both are transmembrane (TM) proteins. The predicted structure contains 12 membrane-spanning domains, with both the N and the C termini (102 and 81 amino acids) located on the cytoplasmic side of the membrane. The extracellular loop between the 7 and 8 TM domains contains a series of conserved proline residues, which are needed for structure stability and transport efficiency (Liang et al., 2001). Other conserved proline residues, putatively located within the TM segments, are thought to be important for determination of protein structure (Liang et al., 2001). A small C-terminus fragment (PICPVFKGFS, amino acids 563–572), encoding a b-turn motif and homologous to other sodium-dependent transporters (Cheng et al., 2002; Sun et al., 2003), is required for targeting SVCT1 to the apical membrane of intestinal cells (Maulen et al., 2003; Subramanian et al., 2004; Boyer et al., 2005).

Reverse transcription-polymerase chain reaction (RT-PCR) analysis showed that SVCT1 is mainly expressed in the intestine; liver and kidney express mainly SVCT1. In addition, it has been described that SVCT2 mRNA levels in kidney are less than 5% of SVCT1 levels (Kuo et al., 2004). SVCT2 has a wider distribution and is present at the membrane of a large number of cell types and is probably involved in entering of AA inside cells. On the one hand, the high capacity of SVCT1 is appropriate for epithelial cells that transport much more ascorbate than required for their own internal use.

The SVCT isoforms appear to function independently of each other because SVCT1 expression and ascorbate concentrations in SVCT1-predominant organs are not affected by SVCT2 deficiency (Kuo et al., 2004). Ascorbate concentrations are lower for ($Slc23a2^{+/-}$) than wild-type ($Slc23a2^{+/+}$) mice in tissues where SVCT2 is the main isoform, such as brain, spleen, and skeletal muscle (Kuo et al., 2004; Sotiriou et al., 2002). Therefore, SVCT2 is a major determinant of ascorbate accumulation in tissues lacking SVCT1. Selective sorting to the apical plasma membrane has been demonstrated for SVCT1 in the human colon adenocarcinoma cell line CaCo-2 (Maulen et al., 2003) and the human nasal epithelial cell line CF15 (Fisher et al., 2004). The evidence for the latter is that, first, cells cultured from human airway epithelia express SVCT1 and SVCT2 mRNA and are capable of sodium-dependent ascorbate uptake, and second, recombinantly expressed SVCT2-enhanced green fluorescent fusion protein is targeted exclusively to the apical membrane pole of the CF15 nasal epithelial cell line (Fisher et al., 2004). Confocal imaging of hSVCT1 truncation mutants expressed in CaCo-2 cells and Madin-Darby canine kidney cells demonstrated that hSVCT1 was expressed at the apical cell surface and also resided in a heterogeneous population of intracellular organelles (Subramanian et al., 2004). Progressive truncation of the cytoplasmic COOH terminal tail of hSVCT1 showed that an embedded ten-amino-acid sequence PICPVFKGFS, in amino acids 563–572, was required for targeting of hSVCT1 (Subramanian et al., 2004).

16.3.4.3 Regulation of Gene Expression

Ascorbate transport is mainly controlled by carriers' availability, and thus it depends on the number of SVCT proteins present in the plasma membrane (which in turn is related to enhanced synthesis, slowed degradation, activation of nonfunctional carriers or cellular redistribution), as well as their substrate affinity. So, transcriptional, translational, and post-translational modifications of SVCTs allow a fine-tuned regulation of AA uptake (Liang et al., 2002).

In vitro and *in vivo* studies have revealed a transcriptional control mechanism for SVCT's activity by hormones and intracellular signaling molecules. For example, SVCT2 mRNA levels are increased by fetal bovine serum and epidermal growth factor in a human trophoblast cell line (Biondi et al., 2007) and by glucocorticoids (Fujita et al., 2001), zinc (Wu et al., 2003b), and calcium phosphate ions (Wu et al., 2003a) in osteoblastic cells. *In vivo*, SVCT2 mRNA is overexpressed following ischemic brain injury (Berger et al., 2003). Moreover, SVCT1 expression decreases in rat hepatocytes during aging (Michels et al., 2003) and SVCT2 expression is down-modulated during differentiation of rat and mouse muscle cells (Savini et al., 2005).

Interestingly, the two Na-dependent transporters are probably regulated by their own substrate. Indeed, elevated levels of ascorbate in the intestinal lumen lead to down-regulation of SVCT1 mRNA in enterocytes (MacDonald et al., 2002), although the exact mechanisms involved are still unknown; down-regulation may be due to transcriptional repression or, alternatively, to decrease in mRNA stability. Ascorbate regulates its bioavailability, controlling not only intestinal SVCT1 expression, but also SVCT2-mediated cellular uptake in other body compartments. Indeed, ascorbate supplementation or deprivation influences its own transport in osteoblasts (Dixon and Wilson, 1992) and astrocytes (Wilson et al., 1990). In addition, it has recently been demonstrated that a substrate-mediated translational control for SVCT2 exists in platelets (Savini et al., 2007). Finally, a feedback mechanism has been demonstrated also in human lung epithelial cells, where a loss of intracellular ascorbate is compensated by active uptake thanks to a marked increase of SCVT2 expression (Karaczyn et al., 2006).

Ascorbate down-regulates the maximal rate of sodium-ascorbate cotransport. This has been shown for absorptive epithelia by feeding excess ascorbate to guinea pigs and afterward measuring the rate of sodium-dependent uptake of AA into the intestinal mucosa (Karasov et al., 1991). Down-regulation of SVCT1 by ascorbate may limit its usefulness for raising intracellular ascorbate concentration. Evidence for down-regulation is that incubation of the human colon adenocarcinoma cell line CaCo-2 TC7 with ascorbate for 24 hours leads to decreases in SVCT1 mRNA level and ascorbate uptake rate (MacDonald et al., 2002). Thus, it appears that the activity of SVCT1 in enterocytes is regulated to adjust for the recent history of ascorbate absorption. Substrate down-regulation of SVCT2 activity has been induced by pre-incubating astrocytes and osteoblasts with ascorbate for 10 to 24 hours and subsequently determining the kinetic properties of the initial rate of [14C] ascorbate uptake (Dixon and Wilson, 1992; Wilson et al., 1990). Changes in SVCT2 activity are rapid and large. The observation that SVCT2 activity varies inversely with intracellular ascorbate concentration is consistent with the hypothesis that this transporter regulates the intracellular concentration of its substrate. Up-regulation of SVCT2 in cells depleted of ascorbate leads to more efficient absorption of extracellular ascorbate and tends to restore the intracellular concentration of the vitamin (Dixon and Wilson, 1992; Wilson et al., 1990). Therefore, in the absence of neurohormonal or paracrine signals, SVCT2 acts to maintain the intracellular ascorbate concentration constant. Experiments with transgenic mice lacking Slc23a2 indicate that SVCT2 normally maintains the high ascorbate concentration found in brain (Kuo et al., 2004; Sotiriou et al., 2002). *In situ* hybridization of mRNA indicated that the ependymal cells of the choroid plexus express SVCT2, where it may be involved in the trans-epithelial transport of ascorbate between the blood and the cerebrospinal fluid (Wilson, 2005). Brain neurons and astrocytes also express SVCT2 (Berger et al., 2003; Castro et al., 2001; Korcok et al., 2000; Tsukaguchi et al., 1999). However, because sodium-ascorbate cotransporters become down-regulated when intracellular ascorbate concentration is high (Wilson et al., 1990), they may not be suitable targets for therapeutic strategies that attempt to raise intracellular ascorbate to supraphysiological levels.

Post-translational modifications represent an additional level of regulation. Different putative regulatory sites are present throughout the amino acid sequence of SVCT1 and SVCT2. The N-terminus contains more than 20% of acidic residues that can act as a regulatory domain deputed to interaction with other proteins (Liang et al., 2001). Regulation based on protein–protein interaction has been demonstrated for the alternatively spliced form of SCVT2 that can act as a dominant negative isoform (Lutsenko et al., 2004). Both carriers also possess putative glycosylation and phosphorylation sites. Conserved N-linked glycosylation sites are located in the second and third extracellular loop (Asn 138, 144, and 230 in human SVCT1; Asn 188 and 196 in human SVCT2) (Faaland et al., 1998; Tsukaguchi et al., 1999; Liang et al., 2001).

In conclusion, much research has been directed to identifying mechanisms that control the ascorbate transport in various cell types. Transport activity may be altered by changes in the affinity of substrate binding, the translocation capacity (i.e., turnover number) of each transporter protein, or the number of transporter proteins present in the plasma membrane. Sodium-ascorbate cotransport may be regulated kinetically by changes in either the concentrations of the transported solutes or membrane potential. The limited capacities of SVCT isoforms and their susceptibility to down-regulation by ascorbate may have influenced the many human clinical trials with oral AA supplements that failed to confer antioxidant protection or clinical benefit. Optimization of the dosing regime may be critical to the success of future intervention studies using AA. An important consequence of substrate regulation of SVCT1 and SVCT2 activities may be more efficient absorption by the intestine, conservation by the kidney, and uptake into target cells of intermittent doses than of continuously ingested doses of ascorbate. A last question regards the possibility that SVCT2 will not be only a transporter but also a receptor of AA, acting as a signaling molecule. This is a pending question that should be asked in the future.

16.4 ASCORBIC ACID AND GENE EXPRESSION

AA has long been considered either as an enzymatic cofactor or as a molecule that acts through the antioxidant properties attributable to its chemical structure. The first indication that it might also modulate gene expression was documented in a series of articles describing the stimulation of procollagen mRNA by ascorbate (Chokjier et al., 1989; Lyons et al., 1984; Murad et al., 1981). The authors clearly demonstrated that this effect was at the transcriptional level. Two other recent papers reported the action of AA on expression of the gene encoding a new form of collagen involved in extracellular matrix formation (Chernousov et al., 1999, 2000). However, no clear explanation has been proposed, except for indirect effects such as hydroxyproline oxidation (Bates et al., 1972).

Therefore, although AA has been shown to regulate the expression of genes encoding extracellular matrix proteins, no additional gene or gene family was proposed until recently. In 2004, Passage et al. demonstrated that treatment of a mouse model of the Charcot-Marie-Tooth 1A disease reverts, at least partly, to the transgenic mouse phenotype. This disease is due to the overexpression of a major myelin gene, *PMP22*, and AA treatment lowers *PMP22* expression. In the same year, using microarrays representing about 6,000 genes, Shin et al. (2004) published a series of genes responding to AA treatment of embryonic stem cells. Most of the overexpressed genes belonged to gene families involved in neurogenesis, maturation, and neurotransmission. More recently, Park et al. (2009) published a proteomic analysis of cancer cells treated with AA. The most relevant effects seem to be overexpression of RKIP and Annexin A5. Finally, in February 2009, Belin et al. (2009) published an article describing changes of gene expression in normal and cancer cells treated with increasing doses of AA. They used pangenomic arrays. The most striking result was that only thirty genes were down-regulated by AA treatment. Among these, twelve belonged to two families, tRNA synthetases and translation initiation factors, involved in cell division. The authors demonstrated that AA stops cell proliferation *in vivo* and *in vitro* and kills the dividing cells by necrosis at higher concentration. To our knowledge, only one mechanism by which AA could modulate gene expression has been proposed, through modulation of the intracellular cAMP pool; we will discuss this later. Are genes expressed and regulated through the cAMP-dependent pathway the only ones affected by AA treatment? That remains an open question.

A general question regarding the control of gene expression by small molecules concerns a potential receptor. Receptors for vitamin A and vitamin D have been well described and characterized (see following), but no classical receptor has been described for AA so far. However, two proteins with high affinity for AA have been described, SVCT1 and SVCT2 (for review, see Wilson, 2005). Both are AA transporters and act through a Na-dependent mechanism. SVCT1 is located in the intestine and kidney and is involved in AA transport into the blood. In contrast, SVCT2 is present in the plasma membranes of different cell types. It has a very specific affinity for L-ascorbic acid and does not recognize closely related molecules (Rumsey et al., 1999). It enables AA to accumulate inside cells, though we do not know exactly where the intracellular AA is located.

Interestingly, the gene encoding a key enzyme in AA biosynthesis, L-gulono-gamma-lactone synthase, accumulates mutations in man and guinea pig (Nishimiki et al., 1992, 1994). In consequence, these two species cannot synthesize AA. In contrast, SVCT2 is one of the most evolutionarily conserved molecules and no species lacks this key protein. We may conclude that it is not AA biosynthesis itself, but the accumulation of AA in cells mediated by SVCT2 that is most important for organisms. The involvement of SVCT2 in AA signal reception will thus be very interesting to investigate.

16.5   ASSOCIATED PATHOPHYSIOLOGY

16.5.1   GENETIC VARIANTS OF TRANSPORTERS

An important question about the influence of variations in AA concentration in blood and cells regards the existence of genetic variants of AA transporters. Eck et al. (2004) characterized the genomic structures of SLC23A1 and SLC23A2. They analyzed variations in sequences of these genes in different populations. In SLC23A1, the majority of single nucleotide polymorphisms (SNPs) are population specific, including three of four nonsynonymous SNPs. In contrast, most SNPs are not population dependent and there are no nonsynonymous SNPs in SLC23A2. These data suggested that nonsynonymous variations could be tolerated in SLC23A1 but not SLC23A2, indicating different selective pressure on sequences of these two genes. A recent publication by Corpe et al. (2010) reports the transport activity of different alleles of SLC23A1. They analyze this activity for four variants, one synonymous SNP and three nonsynonymous. All variants presented a reduction of transport capacity, up to 80% reduction for A722G, a relatively frequent allele in the population (5/5%).

These data raised the question of AA blood levels linked to gene polymorphism in the general population. A recent report (Timpson et al., 2010) demonstrates that a common genetic variant of the SLC23A1 gene locus (rs33972313) is associated with circulating concentrations of L-ascorbic acid. This study included a discovery cohort (the British Women's Heart and Health Study), a series of follow-up cohorts, and meta-analysis (totaling 15,087 participants).

The authors demonstrate that a genetic variant (rs33972313) in the SLC23A1 AA active transporter locus is significantly associated with differences in circulating concentrations of AA in the general population. This finding has implications more generally for the epidemiologic investigation of relations between circulating L-ascorbic acid and health outcomes.

16.5.2 Diabetes

Twenty-three studies asking the question of diabetes and AA concentration in blood have been reported. Seven studies found that blood AA concentrations in persons with diabetes were not significantly lower than concentrations in persons without diabetes (Akkus et al., 1996; Asayama et al., 1993; Jennings et al., 1987; Lysy and Zimmerman, 1992; Owens et al., 1941; Schorah et al., 1988). However, they either did not report dietary intake or illness status, making the conclusions uncertain. In addition, none of the studies examined urinary frequency, a factor we found to be inversely associated with serum AA concentrations.

A paper by Will et al. (1999) reviews the scientific evidence regarding the AA status of people with diabetes mellitus and whether they might have increased dietary AA requirements. They report that English language articles published from 1935 to the present that either compare ascorbic acid concentrations of persons with and without diabetes mellitus or assess the impact of AA supplementation on various health outcomes among persons with diabetes mellitus were examined. Most studies have found people with diabetes mellitus to have at least 30% lower circulating ascorbic acid concentrations than people without diabetes mellitus. AA supplementation had little impact on blood glucose concentrations but was found to lower cellular sorbitol concentrations and to reduce capillary fragility. Much of the past research in this area has been methodologically weak. To further understand the relation of ascorbic acid and diabetes mellitus, randomized clinical trials of ascorbic acid supplementation should be a high priority for research.

Several articles report AA levels in plasma and tissues of diabetic patients and animals have been reported to be low (Will et al., 1996). Supplementation of AA decreases sorbitol levels (Cunningham et al., 1994; Wang et al., 1995) thereby preventing the development of diabetic complications (Brownlee et al., 1988; Gabbay, 1975) and increasing the stability of blood vessels (Cox and Butterfield, 1975; Ting et al., 1996. Therefore, reduced AA levels in diabetes impair the blood vessels and lead to the development of diabetic complications. Thus, it is important to identify the underlying mechanisms of reduced plasma and tissue AA levels in diabetes.

Previous studies have reported impaired urinary excretion of AA in diabetic patients (Seghieri et al., 1994; Yue et al., 1989). Further, the activity of L-gulono-gamma-lactone oxidase (GLO), a terminal enzyme of AA biosynthesis (Majumder et al., 1972) is impaired in diabetic animals, and decreased GLO activity impairs the ability to synthesize AA (Bode et al., 1993). Until recently, however, the molecular properties of the enzymes and transporters involved in AA metabolism have not been characterized, and the precise mechanisms of decreases in AA in plasma and tissues in diabetes have been undetermined.

A recent paper (Dakhale et al., 2011) reports the effect of administration of AA with metformin on fasting (FBS) and postmeal blood glucose (PMBG) as well as glycosylated hemoglobin (HbA1c) in the treatment of type 2 diabetes mellitus (DM). The goal was to examine the effect of oral AA with metformin on FBS, PMBG, HbA1c, and plasma ascorbic acid level (PAA) with type 2 DM. Seventy patients with type 2 DM participated in a prospective, double-blind, placebo-controlled, twelve-week study. The patients with type 2 DM were divided randomly into a placebo and an AA group of thirty-five each. Both groups received the treatment for twelve weeks. Decreased PAA levels were found in patients with type 2 DM. This level was reversed significantly after treatment with AA along with metformin compared to placebo with metformin. FBS, PMBG, and HbA1c levels showed significant improvement after twelve weeks of treatment with AA. In conclusion, oral supplementation of AA with metformin reverses ascorbic acid levels, reduces FBS and PMBG, and improves HbA1c. Hence, both the drugs in combination may be used in the treatment of type 2 DM to maintain good glycemic control.

To conclude, the effect of AA on diabetes is still controversial in humans. However, in an animal model, Kashiba et al. (2002) report that renal expression of SVCTs mRNA was normal in STZ-diabetic rats, but urinary excretion of AA was significantly increased in the short-term study. This could indicate that impaired renal reabsorption of AA is also induced by post-translation modification of the transport proteins. Another possibility is that excessive loss of AA in the urine was simply the result of the increased urine flux due to hyperglycemia and not the disturbance of renal AA transporters. Urinary excretion of NAG, which is released into urine when renal tubules are damaged, was not increased in the STZ-induced diabetic rats of the short-term study, supporting the latter concept that excessive loss of AA in the urine was simply the result of the increased urine flux by hyperglycemia. As a general conclusion, additional studies should be conducted in order to recommend or not recommend supplementation with AA to diabetic regimens.

16.5.3 CANCER

The recommended daily dose of AA intake has varied over time, but it is presently about 75 to 90 mg/day. However, several authors, Linus Pauling among them, have suggested that higher daily doses could have a benefit on health, especially in preventing cancer (Castro et al., 2008). This has been a hotly debated topic in the scientific community.

Different clinical studies, generally methodologically poor, have been conducted, with conflicting results. Reading these data carefully, we could observe that studies concluding that AA has no impact on cancer are generally conducted using oral administration, and studies concluding a benefit have been conducted by IV administration. We know that IV administration, bypassing the SVCT1 transporter, leads to much higher blood concentration (Padayatty et al., 2003). Our personal experience clearly demonstrates that the impact of AA treatment, *in vitro* as well as *in vivo*, is significant only if we could achieve high AA concentrations that could be only obtained *in vivo* by IV administration (Belin et al., 2009). Results of well-conducted clinical trials on cancer treatment with IV injection are thus anticipated with great interest.

Regarding a potential mechanism of AA impact on cancer cells and proliferation, two explanations have been proposed, using animal models. The research group of M. Levine proposes that this effect is mediated by biochemical properties of AA (Du et al., 2010). They demonstrate that there is a time- and dose-dependent increase in measured H_2O_2 production with increased concentrations of ascorbate. Ascorbate decreased clonogenic survival of the pancreatic cancer cell lines, which was reversed by treatment of cells with scavengers of H_2O_2. Treatment with ascorbate induced a caspase-independent cell death that was associated with autophagy. *In vivo*, treatment with ascorbate inhibited tumor growth and prolonged survival.

Based on our experience, we propose that AA inhibits cell division and further promotes necrosis by down-modulating the expression of genes necessary for S-phase progression. We found that actively proliferating but not quiescent cells are susceptible to AA treatment, excluding a nonspecific toxicity of the highest AA concentrations. It is tempting to speculate that the inhibited expression of tRNA synthetases and translation initiation elongation factor subunits leads to the rapid cessation of energy production in proliferating cells, resulting in necrotic cell death. However, the mechanisms underlying the regulation of tRNA synthetase and ieF subunit expression in mammalian cells are still unknown.

Another question regards susceptibility to cancer and AA blood level. Surprisingly, only a few articles report on this subject. However, a recent paper (Kuiper et al., 2010) reports that low levels of ascorbate are significantly associated with tumor aggresivity in endometrial cancers. In addition, a genetic study showed that one intronic SNP (IVS2+1312 G>A), as well as a haplotype that contained the common allele of the IVS3+80 C>T, IVS3+108 A>G, and IVS3+224 T>G markers in *SLC23A2* were significantly inversely associated with gastric cancer risk. Moreover, these authors observed no relation between variants in *SLC23A1* and disease. This is somewhat surprising considering the greater observed diversity in *SLC23A1* compared to *SLC23A2* (Boyer et al., 2005). However, SVCT2 (the protein product of *SLC23A2*) but not SVCT1 (the protein product of *SLC23A1*) has been detected in gastric glands from rats, implicating *SLC23A2* as the primary means of ascorbic acid uptake in this organ (Tsukaguchi et al., 1999) explaining this result. In summary, they show that common variants in *SLC23A2*, a gene that directly regulates active transport of ascorbic acid, can impact gastric cancer risk, without any definitive conclusion.

16.5.4 NEUROPATHY

As mentioned before, we described (Passage et al., 2004) the phenotypic correction of the neuropathic phenotype of a mouse model of CMT1A, created in our laboratory, by high doses of ascorbic acid. We demonstrated that this is likely due to a repressing effect of AA on the expression of PMP22, the gene duplicated in CMT1A (Kaya et al., 2007; Kaya et al., 2008a). This effect is probably due to the impact of AA on cAMP production, as PMP22 expression is under the control of cAMP dependent pathways. PMP22 is one of the major proteins of myelin; it was thus tempting to link AA concentration and myelin formation. This is strengthened by the following:

- Authors have described ascorbic acid's myelination-promoting properties in *in vitro* axon/Schwann cell coculture (Carey et al., 1987; Eldridge et al., 1987).
- Ascorbic acid is systematically added to the culture medium in axon-Schwann cell coculture experiments and is absolutely necessary for *in vitro* myelination (Plant et al., 2002).
- There is evidence of a link between femoral neuropathies and ascorbic acid deficiency (Wilson et al., 1975).

This probable implication of AA in terminal Schwann cell differentiation, involving myelin formation, has been recently confirmed by Gess et al. (2011), demonstrating that heterozygote mice invalidated for *SLC23A2* present a defect in myelination leading to a phenotype similar to CMT1A.

It thus seems very interesting to explore peripheral neuropathic phenotypes that could be associated to a low concentration of AA in blood or tissues and to explore the possibility of a phenotypic reversion by AA administration.

16.5.5 AGE

Although age is not a disorder by itself, it has been associated with several diseases. Old human subjects require more AA in their diet than do young subjects to reach a desired plasma ascorbate concentration (Lykkesfeldt et al., 1998). Animal studies have revealed a diminished expression of SVCT1 mRNA and a decline in the capacity of cells to absorb AA during aging (Michels et al., 2003). The ascorbate concentration in the liver of male rats decreases with age even though the rate of *de novo* ascorbate synthesis does not. When incubated with ascorbate, isolated hepatocytes from old as compared with young rats show decreased maximal rate of ascorbate uptake and lower steady state intracellular ascorbate concentration.

Sodium-free media significantly reduce ascorbate uptake, implicating sodium ascorbate cotransporters. Hepatic SVCT1 mRNA levels decline 45% with age, with no significant changes in SVCT2 mRNA abundance. It thus appears likely that a fall in SVCT1 expression changes hepatic ascorbate concentration. In contrast, Meredith et al. (2011) described differential expression of the SVCT2 cotransporter during development. This finding may indicate a dominant role for plasma membrane transporters in the regulation of ascorbate levels, even in cells capable of synthesizing the vitamin.

Moreover, the deficit in intracellular ascorbate concentration can be overcome by increasing the external supply of the vitamin, which suggests that intervention to increase plasma ascorbate concentrations might be beneficial for hepatic function in elderly subjects (Michels et al., 2003).

16.6 CONCLUSIONS

As a general conclusion, the biochemical function of ascorbic acid has been largely studied and documented. This is also the case of the two transporters. However, it seems to us that the possibility that these proteins, as well as AA itself, could have new, unknown properties. In the future, it will be interesting to envisage the role of AA and transporters (especially SLC23A2) in differentiation and development.

In addition, the impact of AA variations in humans linked to genetic polymorphism or not seems under-studied. It could be important to develop these studies in the future. This will be important not only to unravel pathophysiological mechanisms but also to perform drug development. This development could use AA by itself, new formulations, or administration of AA and development of AA analogues that could perform better than AA by itself.

REFERENCES

Akkus, I.; Kalak, S.; Vural, H.; et al. **1996**. Leukocyte lipid peroxidation, superoxide dismutase, glutathione peroxidase and serum and leukocyte vitamin C levels of patients with type II diabetes mellitus. *Clin. Chim. Acta.* 244: 221– 227.

Asayama, K.; Uchida, N.; Nakane, T.; et al. **1993**. Antioxidants in the serum of children with insulin-dependent diabetes mellitus. *Free Radic. Biol. Med.* 15: 597–602.

Bates, C.; Prynne, C.; Levene, C. **1972**. Ascorbate-dependent differences in the hydroxylation of proline and lysine in collagen synthesized by 3T6 fibroblasts in culture. *Biochim. Biophys. Acta* 278(3): 610–616.

Belin, S.; Kaya, F.; Duisit, G.; Giacometti, S.; Ciccolini, J.; Fontes, M. **2009**. Antiproliferative effect of ascorbic acid is associated with the inhibition of genes necessary to cell cycle progression. *PLoS ONE* 4(2): e4409.

Berger, U. V.; Lu, X. C.; Liu, W.; Tang, Z.; Slusher, B. S.; Hediger, M. A. **2003**. Effect of middle cerebral artery occlusion on mRNA expression for the sodium coupled vitamin C transporter SVCT2 in rat brain. *J. Neurochem.* 86: 896–906.

Bianchi, J.; Rose, R. C. **1985**. Transport of L-ascorbic acid and dehydro-L-ascorbic acid across renal cortical basolateral membrane vesicles. *Biochim. Biophys. Acta* 820(2): 265–273.

Bode, A. M.; Yavarow, C. R. **1993**. Enzymatic basis for altered ascorbic acid and dehydroascorbic acid levels in diabetes. *Biochem. Biophys. Res. Commun.* 191: 1347–1353.

Biondi, C.; Pavan, B.; Dalpiaz, A.; Medici, S.; Lunghi, L.; Vesce, F. 2007. Expression and characterization of vitamin C transporter in the human trophoblast cell line HTR-8/SVneo: Effect of steroids, flavonoids, and NSAIDs. *Mol. Hum. Reprod.* 13(1): 77–83.

Boyer, J. C.; Campbell, C. E.; Sigurdson, W. J.; Kuo, S. M. **2005**. Polarized localization of vitamin C transporters, SVCT1 and SVCT2, in epithelial cells. *Biochem. Biophys. Res. Commun.* 334: 150–156.

Brownlee, M.; Cerami, A.; Vlassara, H. **1988**. Advanced glycosylation end products in tissue and the biochemical basis of diabetic complications. *N. Engl. J. Med.* 318: 1315–1321.

Buffinton, G. D.; Doe, W. F. **1995**. Altered ascorbic acid status in the mucosa from inflammatory bowel disease patients. *Free Radic. Res.* 22: 131–143.

Burns, J. **1956**. Missing step in guinea pigs required for the biosynthesis of L-ascorbic acid. *Science* 124: 1148–1149.

Burns, J. **1957**. Missing step in man, monkey required for the biosynthesis of L-ascorbic acid. *Nature* 180: 553.

Carey, D. J.; Todd, M. S. **1987**. Schwann cell myelination in a chemically defined medium: Demonstration of a requirement for additives that promote Schwann cell extracellular matrix formation. *Brain Res.* 429(1): 95–102.

Castro, M.; Caprile, T.; Astuya, A.; Millan, C.; Reinicke, K.; et al. **2001**. High affinity sodium-vitamin C co-transporters (SVCT) expression in embryonic mouse neurons. *J. Neurochem.* 78: 815–823.

Castro, T.; Low, M.; Salazar, K.; Montecinos, H.; Cifuentes, M.; Yanez, A. J.; Slebe, J. C.; Figueroa, C. D.; Reinicke, K.; De los Angeles Garcia, M.; Henriquez, J. P.; Nualart, F. **2008**. Differential distribution of the sodium-vitamin C cotransporter-1 along the proximal tubule of the mouse and human kidney. *Kidney Int.* 74(10): 1278–1286.

Chatterjee, I. **1973**. Evolution and the biosynthesis of ascorbic acid. *Science* 182: 1271–1272.

Cheng, C.; Glover, G.; Banker, G.; Amara, S. G. **2002**. A novel sorting motif in the glutamate transporter excitatory amino acid transporter 3 directs its targeting in Madin-Darby canine kidney cells and hippocampal neurons. *J. Neurosci.* 22: 10643–10652.

Chernousov, M. A.; Rothblum, K.; Tyler, W. A.; Stahl, R. C.; Carey, D. J. **2000**. Schwann cells synthesize type V collagen that contains a novel alpha 4 chain. Molecular cloning, biochemical characterization and high affinity heparin binding of alpha 4(V) collagen. *J. Biol. Chem.* 275(36): 28208–28215.

Chernousov, M. A.; Scherer, S. S.; Stahl, R. C.; Carey, D. J. **1999**. p200, a collagen secreted by Schwann cells, is expressed in developing nerves and in adult nerves following axotomy. *J. Neurosci. Res.* 56(3): 284–294.

Chojkier, M.; Houglum, K.; Solis-Herruzo, J.; Brenner, D. A. **1989**. Stimulation of collagen gene expression by ascorbic acid in cultured human fibroblasts. A role for lipid peroxidation? *J. Biol. Chem.,* 264(28), 16957–16962.

Clark, A. G.; Rohrbaugh, A. L.; Otterness, I.; Kraus, V. B. **2002**. The effects of ascorbic acid on cartilage metabolism in guinea pig articular cartilage explants. *Matrix. Biol.* 21: 175–184.

Corpe, C. P.; Tu, H.; Eck, P.; Wang, J.; Faulhaber-Walter, R.; Schnermann, J.; Margolis, S.; Padayatty, S.; Sun, H.; Wang, Y.; Nussbaum, R. L.; Espey, M. G.; Levine, M. **2010**. Vitamin C transporter Slc23a1 links renal reabsorption, vitamin C tissue accumulation, and perinatal survival in mice. *J. Clin. Invest.* 120(4): 1069–1083.

Cox, B. D.; Butterfield, W. J. H. **1975**. Vitamin C supplements and diabetic cutaneous capillary fragility. *Br. Med. J.* 3: 205.

Cunningham, J. J.; Mearkle, P. L.; Brown, R. G. **1994**. Vitamin C: An aldose reductase inhibitor that normalizes erythrocyte sorbitol in insulin-dependent diabetes mellitus. *J. Am. Coll. Nutr.* 13: 344–350.

Dakhale, G. N.; Chaudhari, H. V.; Shrivastava, M. **2011**. Supplementation of vitamin C reduces blood glucose and improves glycosylated hemoglobin in type 2 diabetes mellitus: A randomized, double-blind study. *Adv. Pharmacol. Sci.* 2011: 195271.

Daruwala, R.; Song, J.; Koh, W. S.; Rumsey, S. C.; Levine, M. **1999**. Cloning and functional characterization of the human sodium-dependent vitamin C transporters hSVCT1 and hSVCT2. *FEBS Lett.* 460: 480–484.

Dixon, S. J.; Wilson, J. X. **1992**. Adaptive regulation of ascorbate transport in osteoblastic cells. *J. Bone Miner. Res.* 7: 675–681.

Du, J.; Martin, S. M.; Levine, M.; Wagner, B. A.; Buettner, G. R.; Wang, S.; Taghiyev, A. F.; Du, C.; Knudson, C. M.; Cullen, J. J. 2010. Mechanisms of ascorbate-induced cytotoxicity in pancreatic cancer. *Clin. Cancer Res.* 16: 509–520.

Dunn, W.; Rettura, G.; Seifter, E.; England, S. **1984**. Carnitine biosynthesis from gamma-butyrobetaine and from exogenous protein-bound 6-N-trimethyl-L-lysine by the perfused guinea pig liver. Effect of ascorbate deficiency on the in situ activity of gamma-butyrobetaine hydroxylase. *J. Biol. Chem.* 259(17), 10764–10770.

Eck, P.; Erichsen, H. C.; Taylor, J. G.; Yeager, M.; Hughes, A. L.; Levine, M.; Chanock, S. **2004**. Comparison of the genomic structure and variation in the two human sodium-dependent vitamin C transporters, SLC23A1 and SLC23A2. *Hum. Genet.* 115: 285–294.

Eldridge, C. F.; Bunge, M. B.; Bunge, R. P.; Wood, P. M. **1987**. Differentiation of axon-related Schwann cells *in vitro*. Ascorbic acid regulates basal lamina assembly and myelin formation. *J. Cell. Biol.* 105(2): 1023–1034.

Faaland, C. A.; Race, J. E.; Ricken, G.; Warner, F. J.; Williams, W. J.; Holtzman, E. J. 1998. Molecular characterization of two novel transporters from human and mouse kidney and from LLC-PK1 cells reveals a novel conserved family that is homologous to bacterial and Aspergillus nucleobase transporters. *Biochimica Biophysica Acta.* 1442(2–3): 353–360.

Fischer, H.; Schwarzer, C.; Illek, B. **2004**. Vitamin C controls the cystic fibrosis transmembrane conductance regulator chloride channel. *Proc. Natl. Acad. Sci. USA* 101: 3691–96.

Franceschi, R. T.; Wilson, J. X.; Dixon, S. J. **1995**. Requirement for Na+-dependent ascorbic acid transport in osteoblast function. *Am. J. Physiol.* 268: C1430–C1439.

Fujita, I.; Hirano, J.; Itoh, N.; Nakanishi, T.; Tanaka, K. 2001. Dexamethasone induces sodium-dependent vitamin C transporter in a mouse osteoblastic cell line MC3T3-E1. *Br. J. Nutr.* 86(2): 145–149.

Gabbay, K. H. **1975**. Hyperglycemia, polyol metabolism and complications of diabetes mellitus. *Annu. Rev. Med.* 26: 521–536.

Gess, B.; Röhr, D.; Fledrich, R.; Sereda, M. W.; Kleffner, I.; Humberg, A.; Nowitzki, J.; Strecker, J. K.; Halfter, H.; Young, P. **2011**. Sodium-dependent vitamin C transporter 2 deficiency causes hypomyelination and extracellular matrix defects in the peripheral nervous system. *J. Neurosci.* 23; 31(47): 17180–17192.

Hai, M.; Bidichandani, S. I.; Patel, P. I. **2001**. Identification of a positive regulatory element in the myelin-specific promoter of the PMP22 gene. *J. Neurosci. Res.* 65(6): 508–519.

Jennings, P. E.; Chirico, S.; Jones, A. F.; et al. **1987**. Vitamin C metabolites and microangiopathy. *Diabetes Res.* 6: 151–154.

Karaczyn, A.; Ivanov, S.; Reynolds, M.; Zhitkovich, A.; Kasprzak, K. S.; Salnikow, K. **2006**. Ascorbate depletion mediates up-regulation of hypoxia-associated proteins by cell density and nickel. *J. Cell. Biochem.* 97: 1025–1035.

Karasov, W. H.; Darken, B. W.; Bottum, M. C. **1991**. Dietary regulation of intestinal ascorbate uptake in guinea pigs. *Am. J. Physiol.* 260: G108– G118.

Kashiba, M.; Oka, J.; Ichikawa, R.; Kasahara, E.; Inayama, T.; Kageyama, A.; Kageyama, H.; Osaka, T.; Umegaki, K.; Matsumoto, A.; Ishikawa, T.; Nishikimi, M.; Inoue, M.; Inoue, S. **2002**. Impaired ascorbic acid metabolism in streptozotocin-induced diabetic rats. *Free Radic. Biol. Med.* 33(9): 1221–1230.

Kaya, F.; Belin, S.; Bourgeois, P.; Micaleff, J.; Blin, O.; Fontes, M. **2007**. Ascorbic acid inhibits PMP22 expression by reducing cAMP levels. *Neuromuscul. Disord.* 17(3): 248–253.

Kaya, F.; Belin, S.; Micallef, J.; Blin, O.; Fontes, M. **2008a**. Analysis of the benefits of AA cocktails in treating Charcot-Marie-Tooth disease type 1A. *Muscle Nerve* 38(2): 1052–1054.

Kaya, F.; Belin, S.; Diamantidis, G.; Fontes, M. **2008b**. Ascorbic acid is a regulator of the intracellular cAMP concentration: Old molecule, new functions? *FEBS Lett.* 582(25-26): 3614–3618.

Klepper, J.; Wang, D.; Fischbarg, J.; Vera, J. C.; Jarjour, I. T.; et al. **1999**. Defective glucose transport across brain tissue barriers: A newly recognized neurological syndrome. *Neurochem. Res.* 24: 587–594.

Korcok, J.; Dixon, S. J.; Lo, T. C. Y.; Wilson, J. X. **2003**. Differential effects of glucose on dehydroascorbic acid transport and intracellular ascorbate accumulation in astrocytes and skeletal myocytes. *Brain Res.* 993: 201–207.

Korcok, J.; Yan, R.; Siushansian, R.; Dixon, S. J.; Wilson, J. X. **2000**. Sodium-ascorbate cotransport controls intracellular ascorbate concentration in primary astrocyte cultures expressing the SVCT2 transporter. *Brain Res.* 881: 144–151.

Kuiper, C.; Molenaar, I. G.; Dachs, G. U.; Currie, M. J.; Sykes, P. H.; Vissers, MC. **2010**. Low ascorbate levels are associated with increased hypoxia-inducible factor-1 activity and an aggressive tumor phenotype in endometrial cancer. *Cancer Res.* 70(14): 5749–5758.

Kumar, S.; Hedges, S. B. **1998**. A molecular timescale for vertebrate evolution. *Nature.* 392: 917–920.

Kuo, C.; Hata, F.; Yoshida, H.; Yamatodani, A.; Wada, H. **1979**. Effect of ascorbic acid on release of acetylcholine from synaptic vesicles prepared from different species of animals and release of noradrenaline from synaptic vesicles of rat brain. *Life Sci.* 24(10): 911–915.

Kuo, S. M.; MacLean, M. E.; McCormick, K.; Wilson, J. X. **2004**. Gender and sodium ascorbate transporters determine ascorbate concentrations in mice. *J. Nutr.* 134: 2216–2221.

La Du, B. N.; Zannoni, V. G. **1964**. The role of ascorbic acid in tyrosine metabolism. *Ann. N. Y. Acad. Sci.* 92: 175–191.

Lee, J.; Oh, C.; Mun, G.; Kim, J.; Chung, Y.; Hwang, Y.; Shin, D.; Lin, W. **2006**. Immunohistochemical localization of sodium-dependent L-ascorbic acid transporter 1 protein in rat kidney. *Histochem. Cell Biol.* 126(4): 491–494.

Levene, C.; Shoshan, S.; Bates, C. J. **1972**. The effect of ascorbic acid on the crosslinking of collagen during its synthesis by cultured 3 T6 fibroblasts. *Biochim. Biophys. Acta* 257(2): 384–388.

Lévèque, N.; Muret, P.; Mary, S.; Makki, S.; Kantelip, J. P.; Rougier, A.; Humbert, P. **2002**. Decrease in skin ascorbic acid concentration with age. *Eur. J. Dermatol.* 12(4): 21–22.

Liang, W. J.; Johnson, D.; Jarvis, S. M. **2001**. Vitamin C transport systems of mammalian cells. *Mol. Membr. Biol.* 18: 87–95.

Liang, W. J.; Johnson, D.; Ma, L. S.; Jarvis, S. M. **2002**. Regulation of the human vitamin C transporters and expressed in COS-1 cells by protein kinase C. *Am J Physiol Cell Physiol.* 283(6): C1696–C1704.

Linster, C.; Schaftingen, E. **2007**. AA: Biosynthesis, recycling and degradation in mammals. *FEBS J.* 274: 1–22.

Lutsenko, E. A.; Carcamo, J. M.; Golde, D. W. **2004**. A human sodium-dependent vitamin C transporter 2 isoform acts as a dominant-negative inhibitor of ascorbic acid transport. *Mol. Cell. Biol.* 24: 3150–3156.

Lykkesfeldt, J.; Hagen, T. M.; Vinarsky, V.; Ames, B. N. **1998**. Age-associated decline in ascorbic acid concentration, recycling, and biosynthesis in rat hepatocytes—reversal with (R)-alpha-lipoic acid supplementation. *FASEB J.* 12(12): 1183–1189.

Lyons, B. L.; Schwarz, R. I. **1984**. Ascorbate stimulation of PAT cells causes an increase in transcription rates and a decrease in degradation rates of procollagen mRNA. *Nucleic Acids Res.* 12(5): 2569–2579.

Lysy, J.; Zimmerman, J. **1992**. Ascorbic acid status in diabetes mellitus. *Nutr. Res.* 12: 713–720.

MacDonald, L.; Thumser, A. E.; Sharp, P. **2002**. Decreased expression of the vitamin C transporter SVCT1 by ascorbic acid in a human intestinal epithelial cell line. *Br. J. Nutr.* 87: 97–100.

Majumder, P. K.; Banerjee, S. K.; Roy, R. K.; Chatterjee, G. C. **1972**. Effect of insulin on the metabolism of L-ascorbic acid in animals. *Biochem. Pharmacol.* 22: 759–761.

Malo, C.; Wilson, J. X. **2000**. Glucose modulates vitamin C transport in adult human small intestinal brush border membrane vesicles. *J. Nutr.* 130: 63–69.

Maulen, N. P.; Henriquez, E. A.; Kempe, S.; Carcamo, J. G.; Schmid-Kotsas, A.; Bachem, M.; Grunert, A.; Bustamante, M. E.; Nualart, F.; Vera, J. C. **2003**. Upregulation and polarized expression of the sodium-ascorbic acid transporter SVCT1 in post-confluent differentiated CaCo-2 cells. *J. Biol. Chem.* 278: 9035–9041.

Maulen, N. P.; Henriquez, E. A.; Kempe, S.; Carcamo, J. G.; Smid-Kotsas, A.; et al. **2003**. Upregulation and polarized expression of the sodium-ascorbic acid transporter SVCT1 in post-confluent differentiated CaCo-2 cells. *J. Biol. Chem.* 278: 9035–9041.

Meredith, M. E.; Harrison, F. E.; May, J. M. **2011**. Differential regulation of the ascorbic acid transporter SVCT2 during development and in response to ascorbic acid depletion. *Biochem. Biophys. Res. Commun.* 414(4): 737–742.

Michels, A. J.; Joisher, N.; Hagen, T. M. **2003**. Age-related decline of sodium-dependent ascorbic acid transport in isolated rat hepatocytes. *Arch. Biochem. Biophys.* 410(1): 112–120.

Murad, S.; Grove, D.; Lindberg, K.; Reynolds, G.; Sivarajah, A.; Pinnell, S. R. **1981**. Regulation of collagen synthesis by ascorbic acid. *Proc. Natl. Acad. Sci. USA* 78(5): 2879–2882.

Nelson, P.; Pruitt, R.; Henderson, L.; Jennesse, R.; Henderson, L. **1981**. Effect of ascorbic acid deficiency on the in vivo synthesis of carnitine. *Biochim. Biophys. Acta* 672(1): 123–127.

Nishikimi, M.; Fukuyama, R.; Minoshima, S.; Shimizu, N.; Yagi, K. **1994**. Cloning and chromosomal mapping of the human non functional gene for L-gulogamma lactone oxydase, the enzyme for L-ascorbic acid biosynthesis missing in man. *J. Biol. Chem.* 269: 13685–13688.

Nishimiki, M.; Kawai, T.; Yagi, K. **1992**. Guinea pigs possess a highly mutated gene for L-gulono-lactone oxydase. *J. Biol. Chem.* 267: 21967–21972.

Obrenovich, M. E.; Fan, X.; Satake, M.; Jarvis, S. M.; Reneker, L.; Reddan, J. R.; Monnier, V. M. **2006**. Relative suppression of the sodium-dependent vitamin C transport in mouse versus human lens epithelial cells. *Mol. Cell. Biochem.* 293: 53–62.

Oreopoulos, D. G.; Lindeman, R. D.; Vander-Jagt, D. J.; Tzamaloukas, A. H.; Bhagavan, H. N.; Garry, P. J. **1993**. Renal excretion of ascorbic acid: Effect of age and sex. *J. Am. Coll. Nutr.* 12: 537–542.

Orfali, W.; Nicholson, R. N.; Guiot, M. C.; Peterson, A. C.; Snipes, G. J. **2005**. An 8.5-kb segment of the PMP22 promoter responds to loss of axon signals during Wallerian degeneration, but does not respond to specific axonal signals during nerve regeneration. *J. Neurosci. Res.* 80(1): 37–46.

Owens, L. B.; Wright, J.; Brown, E. **1941**. Vitamin C survey in diabetes. *N. Engl. J. Med.* 224: 319–323.

Padayatty, S.; Katz, A.; Wang, Y.; Eck, P.; Kwon, O.; Lee, J.; Chen, S.; Corpe, C.; Dutta, A.; Dutta, S.; Levine, M. **2003**. AA as an antioxidant: evaluation of its role in disease prevention. *J. Am. Coll. Nutr.* 22(1): 18–35.

Park, S.; Ahn, E. S.; Lee, S.; Jung, M.; Park, J. H.; Yi, S. Y.; Yeom, C. H. **2009**. Proteomic analysis reveals upregulation of RKIP in S-180 implanted BALB/C mouse after treatment with ascorbic acid. *J. Cell Biochem.* 106(6): 1136–1145.

Passage, E.; Norreel, J. C.; Noack-Fraissignes, P.; Sanguedolce, V.; Pizant, J.; Thirion, X.; Robaglia-Schlupp, A.; Pellissier, J. F.; Fontes, M. **2004**. Ascorbic acid treatment corrects the phenotype of a mouse model of Charcot-Marie-Tooth disease. *Nat. Med.* 10(4): 396–401.

Plant, G. W.; Currier, P. F.; Cuervo, E. P.; Bates, M. L.; Pressman, Y.; Bunge, M. B.; Wood, P. M. **2002**. Purified adult ensheathing glia fail to myelinate axons under culture conditions that enable Schwann cells to form myelin. *J. Neurosci.* 22(14): 6083–6091.

Qutob, S.; Dixon, S. J.; Wilson, J. X. **1998**. Insulin stimulates vitamin C recycling and ascorbate accumulation in osteoblastic cells. *Endocrinology* 139: 51–56.

Rajan, D. P.; Huang, W.; Dutta, B.; Devoe, L. D.; Leibach, F. H.; Ganapathy, V.; Prasad, P. D. **1999**. Human placental sodium-dependent vitamin C transporter (SVCT2): Molecular cloning and transport function. *Biochem. Biophys. Res. Commun.* 262: 762–768.

Rumsey, S.; Welch, R.; Garrafo, H.; Ping, G.; Lu, S.; Crossman, A.; Kirk, K.; Levine, M. **1999**. Specificity of ascorbate analogs for ascorbate transport. *J. Biol. Chem.* 33: 23215–23222.

Rumsey, S. C.; Daruwala, R.; Al-Hasani, H.; Zarnowski, M. J.; Simpson, I. A.; Levine, M. **2000**. Dehydroascorbic acid transport by GLUT4 in *Xenopus* oocytes and isolated rat adipocytes. *J. Biol. Chem.* 275: 28246–28253.

Rumsey, S. C.; Kwon, O.; Xu, G. W.; Burant, C. F.; Simpson, I.; Levine, M. **1997**. Glucose transporter isoforms GLUT1 and GLUT3 transport dehydroascorbic acid. *J. Biol. Chem.* 272: 18982–18989.

Saberan-Djoneidi, D.; Sanguedolce, V.; Assouline, Z.; Levy, N.; Passage, E.; Fontes, M. **2000**. Molecular dissection of the Schwann cell specific promoter of the PMP22 gene. *Gene.* 248(1–2): 223–231.

Savini, I.; Catani, M. V.; Arnone, R.; Rossi, A.; Frega, G.; Del Principe, D.; Avigliano, L.; **2007**. Translational control of the ascorbic acid transporter SVCT2 in human platelets. *Free Radic. Biol. Med.* 42: 608–616.

Savini, I.; Catani, M. V.; Duranti. G.; Ceci. R.; Sabatini, S.; Avigliano, L. **2005**. Vitamin C homeostasis in skeletal muscle cells. *Free Radic. Biol. Med.* 38(7): 898–907.

Schorah, C. J.; Bishop, N.; Wales, J. K. **1988**. Blood vitamin C concentrations in patients with diabetes mellitus. *Int. J. Vitam. Nutr. Res.* 58: 312–318.

Seghieri, G.; Martinoli, L.; Micelo, M.; Ciuti, M.; D'Alessandri, G.; Gironi, A.; Palmieri, L.; Anichini, R.; Bartolomei, G.; Franconi, F. **1994**. Renal excretion of ascorbic acid in insulin dependent diabetes mellitus. *Int. J. Vit. Nutr. Res.* 64: 119–124.

Shin, D. M.; Ahn, J. I.; Lee, K. H.; Lee, Y. S.; Lee, Y. S. **2004**. Ascorbic acid responsive genes during neuronal differentiation of embryonic stem cells. *Neuroreport* 15(12): 1959–1963.

Smirnoff, N. **2001**. L-Ascorbic acid biosynthesis. *Vitam. Horm.* 61: 241–266.

Sotiriou, S.; Gispert, S.; Cheng, J.; Wang, Y.; Chen, A.; Hoogstraten-Miller, S.; Miller, G. F.; Kwon, O.; Levine, M.; Guttentag, S. H.; Nussbaum, R. L. **2002**. Ascorbic-acid transporter Slc23a1 is essential for vitamin C transport into the brain and for perinatal survival. *Nat. Med.* 8: 514–517.

Subramanian, V. S.; Marchant, J. S.; Boulware, M. J.; Said, H. M. **2004**. A C-terminal region dictates the apical plasma membrane targeting of the human sodium-dependent vitamin C transporter-1 in polarized epithelia. *J. Biol. Chem.* 279: 27719–27728.

Sun, A. Q.; Salkar, R.; Sachchidanand, S.; Xu, S.; Zeng, L.; Zhou, M. M.; Suchy, F. J. **2003**. A 14-amino acid sequence with a beta-turn structure is required for apical membrane sorting of the rat ileal bile acid transporter. *J. Biol Chem.* 278: 4000–4009.

Takahashi, T.; Lord, B.; Schulze, P.; Fryer, R.; Sarang, S.; Gullans, S.; Lee, R. **2003**. Ascorbic acid enhances differentiation of embryonic stem cells into cardiac myocytes. *Circulation* 107(14): 1912–1916.

Timpson, N. J.; Forouhi, N. G.; Brion, M. J.; Harbord, R. M.; Cook, D. G.; Johnson, P.; McConnachie, A.; Morris, R. W.; Rodriguez, S.; Luan, J.; Ebrahim, S.; Padmanabhan, S.; Watt, G.; Bruckdorfer, K. R.; Wareham, N. J.; Whincup, P. H.; Chanock, S.; Sattar, N.; Lawlor, D. A.; Davey Smith, G. **2010**. Genetic variation at the SLC23A1 locus is associated with circulating concentrations of L-ascorbic acid (vitamin C): Evidence from 5 independent studies with >15,000 participants. *Am. J. Clin. Nutr.* 92(2): 375–382.

Ting, H. H.; Timimi, F. K.; Boles, K. S.; Creager, S. J.; Ganz, P.; Creager, M. A. **1996**. Vitamin C improves endothelium-dependent vasodilation in patients with non-insulin-dependent diabetes mellitus. *J. Clin. Invest.* 97: 22–28.

Tsukaguchi, H.; Tokui, T.; Mackenzie, B.; Berger, U. V.; Chen, X. Z.;Wang, Y.; Brubaker, R. F.; Hediger, M. A. **1999**. A family of mammalian Na+-dependent L-ascorbic acid transporters. *Nature* 399: 70–75.

Vera, J. C.; Rivas, C. I.; Zhang, R. H.; Golde, D. W. **1998**. Colony-stimulating factors signal for increased transport of vitamin C in human host defense cells. *Blood* 91: 2536–2546.

Wang, H.; Zhang, Z. B.; Wen, R. R. **1995**. Experimental and clinical studies on the reduction of erythrocyte sorbitol-glucose ratios by ascorbic acid in diabetes mellitus. *Diabetes Res. Clin. Pract.* 28: 1–8.

Will, J. C.; Byer, T. **1996**. Does diabetes mellitus increase the requirement for vitamin C? *Nutr. Rev.* 54: 193–202.

Will, J. C.; Ford, E. S.; Bowman, B. A. **1999**. Serum vitamin C concentrations and diabetes: findings from the third National Health and Nutrition Examination Survey, 1988–1984. *Am. J. Clin. Nutr.* 70: 49–52.

Wilson, J. X. **2005**. Regulation of vitamin C transport. *Ann. Rev. Nutr.* 25: 105–125.

Wilson, J. X.; Dixon, S. J. **1989**. High-affinity sodium-dependent uptake of ascorbic acid by rat osteoblasts. *J. Membr. Biol.* 111: 83–91.

Wilson, J. X.; Dixon, S. J. **1995**. Ascorbate concentration in osteoblastic cells is elevated by transforming growth factor-β. *Am. J. Physiol.* 268: E565– E571.

Wilson, J. X.; Jaworski, E. M.; Dixon, S. J. **1991**. Evidence for electrogenic sodium-dependent ascorbate transport in rat astroglia. *Neurochem. Res.* 16: 73–78.

Wilson, J. X.; Jaworski, E. M.; Kulaga, A.; Dixon, S. J. **1990**. Substrate regulation of ascorbate transport activity in astrocytes. *Neurochem. Res.* 15: 1037–1043.

Wilson, L. **1975**. The clinical definition of scurvy and the discovery of AA. *J. Hist. Med.* 30(1): 40–60.

Wu, X.; Itoh, N.; Taniguchi, T.; Nakanishi, T.; Tanaka, K. **2003a**. Requirement of calcium and phosphate ions in expression of sodium-dependent vitamin C transporter 2 and osteopontin in MC3T3-E1 osteoblastic cells. *Biochem. Biophys.* Acta. 1641(1): 65–70.

Wu, X.; Itoh, N.; Taniguchi, T.; Nakanishi, T.; Tetsu, Y.; Yumoto, N.; Tanaka, K. **2003b**. Zinc-induced sodium-dependent vitamin C transporter 2 expression: potent roles in osteoblast differentiation. *Arch. Biochem. Biophys.* 420 (1): 114–120.

Yue, D. K.; McLennan, S.; Fisher, E.; Hefferman, S.; Capogreco, C.; Ross, G. R.; Turtle, J. R. **1989**. Ascorbic acid metabolism and polyol pathway in diabetes. *Diabetes* 38: 257–261.

Index

A

Acetyl-coenzyme A (CoA), 207
Acetyl-coenzyme A (CoA) carboxylase (ACC), 207, 219–220
Active targeting, 230
Adenosine triphosphate (ATP), 171, 192, 200, 207, 259
   -induced contractile activity, 193–194, 195
Adipogenesis, vitamin A and, 33
Advanced glycation end products (AGEs), 169, 170
   -modified proteins, 170–171
   production, inhibitors of, 172
   and ROS production, 175
   in vascular complications, 170
Advanced lipoxidation end products (ALEs), 169, 170
Afamin, 137
Aging, ascorbic acid and, 270–271
Albumin, 137
Alcohol dehydrogenases (ADHs), 5
Aldose reductase, 175
All-*trans* retinoic acid, 2, 7, 9, 50
α-tocopherol, 133, 137–139
Alzheimer's disease, 169, 177
Angiogenesis, VDUP1 expression and, 64
Anticancer drugs, 229–230
Anticoagulant factors, 158
Antiquitin (ATQ) deficiency, 188
Antisense oligonucleotide, 236
Apoptosis
   glucose-induced, 174–175
   immune cell apoptosis, VDUP1 role in, 59
Ascorbic acid (AA)
   active transport, 261–262
   associated pathophysiology, 257, 266
      age, 270–271
      cancer, 268–270
      diabetes, 267–268
      genetic variants of transporters, 266–267
      neuropathy, 270
   functions of, 258
      biochemical functions, 258–259
      in cell differentiation, 259
   and gene expression, 265–266
   L-ascorbic acid, 133
   passive transport, 260
   SVCT1/SVCT2 cotransporters, 262
      gene structure and expression, 262
      protein structure, 262
      regulation of gene expression, 263–265

   transport, 259, 260, 261, 266
   uptake of dehydroascorbic acid into cells, 260–261
Asthma, vitamin D and, 98
Atorvastatin, 164
Autoimmune diseases, retinoids effect on, 52–54
Avidin-biotin system, 214

B

BAY K 8644, 193
B cells, retinoic acid effects on, 51
Benfotiamine, 171
   chemical structure of, 172
   therapeutic effects of, 174–175, 176, 177
β-methyl-crotonyl-CoA carboxylase (MCC), 207, 220
β,β-carotene, 2, 4
β,β-carotene-9,10-dioxygenase 2 (BCO-II), 4
β,β-carotene-15,15'-monooxygenase 1 (BCO-I), 4
Biotin, non-prosthetic group functions of, 207
   biotin-binding proteins, 214
   biotin-responsive basal ganglia disease, 210
   carbon metabolism genes, starvation effects on, 213–214
   cell differentiation, requirement for, 208–209
   cells in culture, requirement for, 208–209
   glucose metabolism enzymes, 211–212
   neurotrophic factor and, 210
   palatal process development and, 209
   protein synthesis, 211
   and reproductive system, 209
   sensing, 212–213
Biotin-binding proteins, 214
Biotin protein ligase, 212–213
Biotin-responsive basal ganglia disease, 210
Biotin-signaling pathways, 220
   biotinyl-5'-AMP, 223
   calcium, 222
   cGMP, 220
   HLCS-dependent signaling at chromatin level, 223–224
   miR signaling, 223
   NF-κB, 221
   nitric oxide, 223
   receptor tyrosine kinases, 221–222
   Sp1 and Sp3, 221
Biotinyl-5'-AMP, 223